IEE CIRCUITS AND SYSTEMS SERIES 1

Series Editors: Dr. D. Haigh
Dr. R. Soin

GaAs technology and its impact on circuits and systems

GaAs technology and its impact on circuits and systems

Edited by
D Haigh and J Everard

Peter Peregrinus Ltd. on behalf of the Institution of Electrical Engineers

Published by: Peter Peregrinus Ltd., London, United Kingdom

British Library Cataloguing in Publication Data

GaAs technology its impact on
circuits and systems.
1. Gallium arsenide semiconductors
I. Haigh, D. II. Everard, J.
537.6'22

ISBN 0 86341 187 8

Printed in England by Short Run Press Ltd., Exeter

Contents

List of Colour Plates

Foreword

The field of Circuits and Systems has been one of the most active and fastest developing in Electrical Engineering. From early beginnings in Circuit Theory and Filter Synthesis, the subject now spans such topics as Digital Signal Processing and Algorithms for VLSI design, Neural Networks and Systems Theory and, since circuits and systems cannot be divorced from their implementation, Semiconductor Devices and Technological Processes also come within its sphere. Thus the launch of a new series on Circuits and Systems is most timely.

This first volume in the Circuits and Systems series perfectly illustrates the broad scope of the field. The editors have been able to assemble an impressive list of authors covering the disciplines required for larger scale integration of microwave and optoelectronic systems of the future. Many of the authors have made superhuman efforts to produce material much of which has not been previously published.

Randeep Singh Soin
David Haigh
Fareham, July 1989

Preface

This volume arose from an idea by Randeep Soin to attempt to publish the notes for a workshop in Gallium Arsenide Technology and design forming part of the 1989 European Conference on Circuit Theory and Design. Initially, Randeep was met by general lack of enthusiasm and disbelief that the timescales could be met. However, by patient diplomacy, Randeep gradually won people over and the project was launched.

The first letter to the authors was sent in December 1988, manuscripts started to arrive in May and publication is in August 1989. We hope that this accelerated timescale of eight months from concept to publication will help to maximise the topicality, and hence value, of book material.

Both of us were privileged to be part of the design team for a multiproject Gallium Arsenide chip commissioned by the SERC Rutherford Appleton Laboratory and manufactured by the Plessey Three Five Group. Many of the authors became better known to us through this project and therefore the book can be regarded to some extent as a spin–off from the project.

In view of the wide range of necessary topics for Gallium Arsenide design covered in the book, Chapter 1 has been written in the form of a guide to the various chapters. We hope that this book will provide a realistic statement of the current status of Gallium Arsenide technology and design and will help to realise the enormous possibilities which we believe lie ahead.

David Haigh
Jeremy Everard
London, July 1989

Acknowledgements

We would like to thank the authors for their material. In many cases, the scope and quality of the material far exceeded what was originally envisaged and for this hard work and patient effort we are most grateful.

The typesetting has been carried out to a fully professional standard by Pauline Cloudsdale and for this we are most grateful. Pauline has been working a two shift day, seven days a week for several weeks and for her dedication (and patience with the editors) we are also most grateful.

We also acknowledge the generosity of GenRad Ltd for allowing the use of their typesetting facilities and the time and commitment of Randeep Soin for overseeing the typesetting process and for proof reading.

General secretarial support by Bridget Bradley at University College is also gratefully acknowledged.

Finally, we would like to thank our families for tolerating our absence while undertaking this project.

The Editors

List of Contributors

J. Arnold
Plessey Three Five Group Ltd

A. K. Betts
D. G. Haigh
J. E. Midwinter
G. Parry
K. Steptoe
M. Whitehead
Department of Electronic and Electrical Engineering
University College London

D. R. S. Boyd
S. J. Newett
SERC Rutherford Appleton Laboratory

R. W. W. Charlton
P. Saul
J. A. Turner
Plessey Research (Caswell) Ltd

J. K. A. Everard
I. D. Robertson
J. G. Swanson
Department of Electrical and Electronic Engineering
Kings College London

B. W. Flynn
J. Mavor
H. M. Reekie
Department of Electrical Engineering
University of Edinburgh

A. K. Jastrzebski
Electronic Engineering Laboratories
The University of Kent

L. E. Larson
Hughes Research Laboratories (Malibu, California, USA)

R. P. Merrett
British Telecom Research Laboratories

C. Toumazou
Department of Electrical Engineering
Imperial College of Science Technology and Medicine

(All affiliations are UK except where stated)

Chapter 1

Introduction

D. G. Haigh and J. K. A. Everard

1.1 Present Position

The past five years have seen the emergence of Gallium Arsenide technology from the research laboratory to the market place. This has been accompanied by dramatic increases in complexity from early single devices to integrated circuits of considerable complexity for both digital and analogue applications. This volume very much captures the essence of these developments and aims to lay down sound principles for forthcoming developments in which both the speed and level of complexity of GaAs circuits will increase dramatically, meeting needs in such areas as ultra-fast computers, personal communications and new broadcasting technology.

The scope of the Chapters in this book is very broad, ranging for example from electronics to optics, from CAD tools to materials physics and from processing technology to communication systems. Therefore an introductory Chapter cannot possibly provide an introduction in the true sense to the fundamental principles or underlying common elements of the material covered in the book. We shall content ourselves, therefore, with providing a short guide to the content of the book. Table 1.1 shows shortened titles of the various Chapters. The Chapters can be divided into three sections which we now consider in turn.

1.2 The Prerequisites for Design

The first three Chapters (after the Introduction) deal with the three necessary prerequisites for GaAs design, namely processing technology, device modelling and CAD tools. These prerequisites are necessary irrespective of the application, be it analogue or digital, and they are always mandatory, as illustrated in Figure 1.1.

The coming maturity of Gallium Arsenide technology is reflected in the emergence of foundry processes, which allow a designer to translate his requirements into integrated form in a relatively easy and rapid way. The processing technology described in Chapter 2 is such a foundry

Table 1.1 : Chapter Content

Chapter	Topic
2	Process technology
3	Device modelling
4	CAD tools
5	GaAs versus Silicon
6	Digital circuits and systems
7	A/D and D/A converters
8	Low noise oscillators
9	Design of mixers
10	SC filters and op-amps
11	IC's for optical fibre comms
12	Emerging technologies
13	Opto-electronics

process developed by the Plessey Company (UK). The principle active device is the GaAs metal semiconductor field effect transistor (MESFET). Factors relevant to the use and operation of such a process are described including the customer-vendor interface, and the

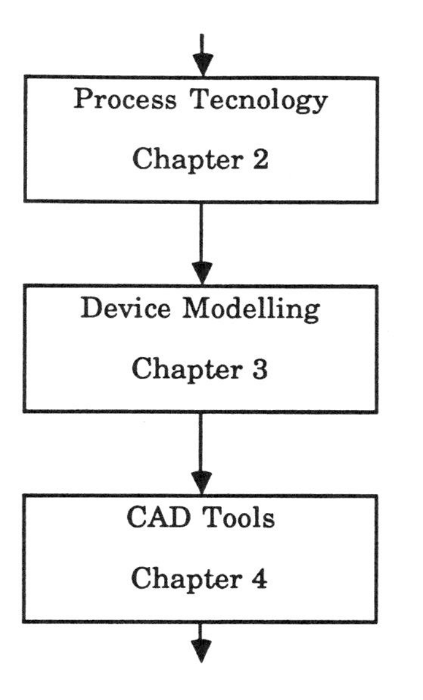

Figure 1.1 Design Prerequisites

characterisation of the semiconductor materials and circuit performance itself. An important aspect of such a process is the maintenance of high yield figures, and the minimum target achieved of 60% illustrates the maturity of present technology. Future developments in terms of processes and devices, as seen from the foundry service perspective, are also covered.

The cost and time involved in committing a circuit, or system, to integration is such that computer simulation at the design stage is mandatory. For the purpose of simulation, semiconductor devices are represented by sets of equations, referred to as device models. The problem of deriving suitable device models in the case of Gallium Arsenide technology has proved to be a complex one and it is only now that a consensus on this subject is emerging. Chapter 3 provides a definitive presentation of this subject, including decisive results from the University of Kent (UK) which have been incorporated in the ANAMIC CAD package. Having derived a suitable model, it remains to determine suitable values for the constants, or parameters, in the model equations for a particular processing technology.This is done by making measurements on sample devices and using optimisation to adjust the model parameters to obtain a good fit. This procedure is termed parameter extraction and is covered also in Chapter 3.

Having obtained suitable device models, with appropriate model parameters for the process to be used, the designer is faced with the task of computer simulation of the design in order to verify it, optimise it and refine it as necessary. There are now a large number of simulation packages available and the SERC Rutherford and Appleton Laboratory (UK) provides an important service of detailed evaluation of such packages for use by the UK academic community. Thus Steve Newett and David Boyd, together with Kevin Steptoe, are in an excellent position to write about CAD tools and this is covered in Chapter 4. Cad Tools also provide the means of translating paper designs into IC layout and mask manufacturing data. This stage cannot be separated from the circuit simulation stage since layout parasitics influence circuit performance. Chapter 4 discusses some common layout editors and compares their distinctive features.

1.3 Design

The central part of the book comprising Chapters 5 to 11 covers design itself according to the overall scheme shown in Figure 1.2. the types of system covered can be classified as analogue, digital or comprising analogue and digital elements.

People working in a particular technology sometimes exaggerate the capabilities of their own technology and do not fully appreciate the

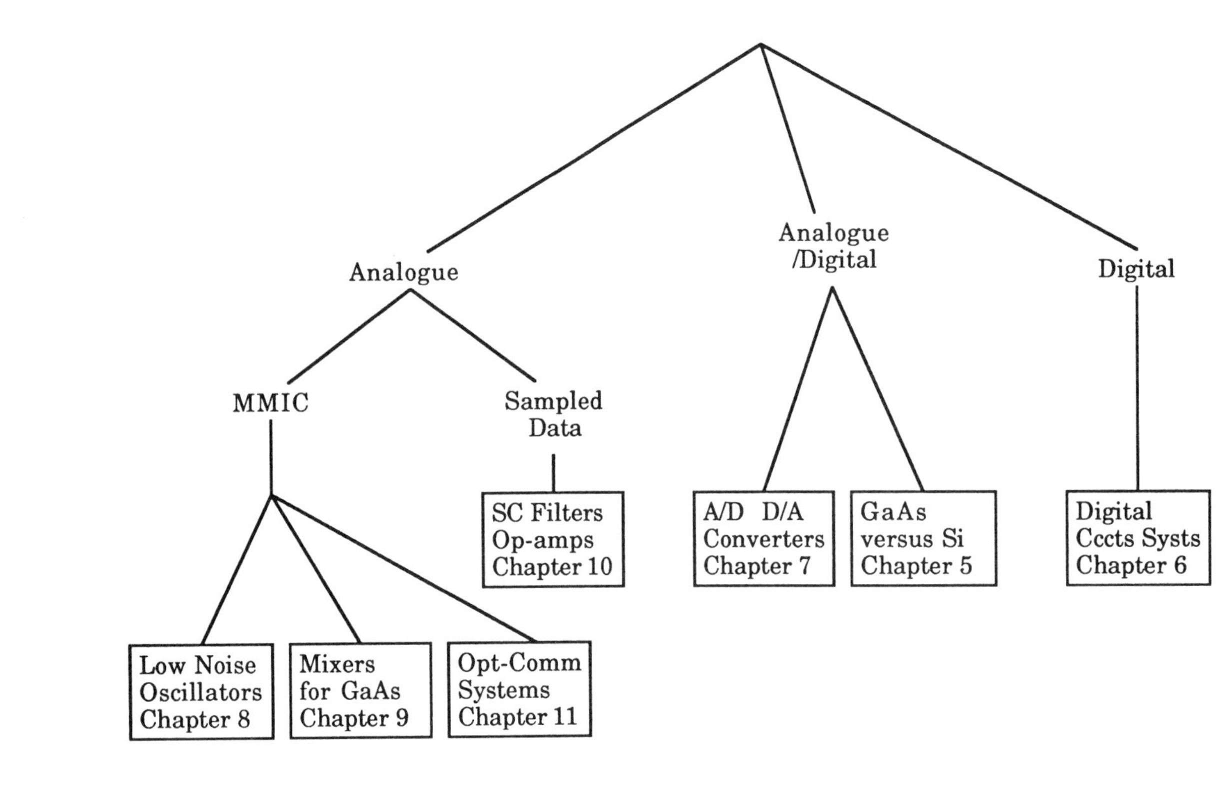

Figure 1.2 GaAs Design

capabilities of competing technologies. Peter Saul at Plessey Research (Caswell, England) is in a unique position to make a valid comparison between GaAs and Silicon across both analogue and digital applications and his assessment is contained in Chapter 5. The comparison is based on evaluating the basic materials and structures from the points of view of physics, engineering, radiation hardness and economics, and the viewpoint is that of the circuit designer.

In the past, the exaggeration of the capabilities of digital GaAS circuits relative to silicon has produced a degree of scepticism in some quarters. Recent work has now established an improvement factor of about 3 in terms of speed–power product for GaAs and Chapter 6 is concerned with the design and capabilities of high speed GaAs logic. This area has certainly received a boost from the development by Gigabit Logic of a supercomputer for Cray using GaAs technology.

The advent of digital circuits and processing has created a major need for the analogue–to–digital (A/D) and digital–to–analogue (D/A) converters, which has hitherto been met using CMOS and bipolar technology. Chapter 7, written by Larry Larson, of Hughes Research Laboratories, is concerned with the application of GaAs technology to this important area with the aim of increased sample rates. Current state of the art using GaAs allows 8–bit resolution at sample rates of 1 gigasample per second, and further improvements are forecast. The best performance in GaAs may well be achieved using heterojunction bipolar transistors (HBT's) which are considered in this Chapter as well as MESFET's.

One of the most exciting possibilities being brought about by GaAs technology is that of the integration of communication system front ends. Perhaps the most critical component of such systems is the low–noise oscillator and Chapter 8 covers the realisation in GaAs of this component. A general theory for low noise oscillator design is presented and this leads to a consideration of a variety of resonator types, including inductor–capacitor and transmission line, which are ideally suited to implementation in GaAs. Some examples of GaAs oscillators conforming with the general theory are presented and finally their performance is critically compared with that of oscillators constructed using Silicon devices. It is concluded that, for frequencies below the maximum limit, Silicon gives lower noise but for the highest oscillation frequencies, GaAs is necessary.

In a communication system, excellent oscillation performance is wasted if the mixer is not also of the highest quality, and Chapter 9 considers the application of GaAs technology to this area. Operating frequencies are currently around 100 GHz, based on the Schottky barrier diode mixer. However, frequencies of 200 GHz are anticipated using high electron mobility transistors (HEMT's). The development of reliable large signal modelling of GaAs devices, as described in Chapter 4, is starting to pay large dividends in the design of high quality mixers with a high degree of confidence.

In CMOS technology, high precision sampled data filters using the switched capacitor circuit techniques have been used very successfully to realise high precision filters in integrated circuit form. Chapter 10 is concerned with application of GaAs technology in this area. Circuit design techniques presented are aiming towards sampling rates as high as 1 GHz and filter midband frequencies up to 100 MHz. A key component in sampled data systems is the operational amplifier, design techniques for which are presented in this Chapter. Such amplifiers are applicable in a broader context, such as in the case of amplifiers in high speed phase locked loops. Also, the high precision switched capacitor integrator, presented as a basic building block of filter synthesis, has a wider sphere of application including that of delta–sigma analogue–to–digital and digital–to–analogue converters.

One case where GaAs has been very successfully applied is to the high speed electronic terminal equipment required for optical fibre communication systems and, in Chapter 11, Bob Merrett of British Telecom Research Laboratories presents this application. The systems covered include computer data links and those for control applications and they are seen in a broad context including new broadcasting technology, such as millimetre wave TV redistribution systems. The development of fully integrated optoelectronic ICs which include the optical sources and detectors , is discussed. the optimum optical wavelength for long distance optical fibre links does not however coincide with that for which GaAs is optimally efficient and therefore optical components using Indium Phosphide are being combined with GaAs. As in other areas, HEMT's and HBT's are likely to have a very important future role in this important system area.

1.4 Next Generation Technologies

The closing two Chapters look to the future and report on topics which are more at an earlier stage at present and which will certainly be very important in the future. Chapter 11 is concerned with new devices which are particularly aimed at very high frequency operation. In this Chapter, Garth Swanson first investigates the basic materials which can be used and highlights their relevant properties. The performance of a device depends not only on the materials used but also on its structure and dimensions and the Chapter goes on to examine the limitations of the relevant semiconductor processing steps and the effect they have on device performance. The conclusion reached is that frequencies well above the present limits around 100 - 200 GHz may be possible by very careful development and sophisticated control of materials and device structure.

The combination of optical and electronic signals, or opto-electronics, is a very rapidly growing field. Chapter 13, written by Gareth Parry, Mark Whitehead and John Midwinter, is concerned with the role of GaAs in optical devices and systems for future applications. It is now generally accepted that all-optical processing cannot be envisaged at present as providing the requirements of high speed and low power simultaneously. This leads to the concept of a judicious blend of optics and electronics and one of the most interesting scenarios is that of using optics to overcome the presently limiting interconnection problem in high speed digital electronic circuits. Such a development depends on a number of issues, including architecture, devices and materials, all of which are considered. Optical modulators based on multi-quantum well devices are one component which is playing a major role in this area. Developments in this area will be rapid, and this Chapter serves to define issues and concepts which will dictate this development.

1.5 A 'Giant Leap' for Technology

Some say, based on early over optimism and exaggerated claims and also on the vagaries and fluctuations of industry and financial manoeuverings, that GaAs technology has somehow failed or is 'on the way out'. This is not our view and nor is it the view of those who are working in the field to produce genuine advances. The facts of the physics of the materials involved cannot be denied and so much has been achieved by so many that the momentum generated cannot possibly stop. We do discern, however, a momentary 'hiccup' or glitch, a time of questioning, of some introspection, about the future role of GaAs technology. But this pause is, we believe, due to the enormity of the next step in the application of GaAs.

The possibilities for large scale integration of systems using GaAs and related compounds are really breathtaking. On one axis, the ability to realise high speed active devices and due to its semi-insulating substrate, high quality reactive components and transmission lines,together with precision filters and analogue-to-digital converters will revolutionise the form of future electronic communication systems. On the other axis, the possibility of large scale optoelectronic integration is equally, if not more, exciting. These developments will require radical changes in engineering practice and education, with the ability to combine, on one chip, circuit technologies which are currently regarded as distinct. It need hardly be mentioned that it will require another quantum leap in the scope and efficiency of modelling techniques and CAD tools, as well as in reliability of processing technology to meet multiple optimisation criteria and using sophisticated material combinations. The super designer engineer of the future will be equally at home with low noise amplifiers, optical modulators, digital shift registers and switched capacitor integrators, to say nothing of the physics of the materials and the devices and also the development and appropriate use of CAD tools. Up till now, the various disciplines have been largely separate and a considerable degree of competence and confidence in these areas, which are represented in the Chapters of this book, have been built up. It is the editors earnest wish that this book should, in demonstrating these achievements, help to bring to realisation the possibilities of large scale system integration.

Chapter 2

An Overview of Processing and Technology

J. Arnold, R. W. W. Charlton and J. A. Turner

2.1 Introduction

The past twenty five years has witnessed radical changes in microwave systems from the point of view of frequency coverage, sophistication of function, complexity, reliability, cost per function and versatility of application. Such radical changes have resulted primarily from Research efforts into solid state devices with the initial postulation of new materials and device structures, improved understanding of the physics and ability to predict performance, and the subsequent realisation of such devices through the development of equipment for the fabrication of material and the implementation of processes. Advances in microwave test equipment and the development of sophisticated software has assisted with the assessment and characterisation of such devices.

Through such Research and Development, the microwave industry has seen the replacement of state of the art devices of the sixties, such as germanium and silicon transistors and negative resistance germanium tunnel diodes operating from a few hundred megahertz to the lower gigahertz frequency ranges to the realisation of heterojunction structures capable of operation to 100 GHz.

The advances in solid state devices has been accompanied by advances in circuit design techniques and materials for realisation of circuits. As a result earlier waveguide systems have been replaced by hybridised circuits combining solid state devices with matching circuits on plastic boards or alumina substrates, and resulting in significantly improved levels of integration and overall size. Although such technologies have lead to major improvements in system realisation, the fabrication, alignment and testing required have resulted in high cost and relatively small volume capabilities. With the realisation of complex processes and the realisation of the concept of the fully integrated microwave circuit, the microwave industry is advancing into a new generation of super-components with reduced size, cost and higher levels of integration.

The advances in solid state devices has primarily developed from a realisation of the advantages offered by materials formed from elements in the Groups III and V of the periodic tables, such as Gallium and Arsenic.

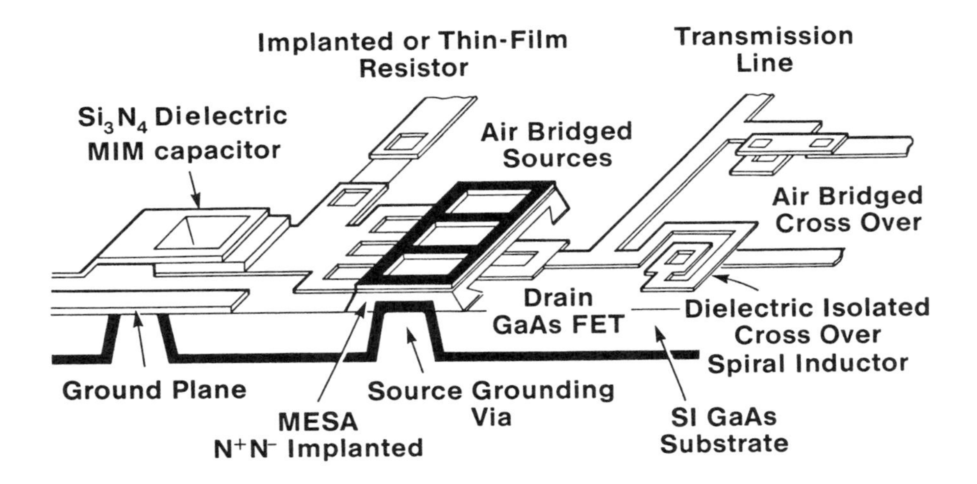

Figure 2.1 Schematic of GaAs MMIC

With the advantages over germanium and silicon of higher mobility, higher peak saturation velocity and greatly simplified device structures, the metal Schottky field effect transistor, MESFET, was destined to establish new performance goals.

Pioneering work by a number of companies worldwide, including Plessey, in the late sixties–early seventies, had demonstrated the potential performance advantages of the gallium arsenide discrete MESFET in hybrid circuits for low noise, wide band and high frequency amplification. The advent of such devices revitalised the microwave industry and led to a substantial growth in a wide variety of applications, both military and commercial, where the reliability of such components were demonstrated.

Even today, the demands for improved system performance requiring lower noise, higher power, improved efficiency and operation into the millimetre frequency bands have continued to sustain research into state of the art performance discrete devices. Improvements in material processing and lithographic techniques have opened the way for new generations of devices based on band gap engineering with the use of heterojunction structures based on a variety of ternary compounds such as AlGaAs/GaAs, AlGaAs/InGaAs and InAlAs/InGaAs on InP. These give rise to a variety of device types such as the high electron mobility transistor, HEMT and the lattice matched HEMT resulting in Fmax up to 170 GHz, noise figures of 0.8 dB at 63 GHz and peripheral power densities of 1.7 W/mm at 59 GHz.

Whilst discrete devices and the associated MIC assembly techniques have become well established, and will continue to form an important facet of microwave system build, it was realised in the early seventies that the semi–insulating properties of the GaAs substrate, namely high resistivity of 10^7 ohm–cm compared to 30 ohm–cm for silicon, would allow the realisation of low parasitic, lumped element matching which could be integrated directly with the active devices. This property coupled with the performance advantages of the GaAs MESFET compared with silicon bipolars resulted in the concept of the fully functional microwave integrated circuit, known as an MMIC. Figure 2.1 shows schematically the physical realisation of key elements such as MESFETs, MIM capacitors, spiral inductors, transmission lines, resistors, airbridges and via holes, all fabricated within a single chip to allow the realisation of a defined microwave function without the need for external components. The potential size reduction, system benefits and the anticipated reduction in system costs from volume manufacture of such an integrated circuit was immediately recognised and resulted in a period of sustained research supported primarily by Government funding for military programmes.

From these simple concepts and the Worlds first GaAs MMIC reported by Plessey in 1974 an impressive variety of circuit functions have been reported by numerous research laboratories. The frequency range from DC to 115 GHz has been researched with examples of small signal amplifiers, both low noise and wideband, power amplifiers, switches, phase shifters, oscillators and mixers. As well as demonstrating single function capability, the integration of multifunctions on a single chip have been demonstrated for applications such as DBS receivers with low noise RF amplifiers, mixers, oscillators and IF amplifiers and T/R modules for phased array radars with low noise and power amplifiers, phase shifters and reciprocal switches.

Table 2.1 identifies some of the key milestones in the history of the GaAs MMIC research and development from 1974 to the present day. Whilst these achievements formed major milestones in the demonstration of the potential of the GaAs MMIC, it was clear, almost without exception, that the objective was to achieve state of the art performance without due regard for yield or manufacturability. It was clear that radical changes of philosophy were required in the areas of design and processing if GaAs MMICs were to become manufacturable, provide high yields and hence become affordable.

In 1984, after a period of 10 years of Research and Development, a number of new companies were formed, either as a subsidiary of a parent company or supported by venture capital, to exploit the new opportunities that had been identified from market surveys.

Many of these new companies reflected the R&D strategy of the parent company, without due consideration for the concept of the GaAs MMIC. This was reflected in the quality of service and products supplied to the

Table 2.1 : Major Achievements in GaAs MMICs 1974 - 1988

X-Band Amplifier	1974	5dB gain/8 to 12 GHz	Plessey
X/J-band Amplifier	1978	6dB gain/6 to 19 GHz/6 dB NF	Plessey
Power Amplifier	1979	10dB gain/1.25 W/9.5 GHz	TI
SPDT Switch	1980	0.5dB loss/40 dB isolation/8 to 11 GHz	Raytheon
RF Generator	1981	Mixer/Oscillator/Amp	HP
Power Amplifier	1981	5.7 to 11 GHz/0.5 W/6dB	Westinghouse
Actively Matched Amp	1981	19dB gain/2 to 4 GHz/4 dB NF	Plessey
Oscillator	1981	13 GHz/8 mW/4% efficient	Plessey
Mixer	1981	31 GHz diode/12dB NF	MIT
Broadband Amplifier	1981	2 GHz BW/low-noise/1st Comrc'l GaAs IC	Siemens
X-Band Switch	1981	FET/0.7dB loss/28 dB isolation	TI
VCO	1982	10 mW/X-J-band	TI
Travelling Wave Amp	1982	5dB gain/1 to 13 GHz	Raytheon
Vector Modulator	1982	S-band/360 deg/analogue	Plessey
4-Bit Phase Shifter	1982	Single-chip/X-band	Raytheon
Multi-chip TR/RX	1982	X-band amp shifter/TR switch	Raytheon
Power Amplifier	1983	10dB gain/8 - 9 GHz/1W	TCSF
RFOW	1983	RFOW	Tektronix
Broadband Amplifier	1983	17dB gain - 4 GHz BW	NEC
Broadband Amplifier	1983	0.6 to 6 GHz/6dB gain/4dB NF	Plessey
Power Amplifier	1984	2W/2 to 8 GHz/10dB gain	Raytheon
Broadband Switch	1984	0-20 GHz/4dB loss/28dB isolation	Raytheon
SATCOMS Amplifier	1984	30dB gain/3dB NF/X-band	Plessey
DBS Receiver	1984	25dB conversion gain/4.5dB NF/2.5x2.0mm	LEP
DBS Receiver	1984	37dB gain/5dB NF/3.5x2.8mm	CISE
Mm Wave Oscillator	1984	69 GHz 0.25 μm/1 mW/1% efficient	Hughes
S-band TR/RX Module	1984	LNA/phase shifter/power amp	Plessey
HEMT Amplifier	1984	X-band low noise	TCSF
20 GHz Transmit Module	1984	Single chip/phase shifter/amp	Rockwell
Transformer Coupled Amp	1984	2 to 4 GHz IR receiver	Honeywell
Dual Gate TWA	1984	2 to 18 GHz with AGC	Avantek
FB Amplifier	1984	6 to 18 GHz	TI
Power Amplifier	1984	6 to 18 GHz/0.5W	TI
Distributed Amp	1984	2 to 30 GHz, 6dB gain	Hughes
Family of VCO's	1984	2 to 18 GHz	TI
Segmented Dual Gate	1984	1-8 GHz phase/amplitude	GI
Planar Mixer	1984	94 GHz	Honeywell
RFOW	1984	RFOW	Plessey
Low Noise Amplifier	1985	30 GHz, 14dB gain, 7dB NF	Hughes
Distributed Amplifier	1985	2 to 40 GHz	Hughes
Mm Wave Oscillator	1985	115 GHz, 0.1 mW	TI
X-Band Power Amplifier	1985	4.5W push-pull, 9.5 GHz	TI
Power Amplifier	1985	2W at 16.5 GHz	TI
Power Amplifier	1985	1W at 28 GHz	Mitsubishi
Phase Shifter	1985	Multibit at 30 GHz	Honeywell
VCO/Divider	1985	11 GHz/VCO anal divider	NEC
Switch Family	1985	Commercial Release	Plessey
Foundry	1985	Commercial Release	Plessey
Power Amplifier	1985	0.5W, 2 to 22 GHz	TI
Low Cost RX	1986	3.7 to 4.3 GHz TVRO	PM
30 GHz Receiver	1986	LNA/Mixer/Phase shifter	Hughes
Distributed Amplifier	1986	Dual gate 2 to 26.5 GHz	HP
Distributed Amplifier	1986	15 to 45 GHz	Raytheon
Cascade Distributed Amp	1986	2 to 18 GHz/10dB stage	Varian
Power Amplifier	1986	135 mW at 41 GHz	TI
Vector Generator	1986	RC Phase Shifter at 4 GHz	GE
Distributed Mixer	1986	1 to 12 GHz, 12dB NF	TI
Converter	1986	4 GHz IR Converter	PM
Levelling Loop	1986	Controlled attenuator	Tektronix
Switch Family/Amps	1986	Commercial Release	Plessey
Regenerative Divider	1986	Anal. at 10 GHz	CCRC
40 GHz Switch	1987	DC - 40 GHz switch	Raytheon
Distributed Mixer	1987	2 - 20 GHz	TI
HEMT Distributed Amp	1987	2 - 20 GHz	Varian
Foundry	1988	Commercial Release, F20	Plessey
HEMT Low Noise Amp	1988	44 GHz	TRW
HEMT Distributed Amp	1988	3 - 40 GHz	Varian
HEMT Low Noise Amp	1988	60 GHz	COMSAT

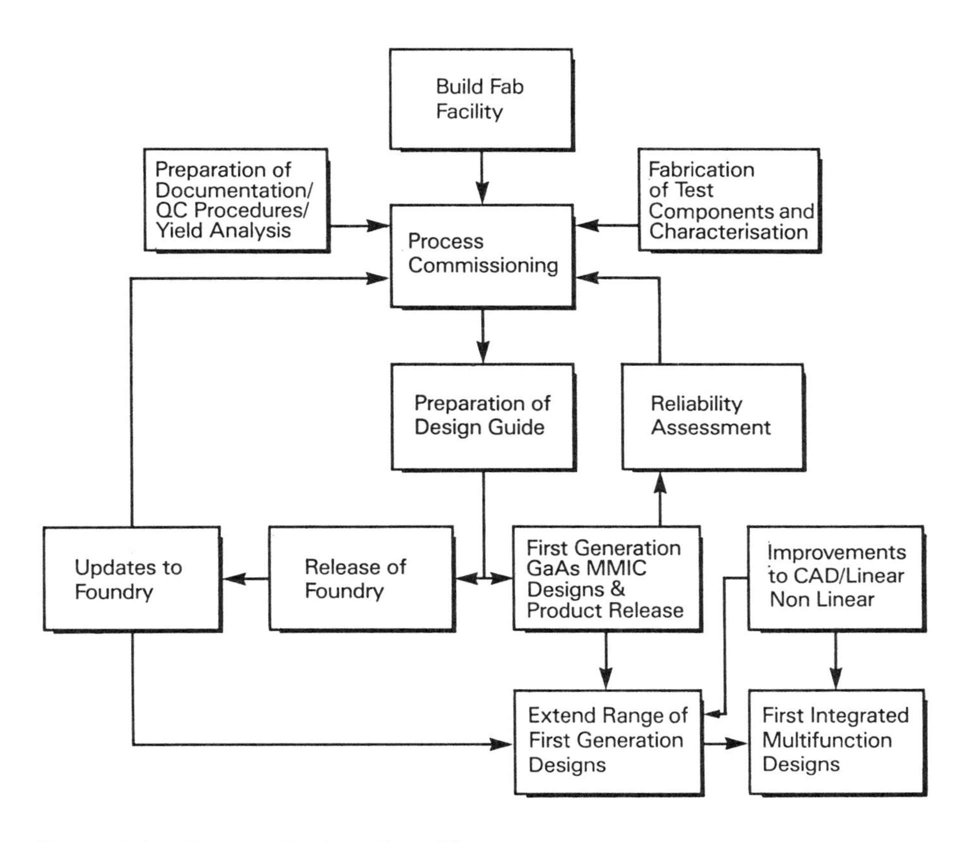

Figure 2.2 Process/Product Flow Chart

customer. A number of companies, such as Plessey Three Five, realised the importance of establishing a fixed reproducible, stable process.

In the case of Plessey, such a process was derived from consideration of the experience of earlier R&D processes and configuring a process to be capable of high yields. The choice of process was determined by a desire to provide enhancements in performance whilst maintaining continuity of manufacturing of products designed on an earlier process. Yield improvements were possible by optimisation and control of material characteristics (ion implanted) and processes and focussing on achievable performance within the process limits rather than with "state of the art" results. Further yield enhancements resulted from the "learning curve" of operating a fixed process. Overall yields rose from a few percent to greater than 60 percent.

The establishing of a stable process was essential to the derivation of the microwave characterisation of all passive and active elements and to the determination of spreads of prime parameters and parasitics within a wafer and from wafer to wafer. The characterisation provided element equivalent circuit models operating over the full process usable frequency range, 14 GHz in the case of the F14 (0.7 μm, non via'd) and 20 GHz in the case of the F20 process (0.5 μm, via'd).

The modelling strategy was selected to be compatible with use in standard commercially available CAD programmes such as TOUCHSTONE or Super–Compact. All process design rules and characterisation were documented into a Design Guide. Figure 2.2 shows a flowchart showing the key steps in establishing and releasing an updated process.

In 1985 the first GaAs MMIC products based on the fixed production process were released and a GaAs Foundry Service was made available to allow customer access to the Plessey process in order to design MMIC components to their own specification.

The GaAs Foundry concept can be summarised as:

(1) A relationship between customer and vendor bonded by a non disclosure agreement.

(2) Provide access, at low cost, to the vendors technology.

(3) No previous knowledge of the technology required.

(4) The vendor provides suitable design tools to enable the customer to transpose specification to chip design.

(5) The vendor undertakes wafer fabrication, test and chip delivery.

In addition any GaAs Foundry should provide:

(1) A stable, reproducible, high yielding process.

(2) Well characterised, toleranced process.

(3) The range of basic elements for circuit realisation together with complete microwave characterisation and models for all passive and active devices.

(4) The design manual with documented characterisation and layout rules and defining quality assurance inspection and control during processing and prior to release of wafers.

(5) Standard cell library.

(6) Design support.

The interface between Foundry customer and Foundry vendor is shown in Figure 2.3

The establishing of the fixed production process and the availability of the Gallium Arsenide Foundry, led to a variety of services being offered

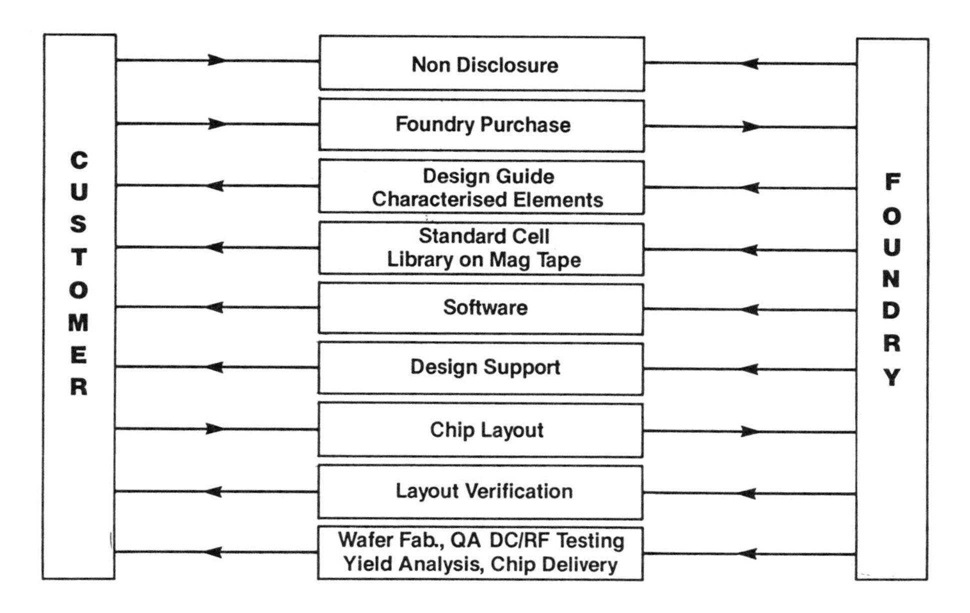

Figure 2.3 Customer/Foundry Interface

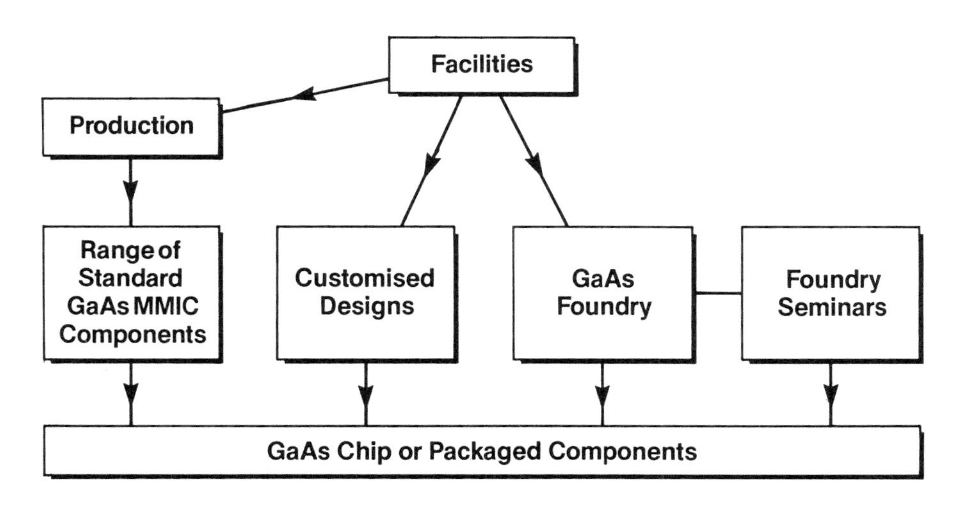

Figure 2.4 Vendor Capability

by vendors. In Figure 2.4, besides standard product and Foundry Service, Foundry seminars were introduced to allow familiarisation with the

Foundry processes and design rules and to enable attendees to design MMIC chips under supervision without prior knowledge of GaAs processing or MMIC design philosophies. Attendees would receive chips to their own designs for subsequent evaluation.

Following the release of initial products and Foundry Services, the market acceptance of GaAs MMIC's was slow, requiring extensive customer education on capability and sampling of products to allow familiarisation with the potential benefits in ease of use and suitability for incorporation into new systems. By 1987, there were clear signs of an improvement in the opportunities for the new MMIC technology, with more Foundry users and greater demands for MMIC products. However the market rate of growth was still considerably slower than predicted by market forecasts in 1983 which had given rise to the formation of the new GaAs MMIC based companies. As a result, some rationalisation was already taking place with a number of start-up companies ceasing trading. Figure 2.5 shows the historical progress from 1974 through 1988. By 1989 a number of companies were forming teaming agreements, the first signs of establishing a common technology for GaAs MMICs.

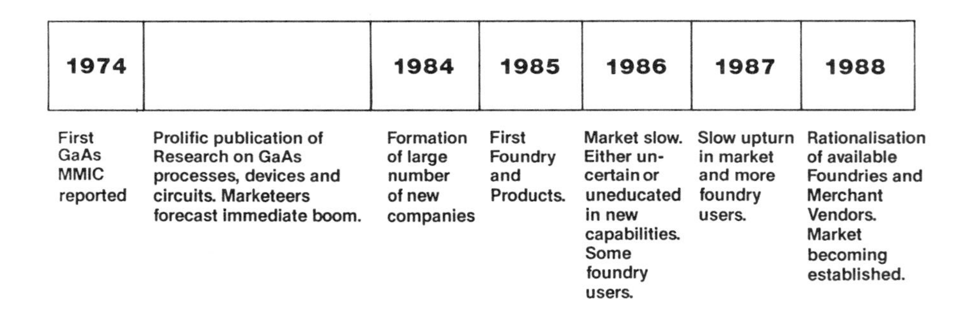

1974
1984
1985
1986
1987
1988
First GaAs MMIC reported
Prolific publication of Research on GaAs processes, devices and circuits. Marketeers forecast immediate boom.
Formation of large number of new companies
First Foundry and Products.
Market slow. Either un-certain or uneducated in new capabilities. Some foundry users.
Slow upturn in market and more foundry users.
Rationalisation of available Foundries and Merchant Vendors. Market becoming established.

Figure 2.5 Historical Profile of GaAs MMICs

During the period 1984 to 1989 GaAs MMIC technology had become established as a viable production technology and a small but increasing range of standard off-the-shelf MMIC components had become available. The majority of companies and products addressed the analogue market from DC to 20 GHz, with far greater uncertainty in applications for high speed or low power digital. The only exception was the development of a "Super computer" by Gigabit Logic for Cray. In general the microwave analogue products remained simple single function building blocks, primarily amplifiers or switches.

In order to increase market penetration, it was important that the GaAs MMIC offered a significant contribution to the system especially in providing a multifunction role. Such "First Generation" products were starting to be offered consisting of hybridised assemblies of single function chips. This concept offered great flexibility to the vendor, being able to configure the chip assembly to provide customer defined functions from the standard building block. The availability of ceramic based multileaded microwave packages increased the level of integration (Figure 2.6).

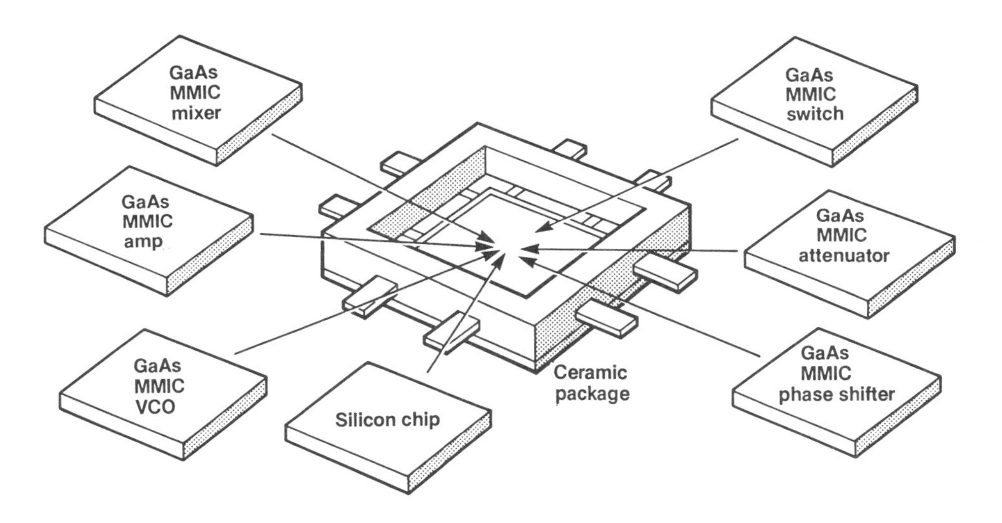

Figure 2.6 **Multifunction Performance from Single Function Components**

The concept of the application specific MMIC, ASMMIC, that is, a chip customised to provide the required multifunctions in a single chip was becoming viable as yields of MMICs improved and applications with volume requirements were identified. In 1986, Pacific Monolithic [1] had developed a complete 3.7–4.2 GHz receiver in a 1 mm^2 chip, and by 1989 companies such as Raytheon and Plessey were demonstrating a capability towards a single chip T/R module for phased array radar. The achievement of cost effective multifunctions on a single chip required further process enhancements, novel circuit concepts, and in particular improvements in computer aided design. The fundamental relationship of chip cost to chip size and the concept of high packing density placed increasing demands on theoretical analysis and prediction of new structures and coupling mechanisms. The concepts of LSI, typical of silicon digital circuitry whilst less demanding at the frequencies of operation and when driven between logic states, demanded accurate theoretical methods to predict effects in microwave linear circuits when wave propagation occurs. The work of R. Jansen [2] on full field analysis pioneered predictions of performance from structural detail and software programmes such as LINMIC were conceived as the core programme for a GaAs work station. Equally important to the multifunction IC was the ability to provide the specific required material profiles for functions such as switches, low noise and power amplifiers, diodes and varactors. The choice of implanted materials, with their superior uniformity characteristics, allowed the selective implantation of the required profiles.

The reliability of the new technologies had been demonstrated by Triquint [3] and Plessey with over 300,000 device hours at 150°C for MMIC switches.

Thus GaAs technology has taken major steps since the early sixties. In addition further significant advances can be predicted with the emergence of heterojunction based technologies into the millimetre wave frequency ranges and with the potential of fully integrated microwave, optic and digital functions on a single chip. These advances are detailed in Section 2.4 of this Chapter and in Chapters 5, 6, 11, 12 and 13.

In the USA, Government support for the six year MIMIC programme is evidence of the system demands for inclusion of the new technology and the provisions of adequate sources of supply of quality material, compatible technology amongst contractors, testability, packaging, CAD/CAM, and the achievement of target costs.

This Chapter briefly details the technologies required for the realisation of the GaAs MMIC, and how these have been implemented in a standard production process in Plessey Three Five. Future trends in applications, new device types and future processing requirements are also addressed.

2.2 Basic Processes

2.2.1 Introduction

The development of high speed process technologies for analogue and digital integrated circuits poses a number of wide ranging challenges for the process engineer. The diversity of frequencies from about 1 GHz to 100 GHz means that a standard technology cannot be adopted for the whole frequency spectrum as the circuit topology required at each end of this frequency range differ widely. At low frequencies, ~1 GHz, traditional distributed (microstrip) style matching elements consume too much surface area of GaAs and so RF techniques using lumped inductors, capacitors and resistors are used. It is only at frequencies in excess of 6–7 GHz that fully distributed circuits are small enough to enable utilisation of this type of impedance matching.

2.2.2 Components and technology Requirements

If we examine the component and technology requirements for circuits in the microwave and millimetre wave frequency band we can draw up a list such as that shown in Table 2.2. By far the most mature technology is based on the MESFET as the active device however, many of the processes and components described in the following sections are equally applicable to heterostructure FETs and bipolar transistors.

Table 2.2 : MMIC Components and Technologies

Low Frequency	High Frequency
MESFETs (1 micron)	MESFETs (submicron)
MIM capacitors ≤10 pf	Transmission lines
Interdigitated capacitors ≤0.5 pf	Series lines
Spiral Inductors (0.5–10 nH)	Short circuit stubs
Resistors, via hole technology	Open circuit stubs
Airbridges, multi level metal	Matching elements
	Resistors, via hole technology
	Airbridges, multi level metal

2.2.3 Circuit Building Blocks

(i) Active Device: The GaAs MESFET

The MESFET is at present the active component of the circuit which has, for the past 14 years, been the 'work horse' for all microwave integrated circuits [4]. Figure 2.7 shows the basic device structure where the width of the gate stripe (the gate length) determines the operating frequency.

One of the major factors affecting MESFET performance is the quality of the material in which the device is made. Presently three major approaches to material supply are being researched. These are:

(1) Vapour phase epitaxy [5].

(2) Molecular beam epitaxy [6]

(3) Ion implantation [7].

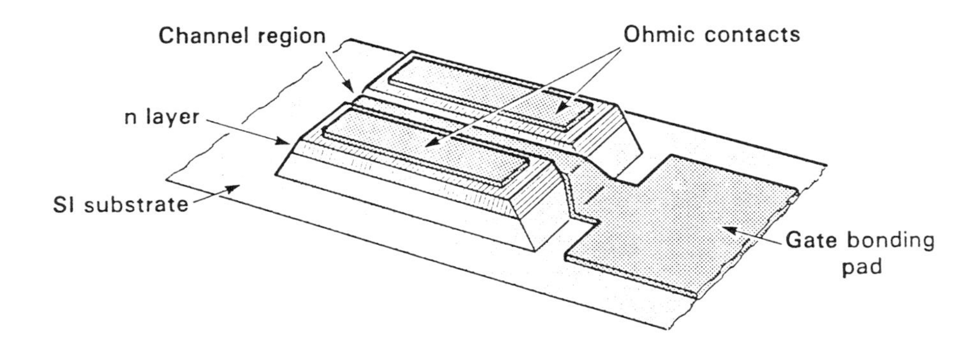

Figure 2.7 Schematic of a Typical MESFET

Table 2.3 : Comparison of Material Processes for MESFET Basic ICs

Material Technology	Profile Control	Electrical Quality	Cost per layer
Vapour phase epitaxy (VPE)	Least control of the three techniques	Best quality material in terms of electrical mobility – used for high quality discrete MESFETs	High-growth rate reasonable but scaling up difficult
Molecular beam epitaxy (MBE)	Monolayer accuracy and very versatile	Good – only slightly worse than VPE	Highest – very slow growth rate – not readily adapted for high throughput
Ion Implantation (II)	Few percent variation across whole wafer	Acceptable for IC fabrication	Low – true 'mass production' process

For MESFET based MMICs the preferred materials formation technique is ion implantation as it offers a technology capable of high throughput (low cost) albeit at a slightly inferior quality to epitaxial growth methods. Table 2.3 shows a comparison of material properties.

For heterojunction devices such as the high electron mobility transistor and the heterojunction bipolar transistor, ion implantation cannot be used and here VPE and MBE techniques are universally utilised.

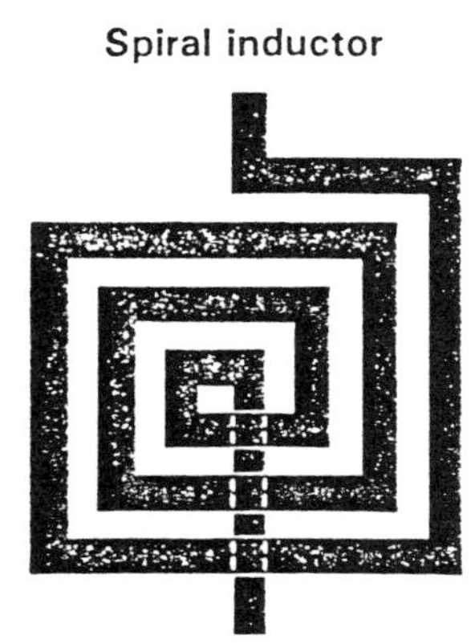

Figure 2.8 Inductor Configurations

(ii) Inductors

Inductors are used particularly in lower frequency circuits as tuning elements between transistors [8,9]. Figure 2.8 shows schematically a common configuration of inductors.

(iii) Capacitors

Capacitors are used within the MMIC for tuning elements, for interstage isolation and for bypass. Two types are currently in use:

(a) Interdigitated for low capacitance (less than 1 pF) [10].

(b) Parallel plate, metal insulator metal for 1–20 pF [9].

Table 2.4 : Capacitor Dielectrics

Material	Dielectric Constant	C/A pf/mm^2	Q	Deposition Technique
Silicon Nitride	6.5	4.80	Good	Plasma assisted CVD
Silicon Dioxide	4.0	340	Good to very good	Low pressure CVD
Polyimide	3–4.5	30	Good to very good	Spun and cured film

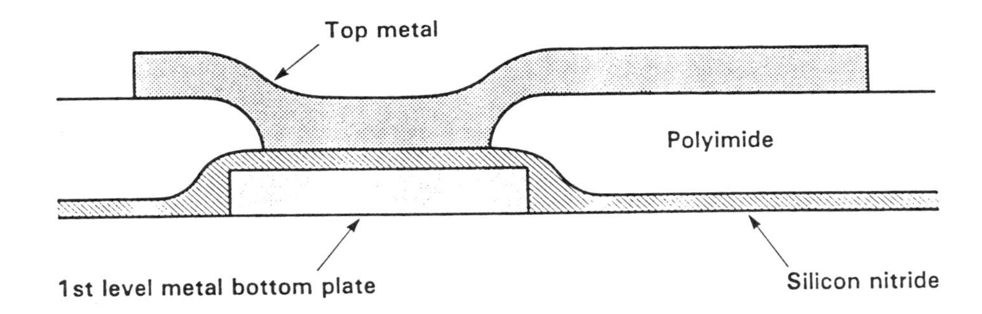

Figure 2.9 Section Through an MIM Capacitor

For the parallel plate capacitor shown in Figure 2.9 a number of different dielectrics are employed. The three major ones are given in Table 2.4. The dielectric thicknesses used have a major outcome on the final yield of the circuit. Generally there are many more capacitors in an MMIC than MESFETs and so the capacitor yield could dominate the yield of the circuit. Generally dielectric thicknesses of 1000–2000 Angstrom are chosen which are sufficient to minimise the pinholes in the film that cause d.c. short circuits while maintaining a reasonable capacitance per unit area.

(iv) Resistors

Resistors are used in the MMIC for determining the bias applied to the transistors, for feedback networks and for terminations in power combiners and Lange couplers [9]. Two families of resistor are commonly used, those made from the GaAs active layers, Figure 2.10, and those produced in deposited resistive films.

Generally, whichever material from Table 2.5 is chosen additional process steps are required in order to incorporate it into an MMIC process – consequently wherever possible GaAs resistors are used.

(v) Airbridges and Dielectric Multilevel Interconnections

In many sections of an MMIC a connection is required between non-adjacent metallised areas. This can be accomplished using either airbridges or dielectric interconnections. As the name implies an airbridge is a bridge of metal running above the surface of the circuit between metal

Table 2.5 : Some resistor materials used in MMICs

Material	TCR ppm/°C	Ω/sq	Manufacturing Technique
TaN	−100	90 typically	Reactive sputtering
CrSi0 (cermet)	−300 to +100	50–500	Sputtering from composite target
Bulk GaAs	+3200	300 for 10^{17} material	Epitaxial or ion implanted
NiCr	200	90 typically	Sputtering

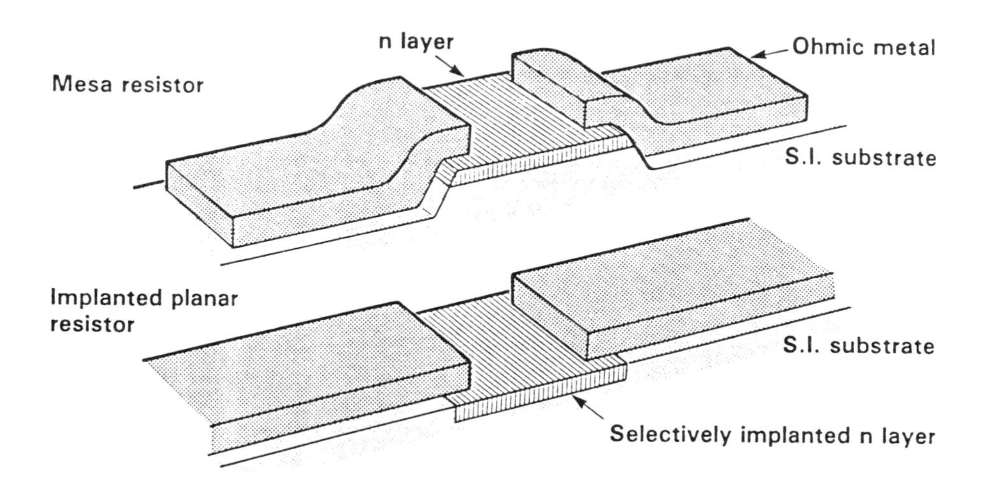

Figure 2.10 Resistor Technologies

areas to be interconnected. They are formed by a photoresist and plating process. An alternative approach is to use a dielectric "cross over" in which a dielectric layer separates the upper and lower metal layers.

(vi) Via Hole Technology

As circuits become more complex it is desirable to remove the constraint of r.f. grounding the circuit around its periphery and processes are being

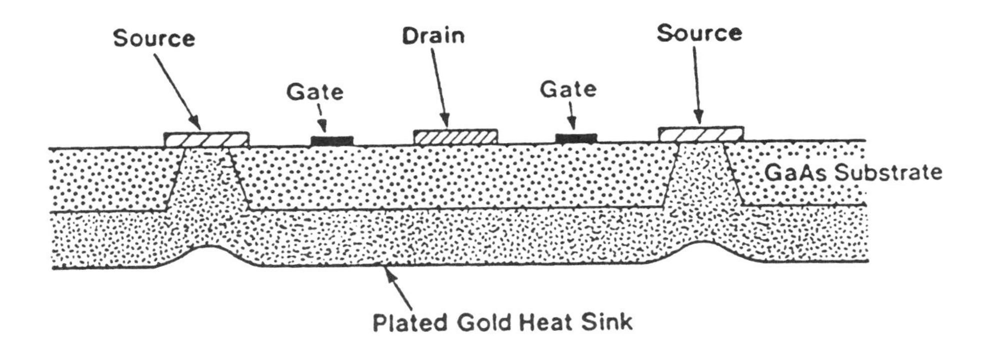

Figure 2.11 Via Hole Process for Individual Component Grounding

developed to directly ground individual components. One way that this can be achieved is by a 'via hole' process in which holes are made from the back of the wafer to contact grounding pads on the front side [11]. This is illustrated in Figure 2.11. Once the holes have been made they can be filled with the appropriate metallisation which makes a direct contact to the front side metal areas requiring to be grounded. The process of mounting the chip onto a suitable substrate automatically grounds the components on the front side of the wafer.

2.2.4 Component and Technology Integration

Defining the components that were described in the preceding sections of this chapter is not sufficient for the realisation of a full integrated circuit process. Its development is an iterative exercise where the requirements of the technologies to produce the components are combined with those needed to interconnect them. Generally many compromises have to be made before a manufacturable process can be finally realised involving not only aspects of the technology, but also satisfying the requirements of circuit design, assembly and packaging. The interaction necessary between circuit fabricators, circuit designers and users cannot be overstressed and all successful processes are the result of constant consultation between all interested parties.

2.2.5 IC Fabrication Processes

Once the active n–layer has been formed, the whole integrated circuit can be processed using three basic fabrication techniques. Various sequences of these three techniques are needed to allow complete circuits to be fabricated. These are:

(a) Lithography

(b) Deposition – metal and dielectric

(b) Etching – metal, dielectric and semiconductor.

(i) Lithography

Lithography forms a vital part of any process technology for all process stages are preceded by a lithography step that defines the pattern for etching or for metal deposition. Basically the lithography process involves defining a pattern in a thin photo sensitive film (photoresist) by a process of optical exposure through a photographic mask and subsequent differential removal by immersion in a resist developer solution. Two techniques are used to transfer the image from the mask into the film of resist. These are illustrated in Figure 2.12. A comparison of these two approaches is given in Table 2.6.

Table 2.6 : Comparison of contact lithography and DSW

Photolith method	Advantages	Disadvantages
Contact lithography	Available at 0.5 μm resolution	Mask procurement difficult Mask wear due to contact with resist Require very flat masks and wafers
Direct step on wafer	No mask wear 5:1 or 10:1 projection system so mask procurement less difficult	Resolution guaranteed only to 1 micron Process conditions need to be very carefully controlled

If higher resolution is required the presently accepted process is to expose the resist with a beam of electrons. This technique allows line dimensions down to 0.1 microns to be achieved and is the most practical way of realising the gate metallisation for millimetre wave MESFETs. Table 2.7 shows the factors affecting the choice of electron beam lithography and from this it can be seen that the main disadvantages are cost of the equipment and the time taken to expose the patterns.

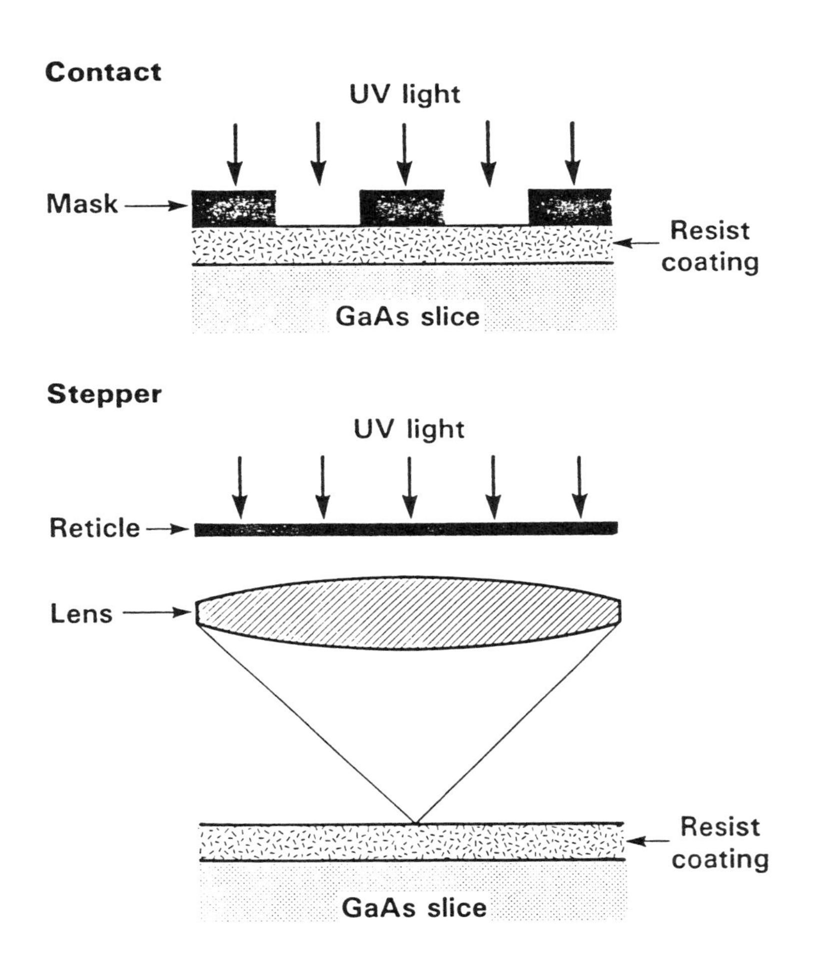

Figure 2.12 Contact Lithography and Direct Substrate Write Configurations

Table 2.7 : Electron Beam Lithography

Resolution to at least 0.2 microns
No masks required
Good yield over a large area
Throughput time limited as each pattern has to be written separately (can mix E–beam and photolighography to partially surmount this problem).
High Machine cost

(ii) Deposition Processes

The fabrication of an MMIC requires the deposition of both metal and dielectric films and this is achieved using some of the different techniques given in Table 2.8.

Table 2.8 : Comparison of Deposition Techniques

	Main Advantages	Main Disadvantages
Filament evaporation	Easy process Relatively low cost equipment	Poor for refractory metals Wafers can reach high temperatures
Electron beam evaporation	Good for refractory metals and lift off processes	Expensive equipment
Plasma assisted chemical vapour deposition	Low temperature process	Produces amorphous films with high etch rate
Sputtering	Good for thick metal, step coverage and alloys	Cannot be easily used for lift off processes

(iii) Etching Processes

Once the metal and dielectric films have been deposited they need to be selectively removed to form the various patterns that build up the integrated circuit. This is carried out by a combination of lithography and etching. There are many etching processes available to the GaAs fabrication technologist. The major ones are listed below:

(a) Ion beam milling [12].

(b) Reactive ion etching and plasma etching [13].

(c) Wet chemical etching.

Of these processes only (c) is not a 'dry' process and the move away from wet chemical etching is becoming an important aspect of GaAs technology. In many instances the dry process technology is being adopted because of its controllability and cost effectiveness. In others it is proving to be the only practical method of removing many metals and dielectrics.

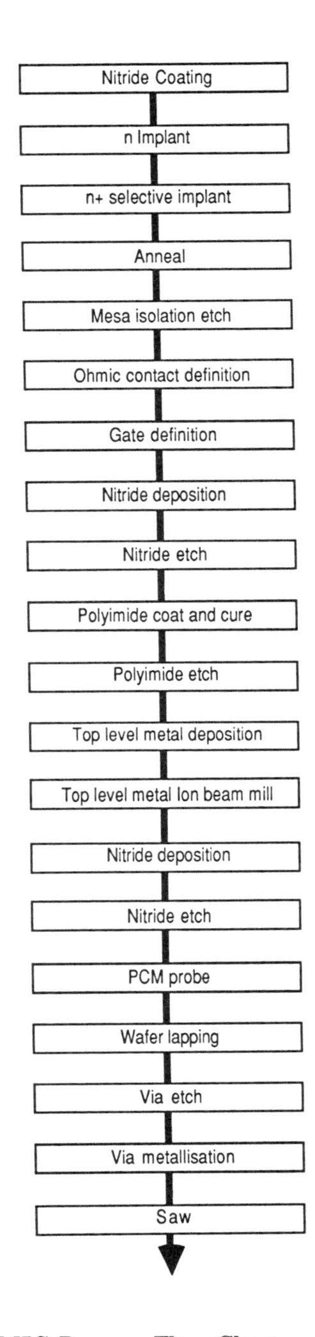

Figure 2.13 GaAs MMIC Process Flow Chart

2.3 Commercial Process

2.3.1 High Yield GaAs MMIC Process

In the preceding section many of the basic semiconductor processes relating to materials, photolithography, plasma processing, deposition and etching were described. In this Section it is shown how these basic processes are combined together into a complete IC chip fabrication process. This is by no means the only possible route but is the process developed by Plessey which is now available as the GaAs MMIC Foundry Process (F20) from Plessey 3–5 Group Ltd.

The entire process is shown in the flow chart in Figure 2.13. The starting material is a cut and polished semi-insulating 100 GaAs substrate of which there are many suppliers world wide. Quality control of the incoming substrates is vital to the overall quality and control of the finished MMIC and this is discussed further in Section 2.3.2.

The first step is to coat the substrates, front and back, with silicon nitride deposited by plasma enhanced chemical vapour deposition (PECVD). The purpose of this coating is to prevent out dissociation of the Arsenic from the substrate surface during the high temperature (850°C) anneal required to activate the implant. Quality of this silicon nitride coating is paramount in controlling the implant characteristics and surface quality following the activation anneal.

The first implantation stage is a 'blanket' implant of Si+29 to form the n–type active layer for the channel of the FET. This implanted layer is also used for mesa resistors. A second high dose Si+29 implant is used to form a n+ layer which reduces contact resistance and parasitic resistances and hence improves MESFET performance. This n+ region is commonly formed by selective implantation as shown in Figure 2.14, where photoresist is used to mask the areas where the n+ layer is not required. The photoresist mask is then stripped and the implants activated either by a high temperature furnace anneal or by 'flash' annealing. More complex, multifunction circuits can be realised by selective implantation of different device implants optimised for alternative device performance, e.g. switch, low noise, power. The appropriate area of the MMIC chip is implanted by repetition of the photoresist mask and implantation process followed by a single, implant activation anneal. Following the anneal stage the silicon nitride implant cap is removed either by plasma etching or by a wet etch.

Isolation of the active devices MESFET's, diodes and mesa resistors from one another is achieved by masking the required device areas with photoresist and etching the exposed GaAs in an ammonia peroxide etch to remove the implanted layer. A mesa to mesa resistance check confirms that adequate device to device isolation has been achieved.

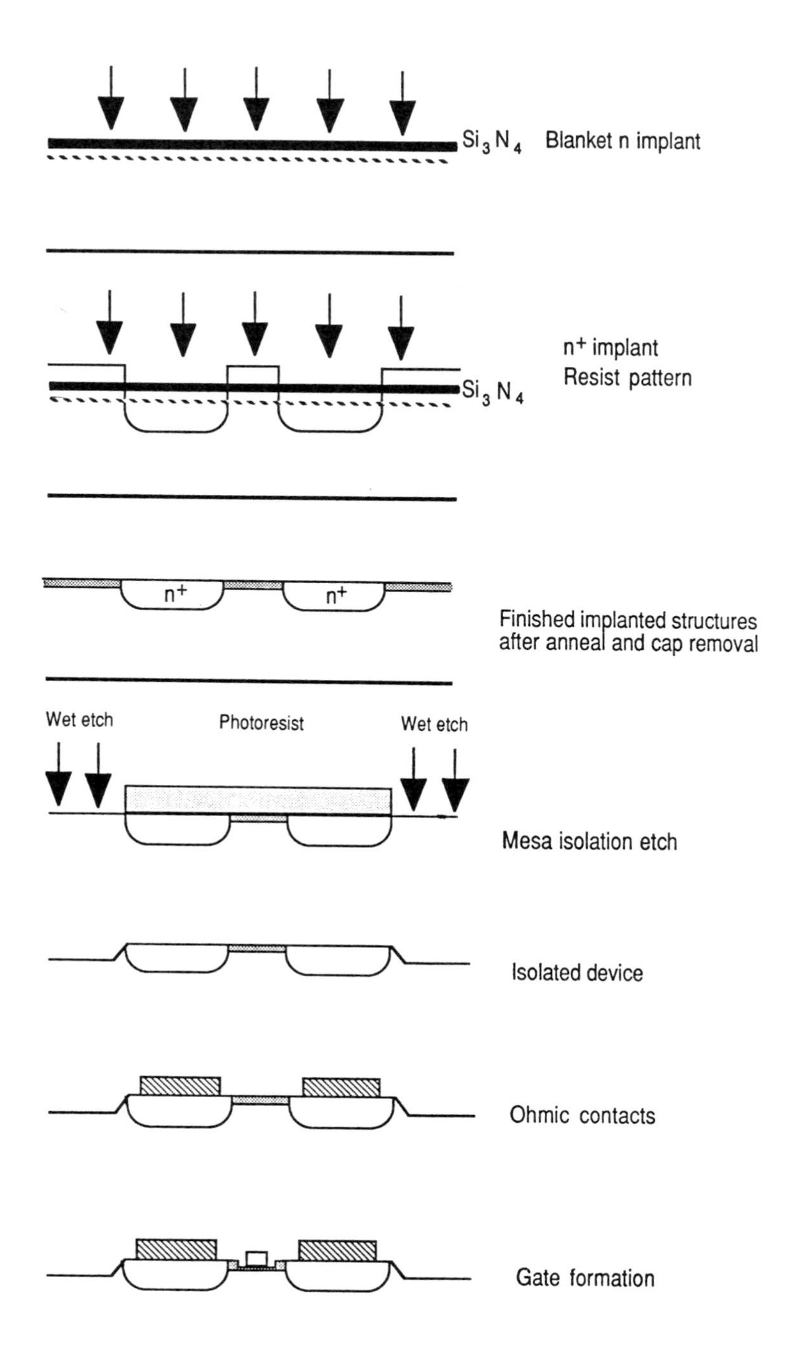

Figure 2.14 Ion Implanted Depletion MESFET Fabrication

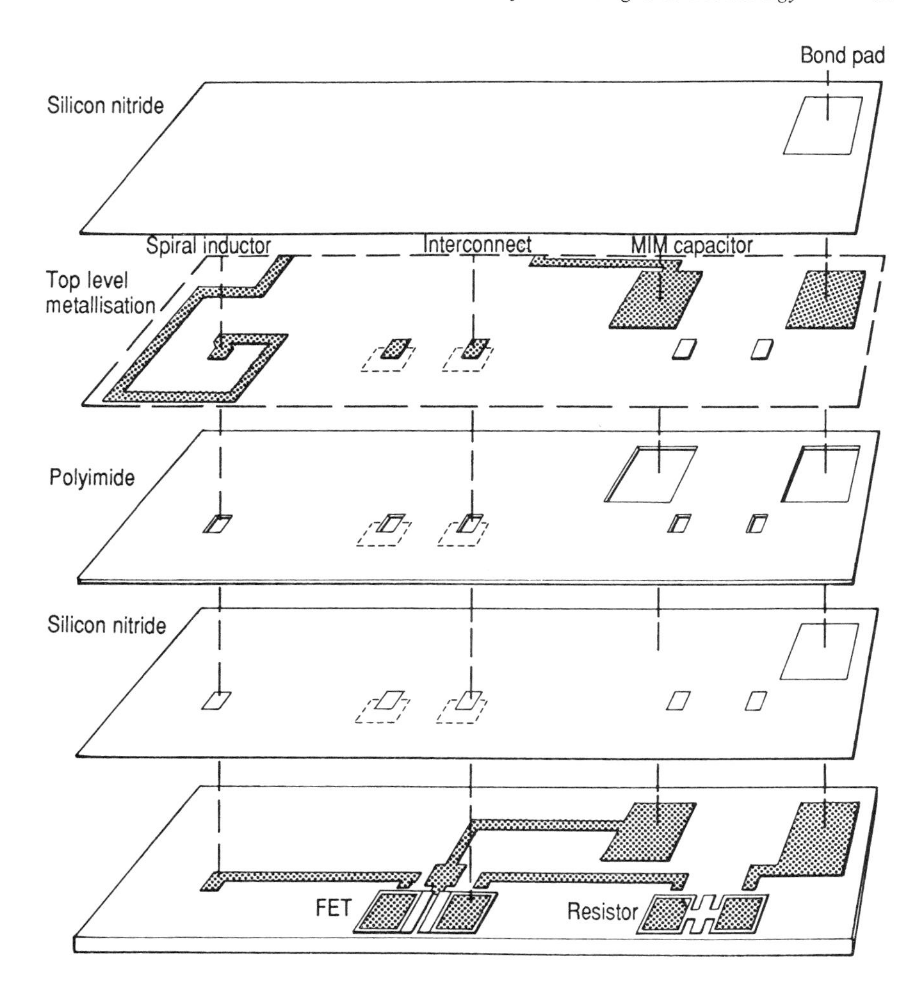

Figure 2.15 Fabrication of Passive Elements and Interconnects

Ohmic contacts must now be formed for the MESFET's and the resistors, these contacts are an alloyed multi–based metallisation, the most common systems being Ni–Ge–Au or In–Ge–Au. The wafer is coated with photoresist and patterned to expose the GaAs in areas where the ohmic contacts are required. The metallisation is then evaporated onto the wafer coating both the photoresist and the exposed GaAs. The photoresist and unwanted metallisation areas are 'floated off' in a suitable solvent. The final stage in the ohmic contact formation is the alloying process which can be achieved either in a tube furnace or by flash annealing. In both cases the time temperature profile is important to achieve a low contact resistance.

The most critical and demanding stage in the MMIC process is the gate formation. The first step is to coat the wafer with photoresist and then expose and develop the required gate pattern to leave exposed GaAs where the metallisation is required. The performance of the MESFET and hence MMIC performance is largely dominated by its gate length; in the F20 Foundry process this is 0.5 μm which is very much at the limit of optical photolithography technology. Control of the photoresist image is critical and is achieved through the use of a scanning electron microscope to inspect sample sites across the wafer and provide accurate linewidth measurements in the sub-micron region. The gate channel recess is etched using the photoresist pattern to give the required FET saturated drain current.

Finally the gate metallisation is deposited by evaporation and 'floated off' to leave the required metal pattern. In the F20 Foundry process the gate metallisation is also used as a first level interconnect, as the lower electrode for metal–insulator–metal (MIM) capacitors, for interdigital capacitors and as the base metallisation for bonding pads.

The first dielectric layer of silicon nitride is deposited using PECVD, passivating the gate channel and ensuring reliable, stable FET characteristics. Windows are etched through the silicon nitride where interconnects to the top metallisation are required, see Figure 2.15. This is achieved by plasma etching using photoresist as the masking material. This silicon nitride layer also acts as the dielectric for MIM capacitors.

Polyimide is spun onto the wafer surface and hotplate cured to complete the imidisation process. The purpose of the polyimide layer is to provide an insulator for metal crossovers. The polyimide also acts as the dielectric for an alternative type of MIM capacitor with a lower capacitance per unit area than silicon nitride. This crossover technique is an alternative to the more common approach of using airbridges. Windows are etched in the polyimide using a photoresist mask and oxygen plasma etching to enable interconnection to be made between metallisation levels.

The top level metallisation, titanium–platinum–gold is rf sputtered onto the wafer surface. RF sputtering is used for its excellent step coverage characteristics which are needed to provide reliable, low resistance interconnections to the lower metallisation levels. The top level metallisation provides a second level of interconnection, forms the upper electrode of MIM capacitors, is used for transmission lines and spiral inductors and provides the final surface for wire bonding. The sputtered metal is patterned by ion beam milling again using photoresist as the masking material.

The second level of silicon nitride which acts as an overall passivation is deposited by PECVD and windows are etched by oxygen plasma etching to expose the gold bonding surfaces. The MMIC front face processing is now complete and the 'process control monitor' cells are autoprobed to

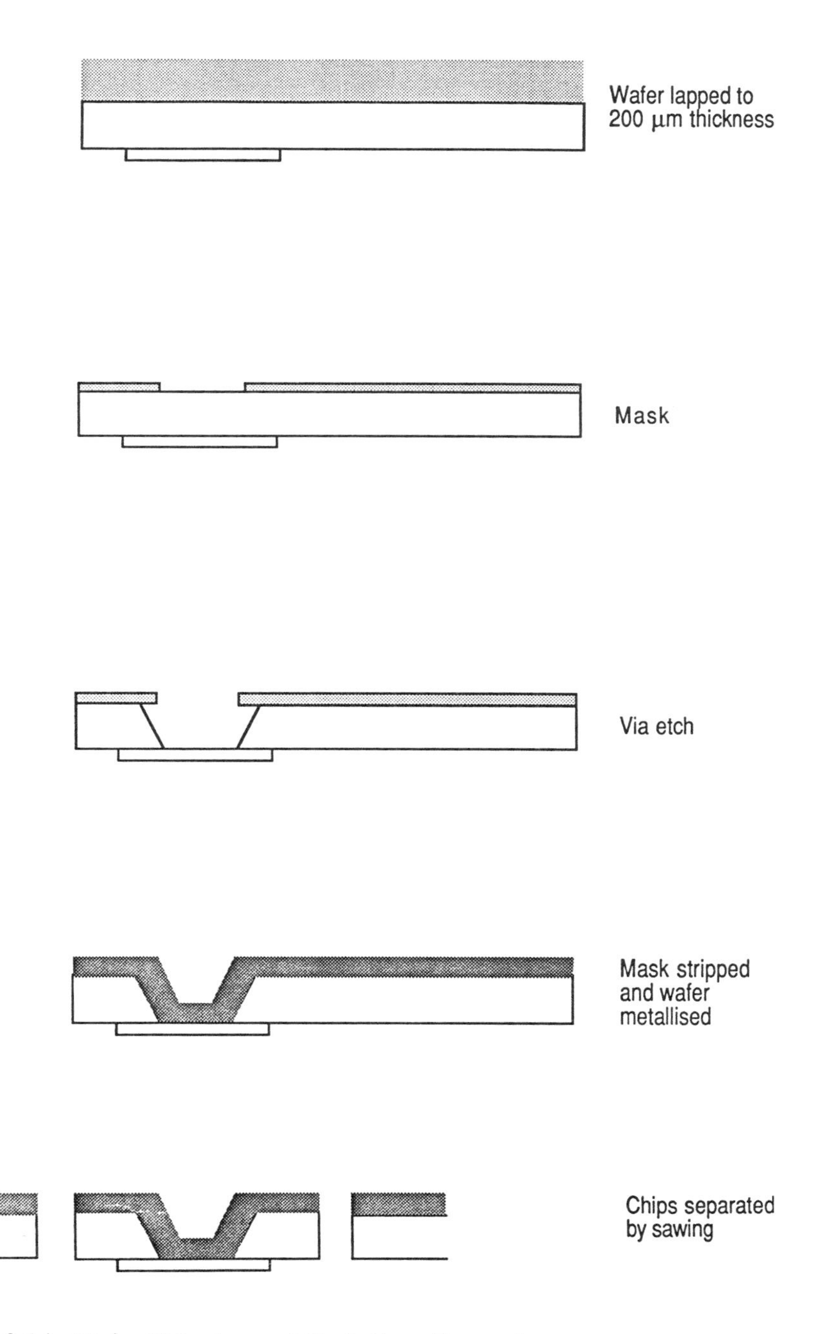

Figure 2.16 Wafer Thinning and Back Face Processing

provide the process control information necessary to release the wafer and maintain the process integrity The PCM cell and other methods of process control are described in Section 2.3.2

The final stages of the MMIC wafer process all involve the back face of the wafer. The wafer thickness for all the front face process stages is 500 μm, at this thickness the wafers are relatively robust. However, from microwave considerations the final MMIC thickness must be considerably less than this, in the case of the F20 Foundry process the final thickness is 200 μm. Many other Foundries use a final thickness of 100 μm which because of the fragile nature of GaAs makes handling extremely difficult. Standard mechanical and chemical lapping techniques are used to reduce the wafer thickness from 500 μm to 200 μm.

The use of 'through GaAs vias' reduces ground bond wire inductance and simplifies layout considerations since ground pads do not have to be adjacent to the chip periphery. The vias are etched from the back face of the wafer using a photoresist mask. The front to back alignment is achieved through the use of an 'infra red' mask aligner, GaAs being transparent in the infra red region.

Two methods of etching vias in GaAs are commonly used, wet chemical etching or plasma reactive ion etching. In either case etch anisotropy is critical to ensure a tapered profile suitable for metallisation. The back metallisation is rf sputter deposited to ensure coverage down to the bottom of the via. See Figure 2.16. As a final step the dies are separated by sawing. An example of a completely processed MMIC chip using selective ion implantation and through GaAs vias shown in Plate I. This is a multifunction chip forming part of a phased array radar chip set and contains a six bit phase shifter complete with interstage amplifiers and transmit receive routing switches. Different implants are used for the amplifier and switch functions.

2.3.2 Process Control

Process control is the key to a reliable, repeatable, characterised, high yielding Foundry Process. Control of the process is exercised through the use of documentation, in-process inspection and measurement and data analysis.

(i) Documentation

Properly controlled documentation is essential to a stable MMIC process, this area can be further sub-divided into the following main categories.

(a) Material Procurement

Material procurement specifications and inspection schedules are used to control the quality of incoming materials. The primary characteristics detailed in the specifications are shown in Table 2.9 for the three key items in the Foundry process, i.e. GaAs substrates, photomasks and process chemicals.

Table 2.9 : Primary Characteristics of Incoming Material

Procured Item	Primary Characteristic
Gallium arsenide substrate	Mobility and resistivity, before and after anneal After test implant: Sheet resistance Hall coefficient Mobility Sheet carrier concentration Total thickness variation Dislocation density Surface defects Contamination
Photomasks	Runout Skew Die to die registration Edge quality Dimensions Defect density
Chemicals, gases, pure metals	Purity

(b) Process Specification

The key document for wafer fabrication is the process specification. All the process steps required to fabricate the wafer are detailed with the process parameters, specified in terms of minimum and maximum limits (where possible control limits are set within these to minimise the spread of results and identify potential out of control situations). Each process step has corresponding operator instructions that describe in further detail the manufacturing method, equipment settings and inspection requirements. The documents help to ensure that the manufacturing method and thus the delivered product are always consistent.

Table 2.10 : Wafer Tests

Stage	Parameter	OL/TP
Implantation	Resistivity	OL
Mesa	Isolation	OL
Ohmic	Isolation	OL
	RSD	OL
	Metal thickness	TP
	S–D gap	OL
	Gate length (resist)	OL
Gate	Metal thicknes	TP
	Channel current	OL
	FE – IDSS, Vp, Vb, gm	OL
	Metal gate length	OL
Nitride 1	Thickness	TP
	Ri	TP
Polymide	Thickness	TP
M3	Thickness	TP
	Resistivity	TP
Nitride 2	Thickness	TP
	Ri	TP
Autoprobe	PCM parameters	OL
Thinning	thickness	OL
Via	–	–
Back metal	thickness	TP
Die separation	WBST	OL
Final inspection	Yield	OL

(c) Process Traveller

This document, or batch card, provides a permanent record of the process conditions and measurements. It is unique to any batch processed and ensures complete traceability.

(ii) In Process Inspection Monitoring

The system established for production monitoring and control is based on the acquisition of data both during and subsequent to manufacture. Analysis of this data is the key element in maintaining a stable process and assuring the quality of the product. The various tests applied to each wafer are listed in Table 2.10. The result obtained for each test is recorded on the batch card. As can be seen from the table the measurements are performed either on the actual wafer (on line OL) or on additional test pieces (TP) which are processed at the same time as the wafers. In addition visual inspection occurs within and after every major stage as part of the operators written instructions. Consistency of the inspection is ensured by providing extensive training and fully detailing the inspection criteria.

An essential part of the in–process measurement is the 'process control monitor' PCM cell which is a standard structure included in any mask set to be processed in the Foundry. The basic structure is shown in Figure 2.17 together with a brief description of each of the components included and the parameters measured in Figure 2.18. The PCM cell can be incorporated onto the mask set in two ways, either by including one PCM cell in each step and repeat array for example as in Figure 2.19 or by building several PCM's into a special array which is then inserted at specific sites in the completed array on the mask, Figure 2.20. Now that rf on wafer testing has reached a mature state it is becoming common practice to include rf structures, e.g. FETs within the PCM. Microwave scattering parameters are measured and an equivalent circuit model derived which provides more detailed process information and control than by measurement of FET dc characterisation.

(iii) Data Analysis

The data obtained from the in–process measurements is analysed using standard statistical process control techniques to ensure control of the process is maintained.

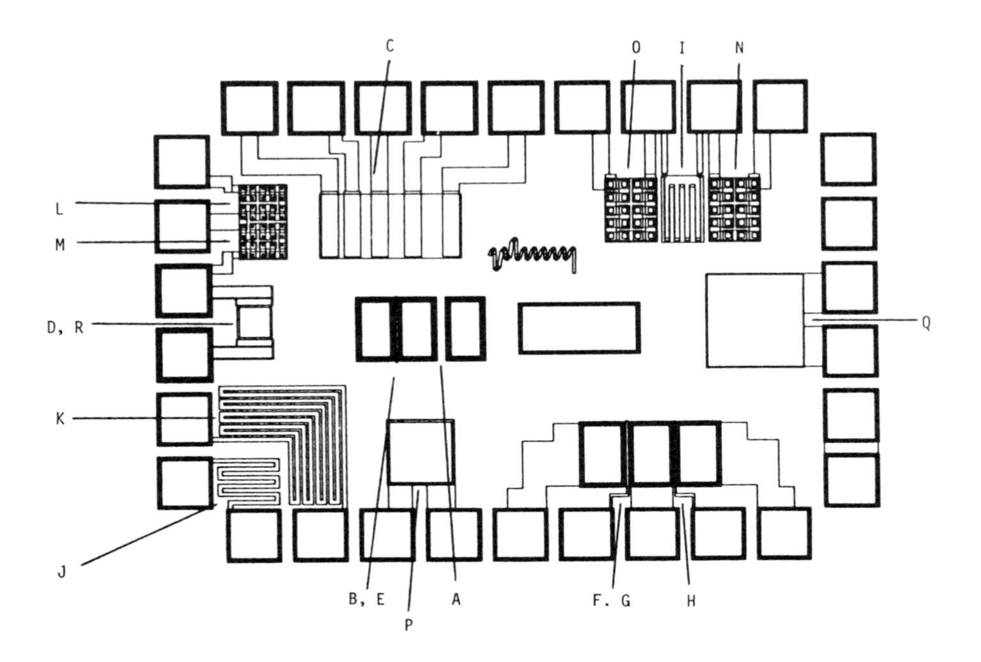

Figure 2.17 Layout of typical PCM cell

There are many forms of control chart which may be utilised. Those routine in the control of the F20 Foundry process are illustrated in the following examples.

(a) Run Charts

In this type of chart the mean of the data obtained for a particular parameter is plotted against the batch number. Two examples, for the MESFET saturated drain current (IDSS) and silicon nitride thickness are shown in Figures 2.21 and 2.22 respectively. The IDSS data is obtained from the PCM autoprobe and each point represents the mean of all the data values obtained for a given wafer. The silicon nitride thickness data is obtained from the average of 5 measurements taken from a test wafer processed in the same deposition run as the GaAs wafer batch. Run charts are used to highlight out of specification data points and indicate obvious process trends requiring corrective action.

	In-Process Test Components	
A	Mesa isolation	20μm x 150μm mesa-mesa gap.
B	Ohmic contact	7μm x 150μm ohmic contact separation, probe to measure Rsd and Isat.
C	Transmission Line	150μm wide, gaps of 12, 24, 36 and 48μm for measurement of GaAs material resistivity and contact resistance.
D	Mesa resistor	330Ω nominal resistor.
E	Gate etch	Provides "windows" in gate resist. Gate dimension .7μm x 150μm.
F	FET	150μm x .7μm for measurement of Idss, gm, Vp, Vb.
	Process Characterisation	
G	FET	150μm x .7μm for measurement of Idss, gm @ LNB, Vp.
H	FET	150μm x 5μm for Rsd and drift mobility measurements.

	Conductivity test strips	Metal	Length	Width
I		M2	1	7
J		M2	1.35	10
K		M3	3.8	10

L	Interconnection resistances	M1-M3	4μm via
M		M2-M3	4μm via
N		M1-M3	12μm via
O		M2-M3	12μm via
P	Nitride capacitor	Nominal 10pF	
Q	Polyimide capacitor	Nominal 1pF	
R	Mesa resistor	Nominal 330Ω	

Figure 2.18 Key to Figure 2.17

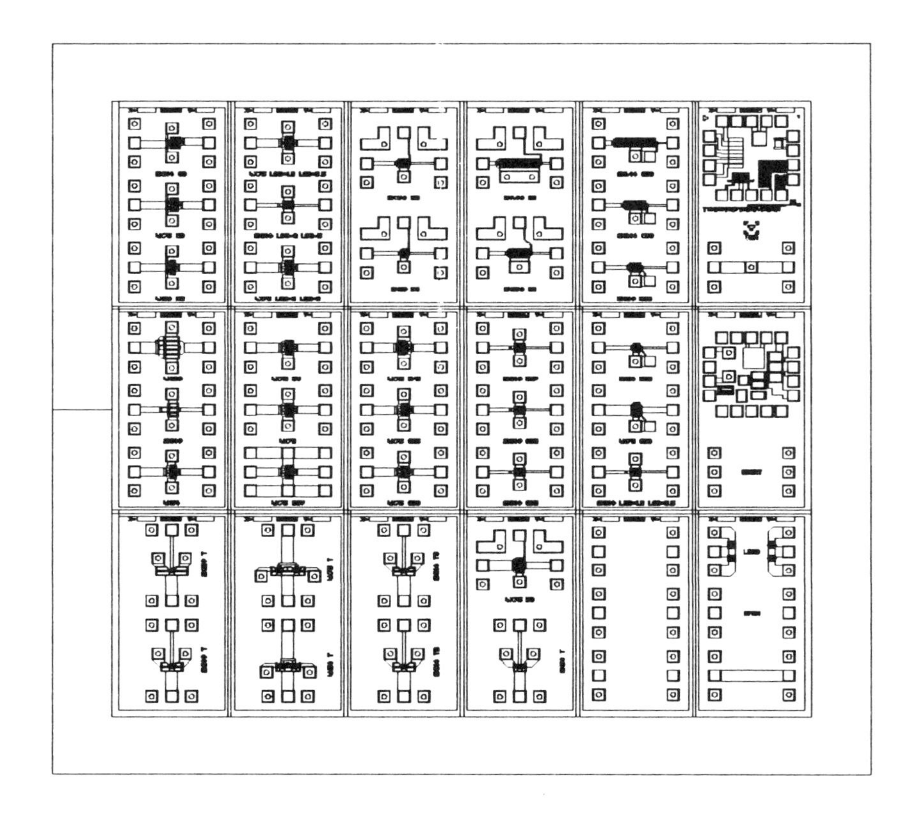

Figure 2.19 PCM Cell in Array

In the IDSS example shown there is an obvious variation from wafer to wafer and several wafers are seen to be out of specification. This is a reflection of the difficulty associated with the gate recess etch which is a manual operation involving an 'etch and measure' iterative process. In the silicon nitride thickness example control of the plasma deposition conditions are much tighter resulting in better control of the nitride film thickness.

(b) Cumulative Sum (CUSUM) Charts

This technique involves plotting the cumulative sum of the deviations from a specified target value against the sample number. An example of this is shown in Figure 2.21 for the saturated drain current of the PCM MESFET. The plot can be divided into three distinct regions, a positive slope between batches 0 – 75, a negative slope between batches 75 and 190

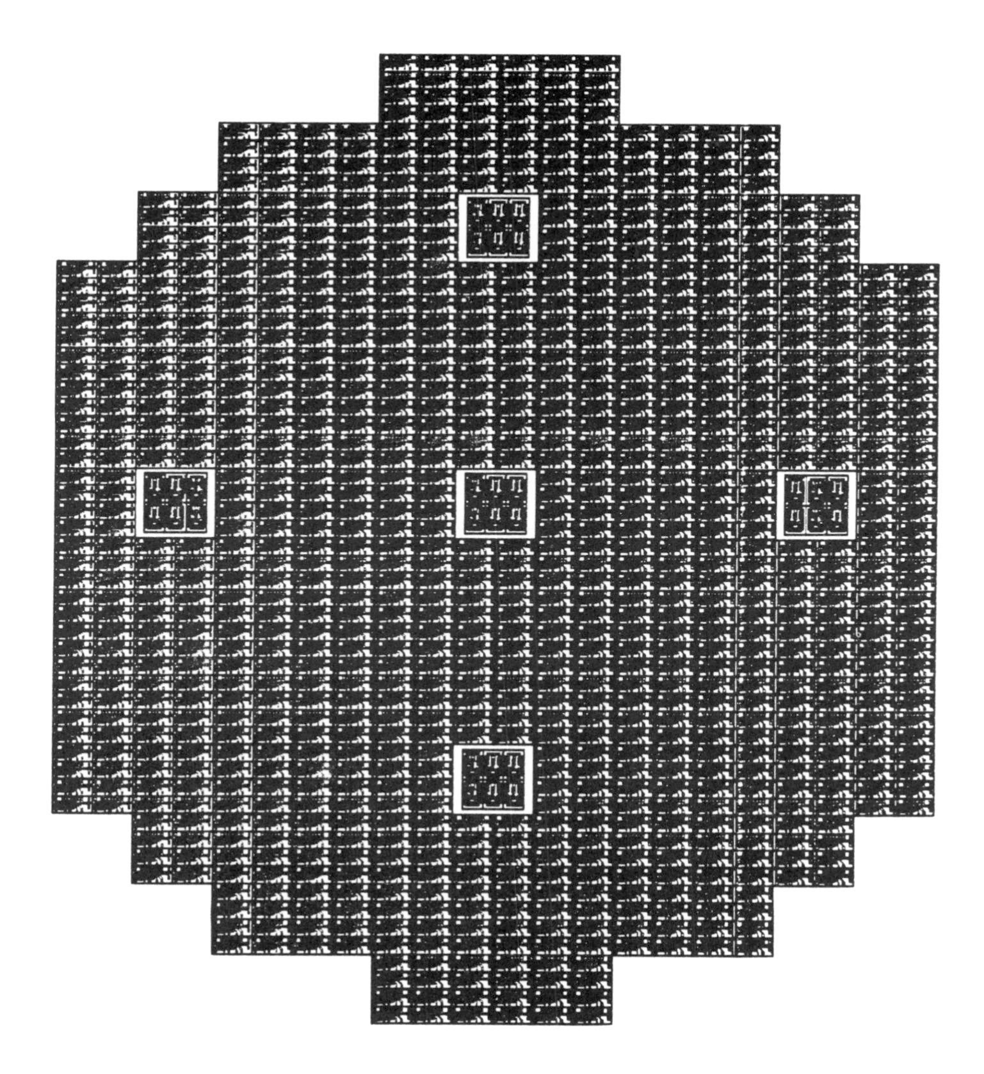

Figure 2.20 Sited PCM Cells on Wafer

and finally another positive slope from 190 to 350. The slope of the CUSUM is related to the deviation of IDSS from the target value, in this instance 22.5 mA. The positive slope means that IDSS is greater than the target value and vice versa. The slope of the line relates to the deviation from target. The CUSUM charts shows that initially the mean IDSS value was about 24.2 mA, at around batch 75 there was a distinct drop to about 21.6 mA which was maintained up to batch 190 when it increased to

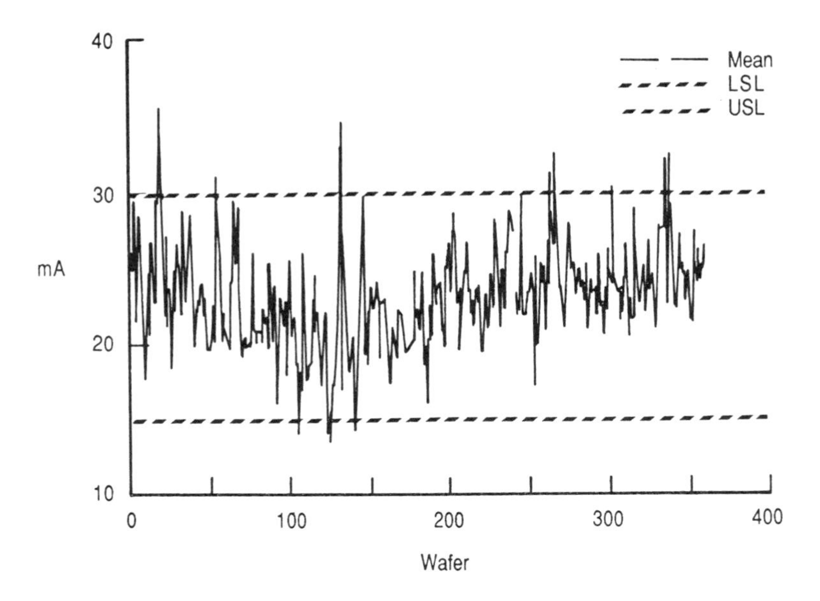

(a) **Trend Chart**

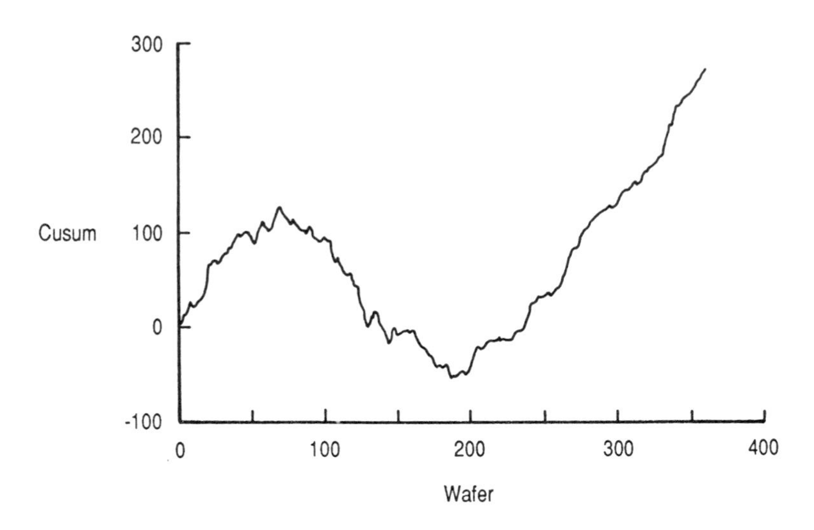

(b) **CUSUM Chart**

Figure 2.21 Run and CUSUM Charts for IDSS

about 23.9 mA. These changes correspond to small changes in the recess etch current target window in an attempt to optimise the MESFET IDSS. These changes in IDSS are embodied in the trend chart of Figure 2.21 but are more easily discernible from the CUSUM chart.

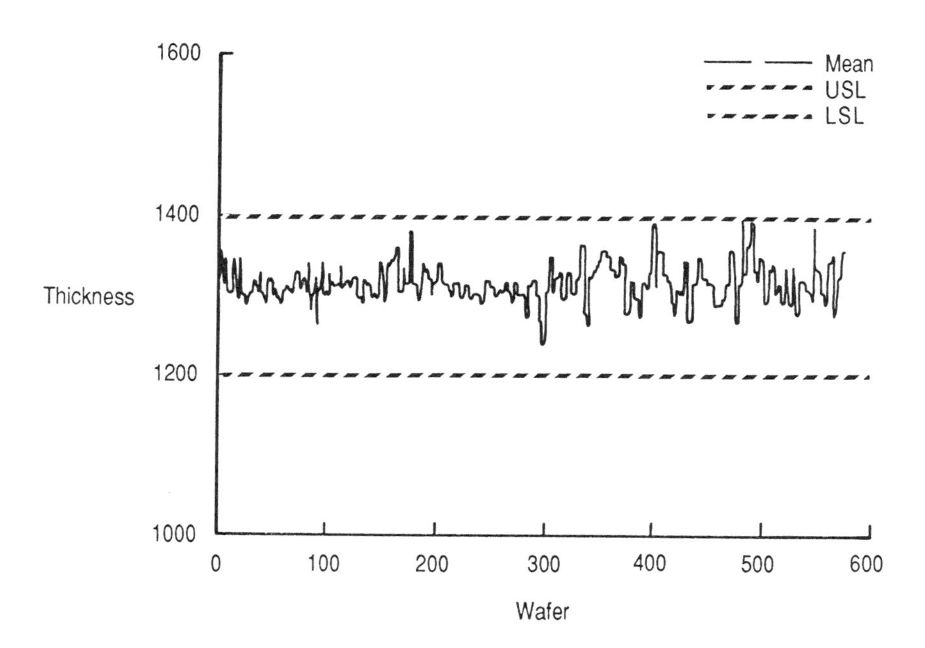

Figure 2.22 Run Chart for Nitride Thickness

(c) Process Capability Chart

Before this type of chart, which predicts the number of out of specification points, can be used it is necessary to first determine that the process is 'in control'. This can be achieved through the use of run charts and CUSUM charts. These are shown in Figures 2.23 and 2.24 for the silicon nitride capacitor combined in the PCM cell. These do not show any underlying trends and therefore indicate that the process is 'in control'. The data can then be used to construct a process capability chart as shown in Figure 2.25. The data is shown in the form of a histogram with an equivalent gaussian distribution super–imposed. The computed mean and standard deviation can then be used to predict the expected number

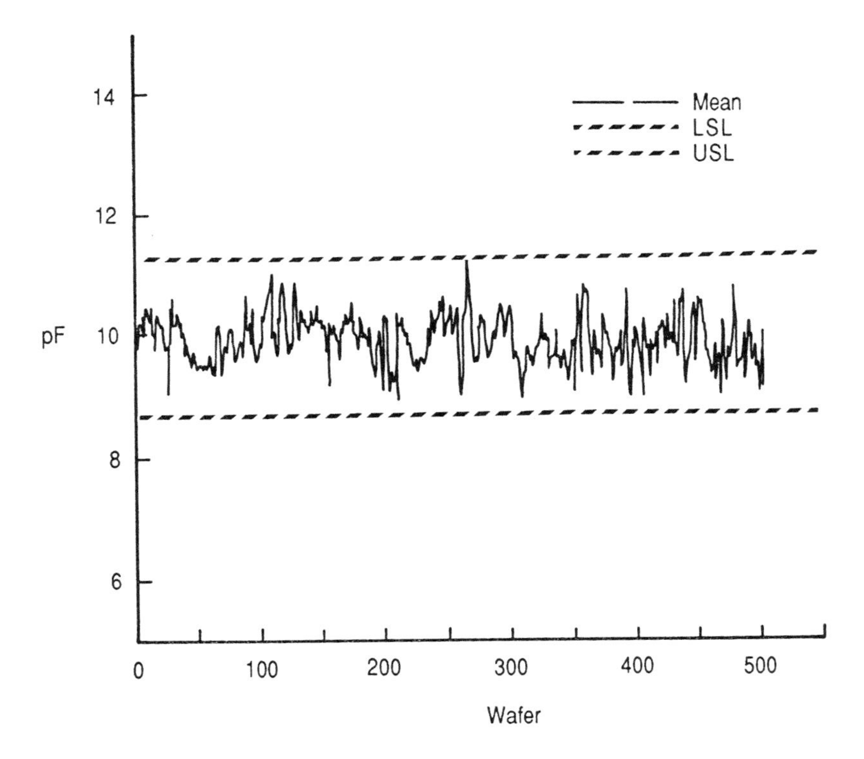

Figure 2.23 Run Chart for Silicon Nitride MIM Capacitor

of out of specification points, in this case 0.17% below the lower spec limit and 0.03% above the upper specification limit.

All the techniques described above help to ensure that the process is maintained in an 'in control' situation. Developing trends can be quickly identified and appropriate corrective action taken to prevent costly scrap material from being produced.

2.3.3 Capability Approval

Capability Approval of a technology such as GaAs MMIC Foundry enables the supplier to release parts to the appropriate standard, in this instance BS 9450. The BS standard is the UK equivalent of the US Mil Std and European CECC 90000 and ensures that the parts released meet a specified quality standard.

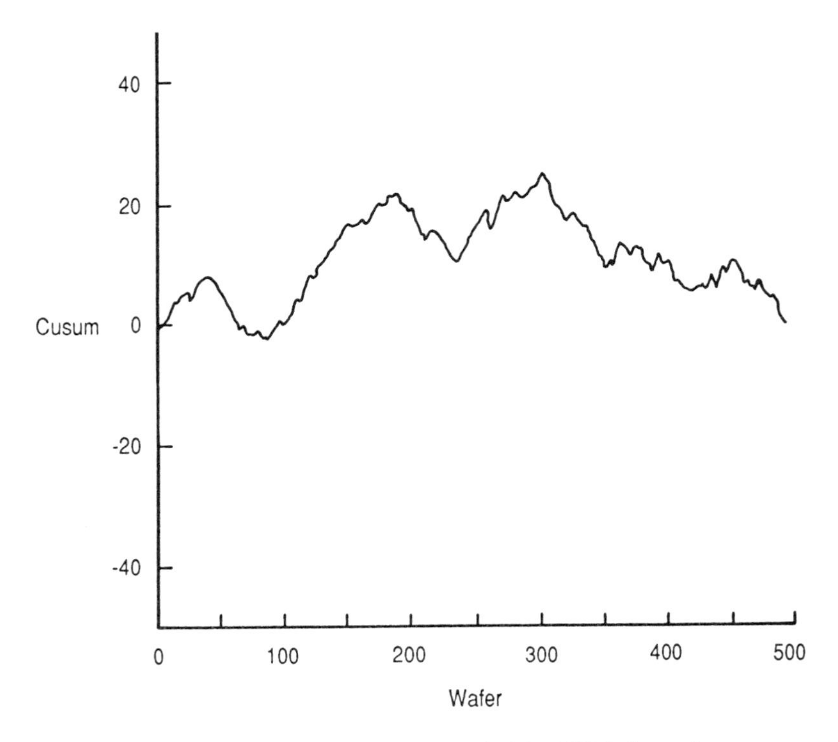

Figure 2.24 CUSUM Chart for Silicon Nitride MIM Capacitor

In order to qualify a particular technology it is necessary to:

(i) Demonstrate control of the process technology.

(ii) Demonstrate ability to meet specified design limits.

(iii) Demonstrate process reliability.

(iv) Prepare comprehensive ability manual detailing process limits and design rules.

Having obtained Capability Approval it is necessary to undertake periodic evaluation to demonstrate continuing compliance with the capability manual. The advantage of qualifying a technology is that it is not product specific and therefore, providing the MMIC chip design is within the process limitation, the MMIC can be released to BS 9450 without further specific approval testing.

The initial phase of approval involves definition of the technology design rules and limits which will ultimately form the core of the capability manual. In the case of a commercial Foundry process the bulk of this has already been defined as part of the Foundry design manual.

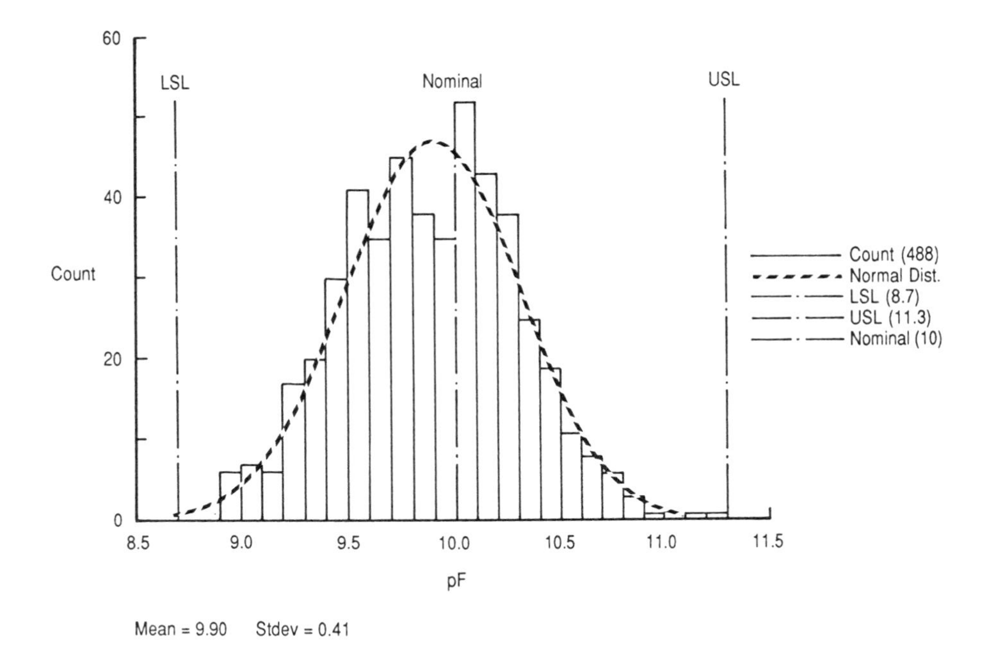

Figure 2.25 Process Capability Study for Silicon Nitride MIM Capacitor

Demonstration of the ability to meet these limits is achieved through the design and processing of capability qualifying circuits (CQC). The CQC that has been designed as part of the programme to qualify the Plessey 3–5 F20 Foundry process is shown in Figure 2.26. The CQC layout contains all the components necessary to demonstrate the design and layout limitations for example:

(i) Minimum track width and spacing.

(ii) Minimum dielectric window dimensions.

(iii) Minimum and maximum resistor dimensions.

(iv) Minimum gate finger spacing.

(iv) Minimum capacitor dimensions and values.

Also included are microwave testable FETs and a simple single stage feedback amplifier which are used for rf characteristic and reliability testing.

A number of batches of the CQC wafer are processed, inspected and characterised to demonstrate conformance to the design rules. In addition

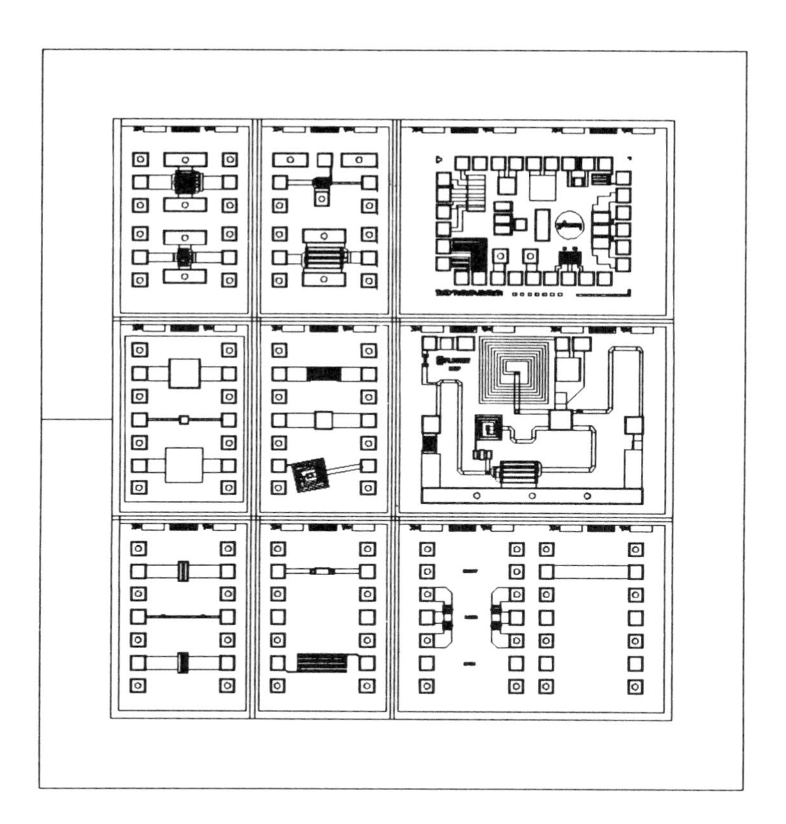

Figure 2.26 Capability Qualifying Circuit Array

circuits are subjected to environmental testing (e.g. damp, heat, temperature cycling) and finally undergo an 8000 hour lifetest.

The Capability Approval programme is monitored throughout by the National Standards Inspectorate (NSI) who are responsible for issuing the final approval.

The Approval is maintained by periodically re-processing the CQC layout and demonstrating the continuing compliance with the design limitations.

2.4 Next Generation

2.4.1 Multifunction ICs with Selective Implantation

As indicated in Section 2.1 of this Chapter the market penetration of the GaAs MMIC will result from the development of Application Specific

MMICs (ASMMICs) with acceptable performance and suitably packaged. At the present time the use of a single sheet implant allows different optimised implants to be used for each different type of single function MMIC, for example, for a switch or low noise amplifier. Whilst optimisation of such profiles will clearly continue (to reduce current consumption and improve the overall dB/mA figure of merit) and characterisation of each implant will be undertaken for a range of device geometries usable for each function, it is imperative that each type of optimised implant can be included within a single chip. Such implantation techniques, known as selective implantation, can be readily adapted with existing processes and will ensure optimum performance of each function rather than compromised performance with the standard sheet implant in a multifunction chip.

The degree of market penetration of the MMIC will depend on a series of complex trade-offs, including size and weight, but also performance, testability, reproducibility and cost. In competing with discrete devices the present 0.5 μm ion implant technology is competing with fully optimised 0.25 μm gate length MESFETs on epitaxial material or heterojunction devices such as 0.25 μm gate length HEMTs. Clearly such devices offer superior noise performance and higher frequency capabilities.

In selecting ion implanted materials for MMICs, the benefits of reproducibility, cost and the ability to selectively implant were traded off against absolute performance. However in some applications the use of epitaxial materials whether for standard MESFETs or heterojunction devices may be preferred. This is particularly true for super low noise or millimetre wave applications. However much will depend on the frequency of application and performance expectation, but in the majority of volume and low cost applications the performance of ion implanted multifunction MMICs will suffice.

In the Authors' opinion, whilst ion implanted devices will always remain inferior to epitaxially fabricated devices, further work on profile optimisation may still reveal further improvements. Clearly the present generation of MMICs could be fabricated with shorter gate lengths through the use of electron beam lithography, whilst sacrificing line throughput. Recent work by G. W. Wang at Ford Microelectronics [14] has claimed performance comparable to state of the art pseudomorphic HEMT devices (based on semi-insulating GaAs) through the use of optimised profile and high carrier concentration with an Si^{29} implant. Whilst the reported results do indicate excellent noise figures (0.8 dB at 16 GHz), the associated gains do not compete with the results of P. M. Smith at General Electric [15] on conventional or pseudomorphic HEMTs based on GaAs substrates or from Umesh K. Mishra of Hughes Research [16] on lattice matched HEMTs on InP substrates.

On the basis of absolute noise and gain performance, the benefits of higher carrier concentration and higher mobility resulting from the use of heterojunction structures (N ~ 10^{18} cm^{-3}) compared with MESFETs on epitaxial or ion implanted material with carrier concentrations of (N ~ 5 x 10^{17} cm^{-3}) will always produce superior performance.

2.4.2 The High Electron Mobility Transistor

(i) The Conventional HEMT

Optimisation of the carrier concentration of the active layer and the minimisation of device structure parasitics, together with reductions in gate length to 0.25 μm through the use of electron beam lithography have produced impressive performance for the GaAs MESFET into the millimetre wave frequency [17,18] with usable performance to 60 GHz. However, further improvements in frequency capability and reductions in noise performance, even at lower frequencies, require increases in carrier concentration and mobility to raise the transconductance and minimise parasitic resistances. Increases of the carrier concentration to the upper 10^{17} cm^{-3} or 10^{18} cm^{-3} are not practical with the conventional MESFET structure, but the concept of band gap engineering through the postulation of the heterojunction has led to rapid improvements in noise performance and usable frequencies to 94 GHz.

The exceptional microwave performance of the high electron mobility transistor, HEMT, is due to the technique of modulation doping which results in the separation of the channel electrons from their parent donor ions by the use of an n^+ AlGaAs/undoped GaAs heterojunction. The band gap of the AlGaAs is larger than GaAs, and can be engineered by the variation of the mole fraction, x ($Al_xGa_{1-x}As$) being typically 0.3. A two dimensional electron gas (2–DEG) is formed at the AlGaAs and undoped GaAs interface and resides in the potential well created by the conduction band discontinuity.

Since the electrons flow in an undoped region (unlike the MESFET), ionised impurity scattering is reduced and high mobility and velocity are obtained. The introduction of a thin spacer layer, typically 20–50 Angstrom thick of undoped AlGaAs has been found to improve mobility further by separating the 2–DEG from the donor ions in the AlGaAs. Similarly the use of a superlattice buffer impedes the migration of substrate impurities to the 2–DEG.

A typical structure of a so–called conventional HEMT is shown in Figure 2.27. The layer sequence is grown by MBE or MOCVD techniques and the structure characterised by very thin layer dimensions requiring accurate control of growth and especially well controlled etching

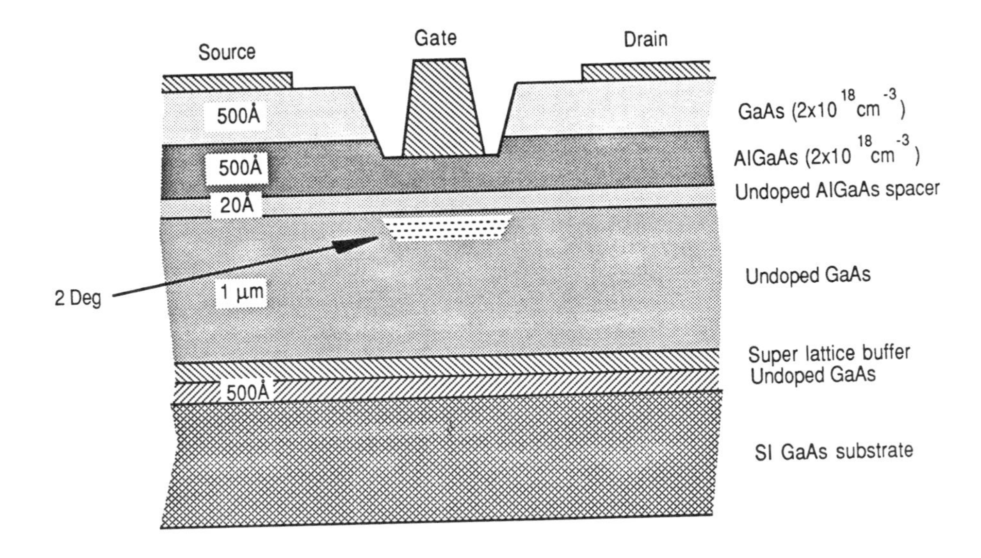

Figure 2.27 Heterojunction Structure

processes. The heterojunction active device is compatible with the MMIC technology, employing boron isolation rf mesa technology.

The work of P. M. Smith at General Electric [15] has indicated the best state of the art results, achieving f_t of 170 GHz from a 0.25 µm gate length with maximum available gains of 9.1 dB and 5.3 dB at 60 and 90 GHz respectively and noise figures of 2.6 dB with 5.7 dB associated gain at 62 GHz. Whilst the performance at millimetre wave frequencies is impressive, the higher associated gains and lower noise figure performance is attractive in some applications at lower frequencies (see Table 2.11).

(ii) The Pseudomorphic HEMT

In addition to the conventional HEMT based on the AlGaAs/GaAs layer structure, a pseudomorphic HEMT has been proposed and demonstrated. This contains an InGaAs channel layer between the AlGaAs spacer and the GaAs buffer. Although lattice mismatched to GaAs, the layer is sufficiently thin that the lattice strain is taken up coherently with the epitaxial layers resulting in dislocation free pseudomorphic material. As a result of the InGaAs layer, higher mobility and increased saturation velocity are obtained compared to GaAs. The work of P. M. Smith [15] has achieved the lowest reported noise figure at 62 GHz of 2.3 dB with

Table 2.11 : Noise performance of HEMTs

	Frequency GHz	Device	Fmin (dB)	Ga (dB)
GE	8	Conventional	0.4	15.2
GE	18	Conventional	0.7	13.2
GE	30	Conventional	1.5	10.0
GE	40	Conventional	1.8	7.5
GE	62	Conventional	2.6	5.7
GE	62	Pseudomorphic	2.3	4.0
Hughes	63	InP	0.8	6.7

11.7 dB and 6.7 dB maximum available gain at 60 and 90 GHz respectively. The devices exhibited f_t's of 200 GHz.

(iii) The AlInAs–GaInAs HEMT

Unlike the conventional lattice matched AlGaAs/GaAs and pseudomorphic AlGaAs–InGaAs, the AlInAs–GaInAs heterojunction reported by Umesh K. Mishra at Hughes [16] has a higher conduction band offset (0.5 eV) and with the higher carrier concentration in AlInAs allows for greater 2–DEG concentration. The GaInAs has a higher electron mobility and peak velocity. Fabricated on an InP substrate and using trilevel resist techniques to fabricate low resistance mushroom gates, devices exhibited the best results to date of any structure at ambient temperature of 0.8 dB noise figure with 6.7 dB associated gain at 63 GHz. See Table 2.11.

(iv) Power Capabilities of HEMT Structures

Due to the low sheet carrier density (10^{12} cm^{-2}) of the undoped GaAs in the HEMT and the reports of low breakdown voltage, the superior low noise HEMT was not expected to offer viable power capabilities. To increase the power, increases in current density were required and as a result devices were fabricated with multiple doped heterojunctions. The work of A. K. Gupta at Rockwell [19] and P. M. Smith [15], although based on small periphery devices have reported reasonable power densities of

0.63 W/mm with 40% power added efficiency at 10 GHz, and 0.43 W/mm with 28% added efficiency at 60 GHz respectively.

2.4.3 The Heterojunction Bipolar Transistor

Heterojunction bipolar transistors (HBT) are currently being developed based on a AlGaAs/GaAs heterojunction for microwave and millimetre wave applications due to their superior high frequency performance over conventional silicon transistors, and the potential of increased power capability over the field effect transistor. Such devices would also be compatible with integration into a MMIC with the advantages of low parasitics on semi-insulating GaAs.

HBTs fabrication on III–V materials offer a number of advantages over silicon:

(1) Due to the wide band gap emitter, a much higher base doping concentration can be used, decreasing base resistance.

(2) The heterojunction at the emitter–base junction suppresses hole injection from the base into the emitter.

(3) Transit times are reduced due to higher mobility.

(4) Reduced emitter doping allows reduction in the base emitter capacitances.

When compared to the FET structure the HBT offers:

(1) A simplified lateral dimensional control since transit times are controlled by epitaxial growth rather than lithography.

(2) A low offset voltage and high breakdown.

(3) High current handling capacity due to conduction of entire emitter area.

(4) High transconductance with low output conductance.

(5) Lower l/f corner frequency being typically 1 MHz compared with 10 MHz for GaAs FETs.

A schematic of an HBT is shown in Figure 2.28. The work reported by P. M. Asbeck of Rockwell [20] has indicated device f_t of 100 GHz from 1.2 μm emitter geometries. The potential of the HBT as a power amplifying device is demonstrated by the results of B. Bayraktaroglu of TI [21] with 4 W/mm of emitter periphery at 10 GHz with 35% power added efficiency under pulsed conditions whilst J. Higgins of Rockwell [22]

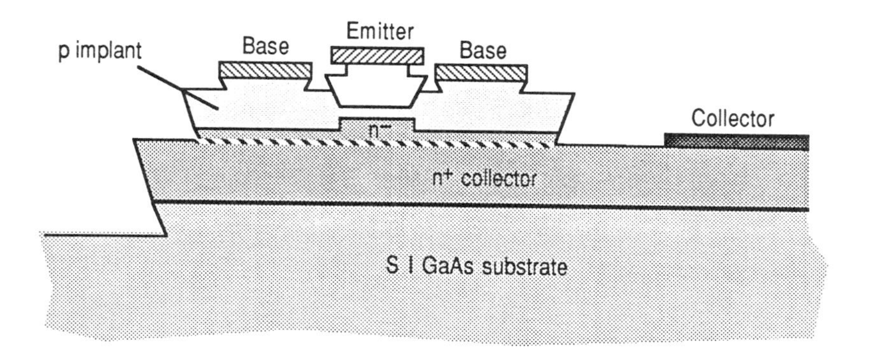

Figure 2.28 Schematic of a HBT

demonstrated 4 W/mm at 10 GHz with 48% added efficiency. Millimetre wave performance was also demonstrated at 59 GHz with 1.7 W/mm at 25% added efficiency. Both of these results were based on small geometry devices when a larger geometry may be limited by the poorer thermal properties of the GaAs compared to silicon.

The potential benefits of the HBT could lead to extensive applications, namely:

(1) High efficiency power amplifiers, even into millimetre frequencies. The high transconductance could provide good class B operation.

(2) Wide band analogue, analogue/digital conversion or high speed digital.

(3) Low noise microwave oscillators.

Future generations of discrete devices and MMICs will be critically dependent on the control of material fabrication through the use of MBE or MOCVD, and on the associated control of processes necessary to provide reproducible and acceptable performance characteristics. For millimetre wave applications especially low noise and high gain the use of the heterojunction structure with short gate length lithography appears inevitable.

However for applications up to Ku band the situation is not so clear. For multifunction circuits where cost is critical, the use of selective ion implantation with improved profile would appear optimum. For improved performance, especially for low noise, the use of an additional device such as a discrete HEMT may suffice. Much will depend on the availability and cost of MBE/MOCVD material. If current trends being observed with commercial AlGaAs/GaAs HEMTs for the DBS market continue we could see 0.5 μm based HEMT ICs in Ku band in the near future.

Already research programmes are identifying the potential of the new HEMT structures when integrated into a MMIC. Examples of such circuits are included in Section 2.5 of this Chapter. Further detailed discussion on new generation device structures are included in Chapter 12 of this book on Emerging Technologies.

2.5 Applications

2.5.1 The Market Requirements

The introduction of GaAs MMIC foundry services and the growth in the availability of a range of standard off the shelf products, within the last five years, has meant that customers are now becoming aware of the benefits of the MMIC from first hand experience. They are also witnessing the beginning of plans for the implementation of system requirements based on this technology. Although the growth in the market has been slow, it should be remembered that the majority of systems require more than five years from concept to production.

Although military applications were responsible for much of the initial research and development, a significant number of commercial applications requiring high volume of GaAs MMICs are now being identified. Table 2.12 shows a number of identified applications for analogue GaAs MMICs.

The types of functions required within these applications is diverse in terms of performance, power consumption, frequency and cost. In general the functions can be categories into:

(1) Switches (active and passive)

(2) Amplifiers (small signal, low noise, wideband and power).

(3) Frequency converters.

(4) Voltage controlled oscillators.

(5) Control for amplitude and phase.

(6) Analogue/digital control interfacing.

(7) Signal detection.

(8) Active splitting and combining.

In the majority of the applications the customer requires either an application specific assembly of MMIC chips or an application specific

Table 2.12 : Applications for Analogue MMICs

Military	Industrial	Communications	Consumer
Phased array radar	Instrumentation	Fibre optics	DBS
EW	Anti collision	Satellite transponders	Cordless and cellular telephones
Decoys Seekers	Radiometers	Mobile communications (including steerable arrays)	
			Automotive anti–collision
Smart munitions		VSAT	Speedmeters
Instrumen–tation		EPIRBS	HDTV
ATE (BITE)		Search and rescue transponders	MDTV detectors
		Position indicating, position reporting, and system manage–ment updates	MDTV

fully integrated single chip, ASMMIC, assembled into an hermetic or low cost package for integration compatibility with the other system functions. Figure 2.6 shows an example of an ASMMIC assembly. These requirements place a diverse range of demands on material characteristics, the maturity state of the MMIC process, packaging, testing and on design capabilities. At the present time the majority of the applications are below 18 GHz, although opportunities do exist into millimetre wave frequency bands.

2.5.2 Process Characterisation

In order to be capable of minimising design cycle times and to be assured of a high probability of first time design success, as well as achieving

Table 2.13 : Equivalent circuit netlist parameters and spreads for F20 FET1 (4x75) at IDSS/2

```
DIM
    FREQ GHZ
    RES OH
    IND NH
    CAP PF
    ANG DEG
    TIME PS
    LNG MM

            DC DATA

            IDDS      =  52.74 mA
            Std Dev   =  4.74 mA      (wafer to wafer)
            Std Dev   =  2.49 mA      (within wafer)

VAR
```

BIAS CONDITION = IDDS/2 9Vds–5V)

	VALUE	STD DEV	MIN	MAX
LG =	0.02451	0.00179	0.02223	0.02887
LD =	0.01553	0.00165	0.01244	0.01819
LS =	0.01005	0.00118	0.00788	0.01224
RG =	1.60519	0.50050	1.00837	2.40766
RD =	1.63315	0.12220	1.42948	1.87813
RS =	2.96053	0.16015	2.69220	3.24495
RDG =	4.21263	0.24377	3.84302	4.67408
RDS =	283	16	251	319
RI =	2.00587	0.83297	0.80147	3.27671
CDG =	0.02768	0.00167	0.02552	0.03139
CGS =	0.28524	0.00957	0.26788	0.30272
CDC -	0.01541	0.00195	0.01247	0.01875
CDS =	0.04945	0.00452	0.04190	0.05737
GM =	0.03552	0.00176	0.03308	0.03794
Fl =	0	0	0	0
TAU =	2.83	0.16378	2.51	3.08

```
CKT
    IND  1  2  L^LG
    RES  2  3  R^RG
    CAP  3  4  C^CDG
    RES  4  5  R^RDG
    CAP  3  8  C^CGS
    CAP  8  5  C^CDC
    RES  8  9  R^RI
    VCCS  3  5  8  9  M^GM A=0 R1=0 R2^RDS F^F1 T^TAU
    CAP  5  9  C^CDS
    RES  5  6  R^RD
    IND  6  7  L^LD
    RES  9  10  R^RS
    IND  10  0  L^LS
    DEF2P  1  7  FETMOD

FREQ
    SWEEP  1  20  1

OUT

    FETMOD MAG[S11]
```

consistently high yields of the product against a defined specification, it is essential that a stable, reproducible production process has been established. Vital to the establishing and maintaining of such a process is the preparation and rigid implementation of process procedures and quality control, and the assessment of the results from monitoring the process control monitor, PCM. The range of elements within the PCM cell, Figure 2.18. with testing at DC or 1 MHz provides valuable data which can be correlated to the microwave characteristics. However, the design of GaAs MMICs requires characterisation of each element type over a sufficient range of element values and over the full frequency range of the process (20 GHz for F20 process).

During the initial process commissioning phases a wide range of test structures for each type of element was designed and realised on a mask set allowing a large number of wafers to be processed and the microwave performance assessed. Measurement of such a diverse range of components allowed broadband equivalent circuit models to be derived for each type of element and the tolerance of both the prime element values and the parasitic elements to be established.

The accuracy of such models has been critically dependent on the development of parameter extract software and improvements in microwave network analysers, such as the Hewlett Packard 8510. More recent advances in the development of microwave coplanar probes, by companies such as Cascade Microwave (now usable up to 50 GHz) and the establishing of suitable calibration standards, has enabled rapid and accurate microwave characterisation of elements on wafer by the use of RFOW techniques. The CQC array shown in Figure 2.26 features calibration standards for short circuit, open circuit, 50 ohm termination and a through line established on the wafer rather than the use of alumina calibration standards. This concept allows more accurate de-embedding of the elements through a properly structured set of assembled elements on the wafer.

Table 2.13 shows the data derived for an F20 4 x 75 μm ion implanted MESFET, when represented by an equivalent circuit model. The data indicates the nominal value for each element together with the standard deviation and maximum and minimum range derived across a wafer and from wafer to wafer. Data derived by such techniques overcomes many of the earlier difficulties of repeatability experienced by the mounting of individual chips into customised r.f. test fixtures. Extension of RFOW techniques to noise characterisation are now becoming possible through the availability of software controlled mismatch standards and data extraction software from Automatic Testing and Networking Inc.

The derivation of equivalent circuit models to represent the performance of passive elements over the full microwave frequency range, and ideally over the usable range of element values, has been possible

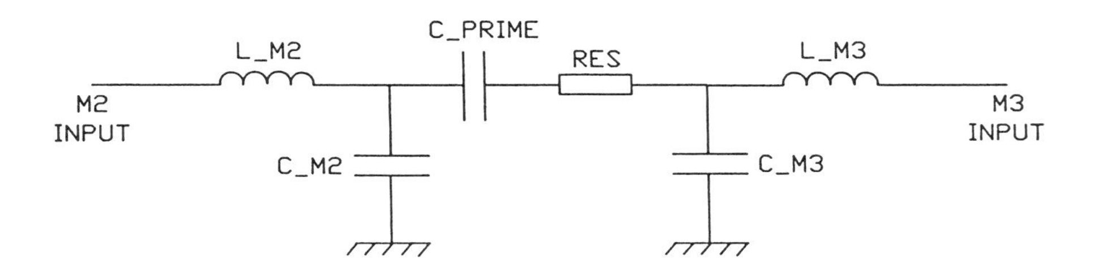

Figure 2.29 Overlay Capacitor Model

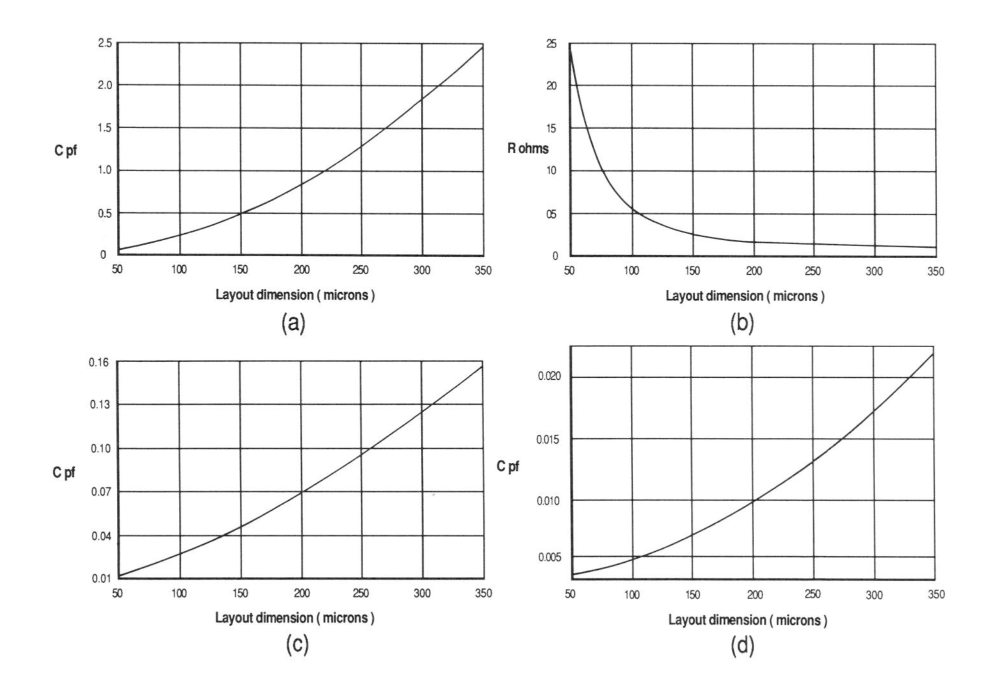

Figure 2.30 Effect of Layout Dimensions on Component Parameters
(a) Prime Capacitance of Polyimide Capacitors
(b) Parasitic Resistance of Polyimide Capacitors
(c) M2 Input Parasitic Capacitance (CAP–M2) for Polyimide Capacitors
(d) M3 Input Parasitic Capacitance (CAP–M3) for Polyimide Capacitors

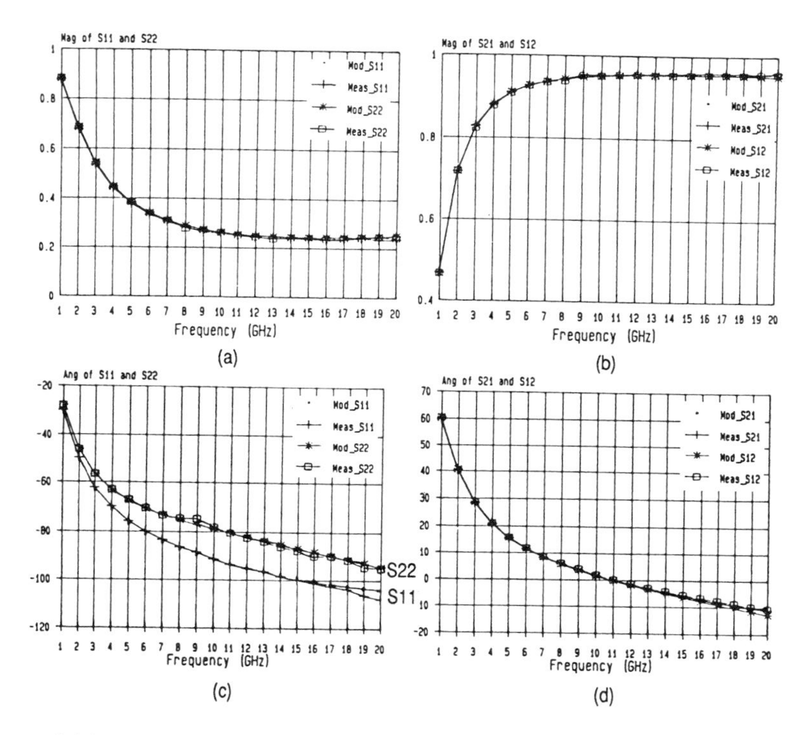

Figure 2.31 Polyimide MIM Model/Measure Fit
(a) Modulus S11 and S22 **(b) Modulus S21 and S12**
(c) Arg S11 and S22 **(d) Arg S21 and S12**

through the availability of good quality measured data and optimisation routines such as those included in TOUCHSTONE. Although earlier models were dependent on the use of look–up tables or graphs, the availability of equation blocks built into the circuit files of TOUCHSTONE have led to more sophisticated use of the model. Such models can be based on prime element capacitance values or on the structure layout dimensions.

Figure 2.29 shows an equivalent circuit model derived for an MIM capacitor usable to 20 GHz. Figure 2.30(a), (b), (c) and (d) shows the quality of fitting of the closed form expressions to the measured data (shown x) over a wide range of MIM capacitor plate dimensions whilst Figure 2.31(a), (b), (c) and (d) shows the quality of model fit to the complete range of measured S–parameters. Similar models have been derived for all the passive elements realisable with the standard process.

The use of equation blocks and closed form expressions have overcome the limitation of programmes such as TOUCHSTONE or Super–Compact to analyse transmission lines fabricated on a complex structure of multiple

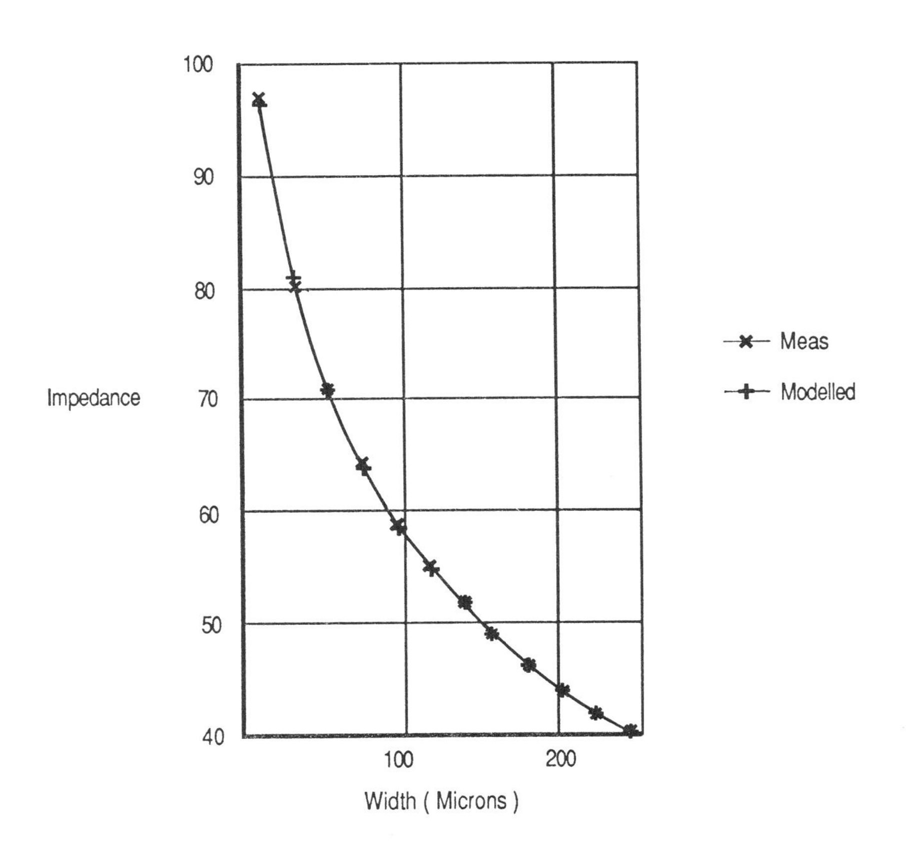

Figure 2.32 A Plot of Modelled and Measured Impedance vs Width for an M3 Line

dielectrics. Figure 2.32 shows the agreement of line impedance derived from measurements of effective dielectric constant and the predictions from closed form expressions. The introduction of RFOW testing has allowed additional monitoring of the RF characteristics of the test FET structures within the PCM cell and ensures compliance against the assigned limit values for each processed wafer. All characterisation data and layout rules together with process limits are included in the Design Guide issued to Foundry customers.

2.5.3 Design Procedures for MMICs

Unlike a hybrid MIC circuit designer, the MMIC designer cannot resort to tuning of his circuit to achieve the design goals. It is important, therefore, that the quality of the design information and the range of variability of the prime elements and parasitic elements is known. Clearly, the need to accept process variations in the design places limitations on the approaches to designs to minimise the effect, and on the final specification. Figure 2.33(a) and (b) define the basic design procedure and layout flow chart to be followed with any GaAs MMIC design. These procedures have been derived from many years of design experience within Plessey. A critical element of circuit design lies in the engineer's choice of initial circuit topology, and the subsequent effects of tolerancing and parasitic elements on the target specification. The introduction of closed form expressions has simplified the working relationship of prime element tolerancing and the resulting parasitic variations. Conversion of a theoretical circuit to layout will involve further elements being defined and analysed. The use of Calma and standard cells greatly assists with circuit layout and ensures that design rule checking errors are minimised.

Although careful characterisation of individual elements within the process has been undertaken, the layout of the final circuit can be crucial to the ultimate performance. At the current status of theoretical analysis of structures [2] it has not been possible until recently to provide anything other than general guidelines. The availability of LINMIC is allowing the analysis of coupled structures, and will eventually result in a complete work station capability enabling final physical layouts to be analysed for microwave performance.

The use of standard cells and algorithms built into software has greatly enhanced the specification and layout of GaAs MMICs and the next few years will see great strides in the development of the complete GaAs work station from specification to mask layout. The reader is recommended to read Chapter 4 for further details of design procedures and CAD tools.

2.5.4 State of the Art Results

The result of Podell in 1986 [1] is still outstanding in terms of the realisation of a multifunction GaAs MMIC based on novel circuit approaches, especially based on the concepts of push–pull to provide virtual earths and hence dispensing with r.f. ground capacitors, and the use of miniature coupled spirals. However, the chip does require extensive off–chip circuitry to provide full functionality.

The work of Plessey in T/R chips for phased array applications have combined the advantages of novel circuit techniques, particularly in the

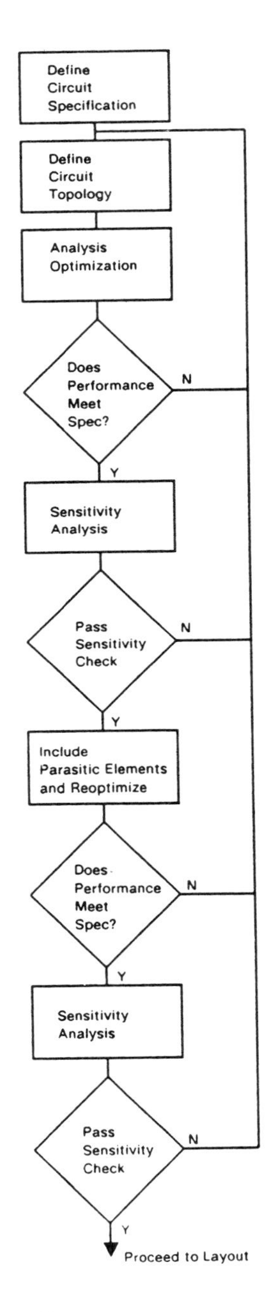

Figure 2.33 (a) **Design Flow Charts**
GaAs MMIC Design Procedure

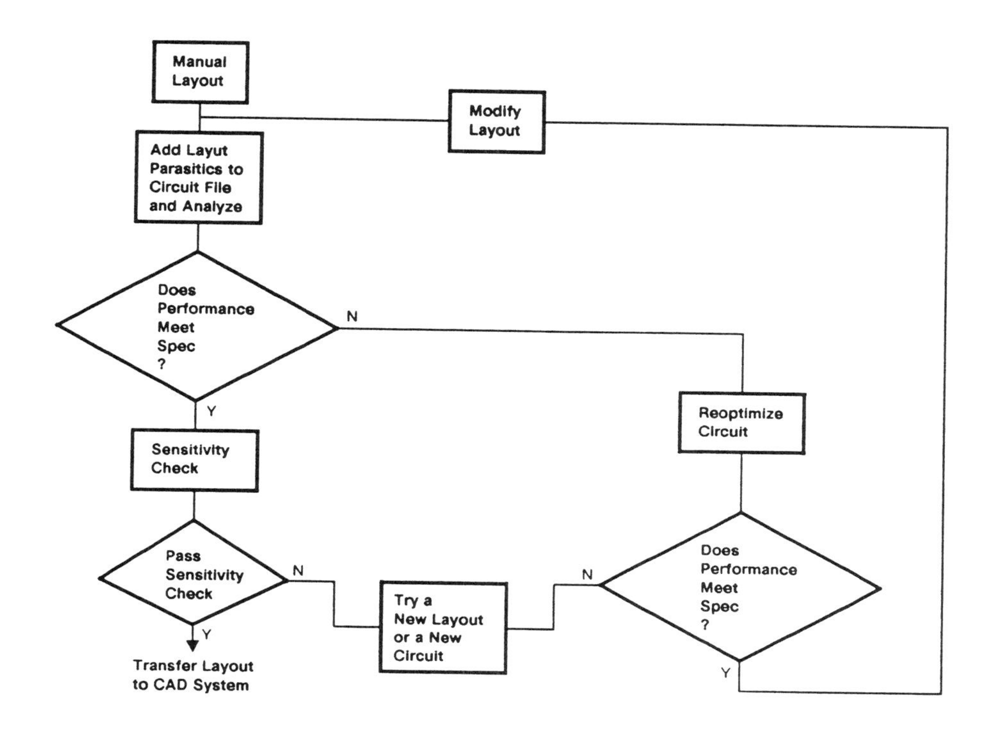

Figure 2.33 (Contd)
(b) Layout of MMIC

area of phase shift elements [23] with the advantages of selective implant processing, and produced the integrated phase shifters, amplifiers and switches shown in Plate I. This is based on the Plessey standard production process.

Clearly as applications defined in Section 2.5.1 mature into circuit realisations, these concepts will become more standard. However much will depend on the advances in CAD to enhance the capability and integration levels on the standard, preferred, volume production processes. In the meantime, research is reporting major strides in MMIC performance based on the benefits of the new material and structure technologies outlined in Section 2.4. The benefit gained from the heterojunction HEMT structure based on AlGaAs/GaAs, namely super low noise, high gain, wide bandwidth and high frequency have been reported. N. Ayaki et al of Mitsubishi [24] has realised a 1.7 dB noise figure MMIC over the 11.7–12.7 GHz band for a two stage amplifier based on 0.5 μm lithography. Wideband, high gain per stage amplifiers have been realised by C. Nishimoto [25] at 2–20 GHz with 11 ± 0.5 dB gain and C. Yuen [26] at 3–40 GHz with 8 dB gain and a noise figure of 3 dB from 10–30 GHz. In the millimetre wave frequency band, low noise, narrow bandwidth amplifiers have been demonstrated by J. Berenz [27] at 44 GHz with 5 dB noise figure, and H. Hung [28] at 60 GHz with 6.4 dB noise figure. Clearly significant work is required to demonstrate the results reported by P. M. Smith on discrete devices but rapid progress is being made in demonstrating the potential of both super low noise and high frequency MMICs.

2.6 Future Trends

It is never easy to predict the future in a rapidly advancing area such as GaAs based IC technology. However, in order to meet the requirements of the systems engineer in high speed digital and analogue circuitry a number of new techniques are being researched.

2.6.1 Materials

A major concern of research at present is in substrate quality whether for homo or heterojunction III–V based material. Ingot annealing is currently being pursued to improve the general quality of the substrate and surface preparation needs to be improved if the fine line lithography required for the highest frequency circuit is not to be a yield limiting factor.

In homojunction devices such as the GaAs MESFET, ion implantation is now generally accepted as being totally adequate and recently has demonstrated excellent performance at millimetre wave frequencies [16]. However, for heterojunction devices further work on improving the epitaxial growth techniques whether it be MBE, MOCVD or MOMBE is required particularly regarding reproducibility and throughput.

2.6.2 Lithography

Present trends favour electron beam lithography as a means to achieve sub half micron geometries. However, throughput is a major block and consequently alternative approaches need to be addressed. To this end X–ray lithography is a potential technique. It is capable of sub half micron definition in thick resists but suitable X–ray sources need to be developed and the manufacture of the masking membranes needs to be refined. Optical stepping is also progressing apace at sub micron geometries with the introduction of improved lenses and laser light.

2.6.3 Etching and Deposition

There is a move away from wet etching processes with consequent improvement in yield and uniformity. Dry processing however is not without its problems as the energetic plasmas used can damage the underlying semiconductor. New techniques involving microwave plasmas and high magnetic fields (electron cyclotron resonance) are being developed in which ion energy can be independently controlled leading to reduced damage. Modification of this technique can also be applied to deposition processes.

2.6.4 Rapid Thermal Processing

With thinner and thinner layers both of semiconductor (HEMT 2D electron gas and base width of heterojunction bipolar) and dielectric required for higher frequency performance, processes involving prolonged heating are not acceptable as the integrity of the interfaces can be destroyed. Rapid thermal processes in which short high temperature pulses can be applied during the annealing or deposition process to reduce the overall thermal exposure are emerging as a solution to this problem.

2.6.5 Laser Processing

The use of lasers in compound semiconductor processing is opening up new areas for technological improvement. They are already being used in research laboratories to condition the surface of the substrate prior to epitaxial growth by removing traces of unwanted oxide. They can be used for selective heat treatment for contact sintering and as an alternative to reactive ion etching for via hole drilling.

All of these approaches can be applied equally to homo and hetero–junction devices and circuits and are all aimed at improving their manufacturing and reliability.

2.7 References

1 Allen F. Podell et al: High volume, low cost, MMIC receiver front end, IEEE Microwave and Millimetre Wave Circuit Digest 1986, pp57–59

2 Rolf H. Jansen et al: A comprehensive CAD approach to the design of MMICs up to mm–wave frequencies, IEEE Transactions, MTT Vol 36, No. 2, February 1988, pp208–219

3 W. J. Roesch et al: Depletion mode GaAs IC reliability, GaAs IC Symposium Digest, pp27–30, Portland, 19087

4 Pengelly, R. S., Turner J. A., (1976): Monolithic broadband GaAs FET amplifiers, Electronics Letters Vol. 12, No. 10, 251–252

5 Knight, J. R., Effer D., Evans P.R. (1965): The preparation of high purity gallium arsenide by vapour phase epitaxial growth, Solid State Electronics, 8, 178

6 Pamplin, B. R. (1980): Molecular Beam Epitaxy, Pergamon Press (Oxford) England

7 Gibbons J. F., Johnson W. S., Mylroie S. W. (1975): Projected range statistics, semiconductors and related materials, 2nd Edition Stroudsburg, PA: Dowden, Hutchinson and Ross

8 Pucel R. A (1981): Design considerations for monolithic microwave circuits, IEEE Trans. Microwave Theory Tech., 29, 513

9 Pengelly R. S (1982): Microwave field effect transistors – theory, design and applications, New York: Wiley

10 Esfandiari R., Maki D. W., Siracusa M. (1983): Design of interdigitated capacitors and their application to GaAs monolithic filters, IEEE Trans. Microwave Theory Tech., 31, 57

11 D'Asaro L. A., DiLorenso J. W., Fukui H. (1978): Improved performance of GaAs microwave field effect transistors with low inductance via-connections through the substrate, IEEE Trans. Electron Devices, 25, 1218

12 Spencer E. G., Schmidt P. H (1971): J. Vac. Sci. Technol, 8, S52

13 Coburn J. W (1982): Plasma Chemistry and Plasma Proc. 2, 1

14 G. W. Wang et al: High performance millimetre wave ion implanted GaAs FETs, IEEE Electron Device Letters, Vol. 10, No. 2, Feb. 1989, pp95–97

15 P. M. Smith et al: Advances in HEMT Technology and applications, IEEE MTT–S Symposium Digest, 1975, Vol. II, pp749–752

16 Umesh K. Mishra et al: InGaAs/AlInAs HEMT technology for millimetre wave applications, GaAs IC Symposium 1988, pp97–100

17 J. Arnold et al: Extended frequency range GaAs MESFETs using 0.3 μm gate lengths, EUMC Amsterdam 1981

18 E. T. Watkins et al: A 60 GHz FET amplifier, IEEE MTT–S Digest 1983, pp145–147

19 A. K. Gupta et al: Power saturation characteristics of GaAs/AlGaAs high electron mobility transistors, IEEE Microwave and Millimetre Wave Monolithic Circuits Symposium, 1985, pp50–54

20 P. M. Asbeck: Heterojunction bipolar transistors for microwave and millimetre wave integrated circuits, IEEE Microwave and Millimetre Wave Monolithic Circuits Symposium 1987, pp1–5

21 B. Bayraktaroglu et al: AlGaAs/GaAs heterojunction bipolar transistors with 4 W/mm power density at X band, IEEE MTT–S Digest 1987, Vol. 2, pp969–972

22 J. A Higgins: Heterojunction bipolar transistors for high efficiency power amplifiers, GaAs IC Symposium 1988, pp33–36

23 A. A. Lane et al: Novel GaAs MMIC components and multifunction circuits, IEEE Colloquium on Microwave and Millimetre Wave ICs, Nov 1988, pp61–65

24 N. Ayaki et al: A 12 GHz band monolithic HEMT low noise amplifier, GaAs IC Symposium 1988, pp101–104

25 C. Nishimoto et al: A 2–20 GHz high gain, monolithic HEMT distributed amplifier, IEEE Microwave and Millimetre Wave Monolithic Circuits Symposium, 1987, pp109–112

26 C. Yuen et al: A monolithic 3–40 GHz HEMT distributed amplifier, GaAs IC Symposium 1988, pp105–108

27 J. Beronz et al: 44 GHz monolithic low noise amplifier, IEEE Microwave and Millimetre Wave Monolithic Circuit Symposium 1987, pp15–18

28 H. Hung et al: A 60 GHz GaAs MMIC low noise amplifier, IEEE Microwave and Millimetre Wave Monolithic Circuits Symposium 1988, pp87–90

Chapter 3

MESFET Modelling and Parameter Extraction

Adam K. Jastrzebski

3.1 Introduction

Successfully designing microwave circuits, and in particular MMICs, critically depends upon the accuracy of the models used in CAD programs. All models are only an approximation of physical reality and a compromise has to be achieved between model complexity and accuracy. Models should predict correctly all important aspects of the device behaviour, but at the same time, they must be computer–time efficient. Therefore, it is very important to understand what are the limitations of the models used and under which conditions they will break down. Suitability of a model depends on the type of circuit to be simulated and there are usually different requirements for different applications. For example, in designing microwave amplifiers matching is very important and the non–linear model of a MESFET must accurately predict both small– and large–signal terminal impedances of the device. Therefore, the drain–source conductance $\delta Ids/\delta Vds$ (i.e. the slope of the output I–V characteristics for constant gate–source voltage Vgs) needs to be accurate. Precise values of the channel voltage Vds and the channel current Ids are not so critical in this case. This fact will be used later in modelling of MESFETs with frequency dispersion, where the I–V characteristics of a MESFET are optimised for correct values of conductance and transconductance at microwave frequencies, rather than to match the DC behaviour.

For digital applications, symmetric models are often required, i.e. models capable of simulating both normal (when the drain potential Vd is higher then the source potential Vs) and inverse (when $Vd<Vs$) modes of the device operation. This leads to much more complicated expressions for the gate–source and gate–drain capacitances than is necessary in most microwave applications. However, accuracy of simulated terminal impedances is much less important and the DC characteristics can usually be used directly to model non–linear behaviour of the device.

Apart from limitations arising from the inherent approximations in MESFET models, accuracy can also be severely limited by the parameter extraction process. Most of the parameters of the nonlinear MESFET model cannot be obtained by direct measurements or calculated from technological data, but have to be extracted from the small-signal S-parameters and DC (or pulsed) I–V curves. Even in the simple case of a linear model at one bias point, the parameter extraction problem is difficult and complicated. This is mainly due to the large number of unknowns (typically about 12). Also measurement errors have pronounced effects on the calculated parameter values. Because existing circuit MESFET models do not accurately describe the internal working of a device, they are neither inherently consistent nor uniquely defined. Even relatively small measurement errors can cause large deviation of the extracted model parameters. Therefore, it is very important to accurately de-embed S-parameters used for modelling up to the reference planes of a device itself, and to take into account all extrinsic parasitic components such as bond wire inductances, pad and end-effect capacitances.

In the second Section of this Chapter various non-linear MESFET circuit models and their properties will be discussed. In particular, popular models used in commercial CAD programs will be analysed from the point of view of their limitations. In the third Section, an improved non-linear MESFET model will be proposed, which combines simplicity with sufficient accuracy for most microwave applications. Various aspects of parameter extraction for non-linear MESFET models are discussed in the next two Sections. An algorithm based on simultaneous optimisation to measured S-parameters at several bias points will be presented. The procedure for non-linear MESFET modelling utilising this algorithm for both quasi-static and dispersion modelling will be shown, together with a supporting example. In the final Section, some recent results in modelling of MESFETs will be presented and further developments suggested.

The subject of MESFET modelling and parameter extraction spans through many interdisciplinary fields, and it is impossible to cover in one chapter all the important issues. A thorough treatment of the subject would need to include such additional aspects as: device physics; device characterisation and measurements; dependence and sensitivities of model parameters on technological processes; reverse modelling; noise modelling; thermal effects; numerical methods used in parameter extraction; specific problems related to implementation of device models in different types of simulators; computer-time efficiency of models.
At the moment, there is no single publication covering all the above issues, but books by Shur [33], Pengelly [32], Soares (ed.) [34] and Ladbrook [35] are helpful in studying some of these matters.

3.2 MESFET MODELS

MESFET models can be divided into the following categories:

(1) Equivalent circuit models

  (a) Empirical and semi–empirical

  (b) Analytical, physics based

(2) Numerical models, physics based

Empirical and semi–empirical circuit models are most commonly used in CAD programs. Equations describing these models are usually modifications of some analytical formulae for the device, but constants in the equations and values of model elements are extracted from the DC, pulsed and small–signal S–parameter measurements. Such models can be made much more accurate than the corresponding analytical models, because their parameters are adjusted to match the real terminal behaviour of the device. They are also computer–time efficient and are usually easy to implement in general CAD programs. The major disadvantages are the lack of direct correlation between model parameters and physical parameters, and the need for complicated parameter extraction procedures.

Numerical models are based on the computer solution of four non–linear, coupled partial differential equations describing device physics [13,16]. These models are of primary importance for the understanding of the device operation and for designing new types of devices. However, they are very complicated, take very long computer times and require detailed information on the material properties and device geometry. Accuracy of these models is often compromised in order to increase speed of computations. In practice, these models are mainly used by device physicists although they were also applied to the design of MMICs on the superfast computer [13] and to the derivation of MESFET circuit models [16].

Analytical models are also based on the device physics but are derived by analytical solution of the device equations under some simplified assumptions [14,15]. Similarly, as for numerical models, device properties can be related to physical parameters and the fabrication process. However, operation of modern MESFETs with submicrometer gates is difficult to describe analytically, so these models are less accurate than the numerical and are most useful for a qualitative analysis of the device. Even very complicated analytical models, which require a computer to solve the equations, often have rather poor accuracy comparing with empirical models. Nevertheless, some simple properties, like for example power saturation of an amplifier can be predicted quite correctly [15].

Although the accuracy of all types of models can be only as good as the accuracy of the measurements, the empirical models are particularly sensitive to measurement errors. This is because the optimisation or curve fitting processes by which model parameters are identified are often ill-conditioned and small relative errors in measurements may give large errors in extracted model parameters.

3.2.1 General MESFET Circuit Model

Most MESFET circuit models have a basic structure which can be derived from the one shown in Figure 3.1a. From the point of view of both the circuit topology and the general form of the describing equations, this is an asymmetric model, which is only valid for the normal mode of the device operation, i.e. when the channel voltage Vd is greater than zero. The general model in Figure 3.1a consists of seven linear and six non-linear elements.

The linear elements in the model, i.e. those not changing with the bias are:

(1) Series lead inductances Ls, Ld, Lg;

(2) Series resistances Rs, Rd, Rg;

(3) Drain-source fringing capacitance Cds.

Non-linear elements in the circuit model depend on three internal voltages: Vg, Vd, and Vgd (see Figure 3.1a). The DC I-V characteristics of a MESFET are described by the non-linear current source Jds(Vd,Vg). This is the main non-linear element in the MESFET model, simulating the control of the channel current by voltages applied to the device terminals. The non-linear capacitors Cgs(Vg) and Cgd(Vgd) represent the depletion layer and fringing capacitances associated with the source and drain ends of the channel, respectively. The parametric resistor Ri(Vg) corresponds to the undepleted part of the channel under the gate, through which the gate-source capacitor is charged. Components Dgs(Vg) and Dgd(Vgd,Jds) represent Schottky-barrier junctions at the source and drain ends of the channel, respectively.

The non-linear model of Figure 3.1a can be linearised at a given bias point (Vgo,Vdo,Vgdo). Assuming typical bias conditions, i.e. $Vgo \leq 0$, $Vd > 0$, $Vgd < 0$ and Vgd greater than the breakdown voltage, we can neglect reverse-biased diodes Dgs and Dgd, which results in the small-signal model shown in Figure 3.1b. Here, the non-linear current source Jds is replaced by the linear controlled source Jds of the transconductance $g_m = G_{mo} \exp(-j\omega\tau)$ in parallel with the output resistance Rdo.

A modified structure for the non-linear MESFET model with the added branch Rf,Cf is shown in Figure 3.1c. This branch decouples AC and DC models for devices with frequency dispersion, decreasing output resistance at RF in comparison with its value at DC [6,10,19]. For non-

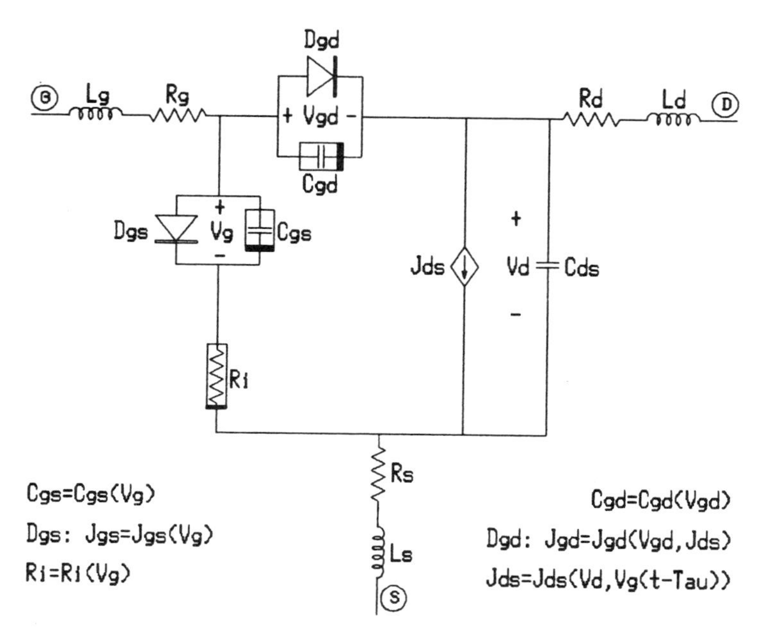

(a) Basic structure of non-linear model;

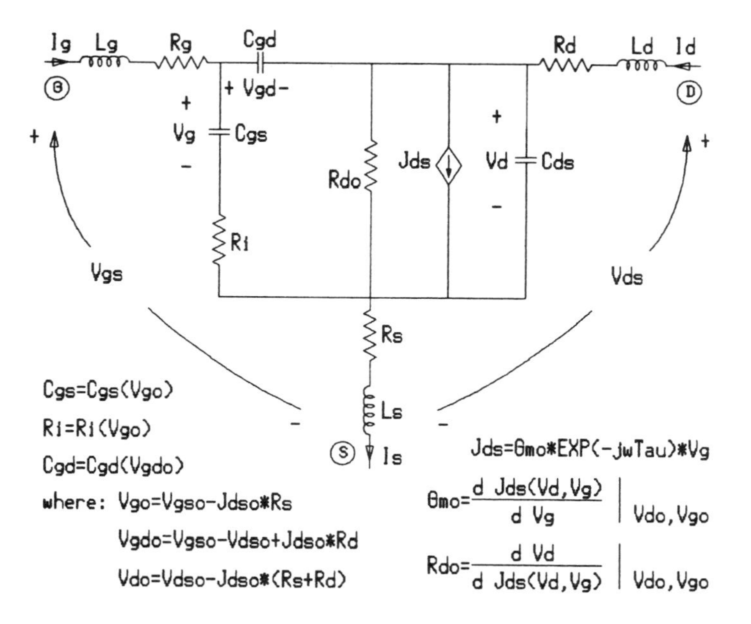

(b) Basic structure of linearised small-signal model
Figure 3.1 General MESFET circuit models.

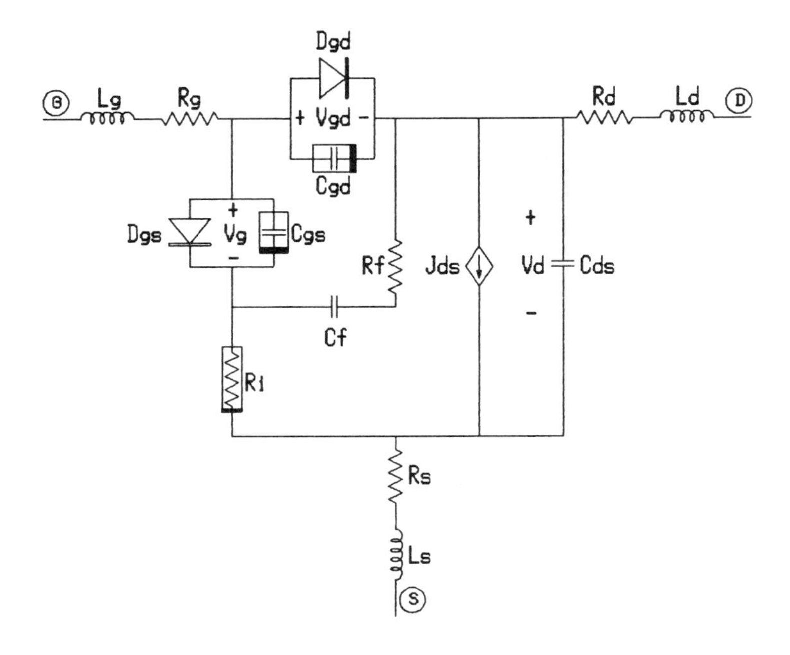

(c) Model with AC–DC decoupling branch Rf, Cf (alternatively, Rf,Cf can be connected in parallel with Jds)

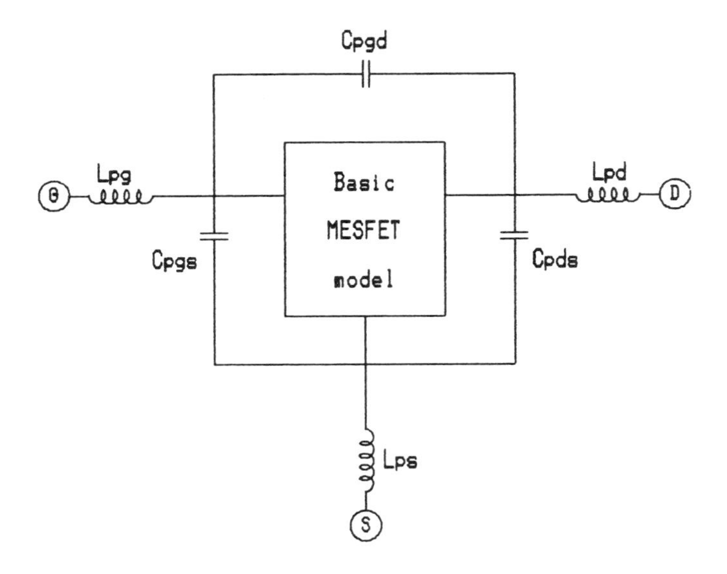

(d) Model with extrinsic parasitics

Figure 3.1 General MESFET circuit models (Contd)

dispersive devices, the Rf, Cf branch can be used to simulate the dipole layer formed at the drain end of the channel. A variation of this structure is often used, where the Rf, Cf branch is connected in parallel with the current source Jds [26].

For packaged devices or devices mounted on a carrier, additional parasitic elements should be added to the basic MESFET model. Figure 3.1d shows extrinsic parasitics associated with carrier mounted chip devices. The same equivalent circuit can also be used for most of the packaged devices.

3.2.2 Overview of Some Non–linear MESFET Models

From the point of view of applications in the design of large–signal analogue microwave circuits, the most important properties of MESFETs, which have to be accurately simulated are:

- Non–linearities of the drain–source current characteristics Ids(Vgs,Vds), and in particular pinch–off and saturation effects;
- Transconductance and drain–source conductance at microwave frequencies. At these frequencies, these parameters may be different than at DC due to frequency dispersion;
- Terminal impedances over a range of useful frequencies;
- Forward current of the gate–source Schottky–barrier junction;
- Gate–drain breakdown current.

In attempts to simulate above properties more and more accurately, a large number of non–linear empirical MESFET circuit models have been reported in literature. Here, only five of the most popular models used in general CAD programs are discussed and their properties compared.

A. Curtice Quadratic Model [1] (1980)

This is a very simple model, which has been implemented in most CAD programs. Unfortunately, it is not very accurate for large–signal microwave circuit simulation, particularly for devices with large pinch–off voltage. It's main characteristic is a square–law relationship between the drain–source current Ids and the gate–source voltage Vgs. The tanh function is used to describe the current saturation. Referring to Figure 3.1a, the model assumes a constant value for Cgd capacitance and elements Ri and Dgd are not present. Expressions for the remaining model elements are listed below:

$$Jds(Vd,Vg) = \beta(Vgg+Vt)^2\ (1+\lambda Vd)\ tanh(\alpha Vd) \tag{1}$$

$$Cgs(Vg) = Cgso/\sqrt{1-Vg/Vbi} \tag{2}$$

$$Jgs = Jo[exp(Vg/VT)-1] \tag{3}$$

$$Vgg(t) = Vg(t-\tau) \tag{4}$$

$$VT = nkT/q = n\ 0.026\ V \text{ for } T = 300°K \tag{5}$$

Where:

β, λ, α – DC model parameters;

Vt – threshold voltage;

τ – transit time (channel delay);

Cgso – depletion capacitance at zero bias;

Vbi – built–in voltage;

Jo – zero current of the Schottky diode;

n – ideality factor of the diode.

Modifications to the above model, proposed in [6], include a general power relationship between Ids and Vgs, saturation voltage control, and an additional component to simulate negative conductance of the dipole layer which forms at the drain end of the channel (Gunn domain effect).

A recently proposed MESFET model in [11] has essentially the same form of expression for Jds as the original Curtice model, but several empirical terms were added to increase accuracy and account for various second–order effects, including dispersion of the output conductance. However, the resulting formulae are very complicated and most of the parameters have no physical meaning, which makes model identification extremely difficult.

B. Tajima models [2] (1981) and [3] (1984)

Two slightly different versions of the same model were proposed and a further small modification was suggested in [30]. The model provides a very good fit to the DC I–V characteristics, includes breakdown current simulation, and is very suitable for microwave applications. However, the complicated expression for Jds(Vd,Vg) is purely empirical rather than physics based and therefore, identification of parameters is difficult.

The gate–drain breakdown is simulated in this model by the reverse current in the diode Dgd (see Figure 3.1a), using the following equation:

$$Jbgd(Vgd,Jds) = \begin{cases} (Vgd+Vb)/Rb & \text{for } Vgd<-Vb \\ 0 & \text{for } Vgd\geq-Vb \end{cases} \tag{6a}$$

$$Vb = Vbo+R1\ Jds \tag{6b}$$

$$Rb = Rbo+R2\ Jds/Jdss \tag{6c}$$

Where:

Vb – effective breakdown voltage;

Vbo – breakdown voltage, when Jds=0;

R1 – resistance relating breakdown voltage to the channel current;

Rb – effective breakdown resistance;

Rbo – breakdown resistance, when Jds=0;

R2 – increase of the breakdown resistance between Jds=0 and Jds=Jdss.

It can be seen from the above expressions, that the breakdown current depends on both the gate–drain voltage Vgd and on the channel current Jds, decreasing for larger values of Jds.

C. Materka & Kacprzak model [4] *(1983) and* [5] *(1985)*

This model uses a very simple but quite accurate formula to describe the DC I–V characteristics of a device:

$$Jds(Vd,Vg) = Jdss\ (1+Vg/Vt)^2\ \tanh(\alpha Vd/(Vg + Vt)) \tag{7}$$

$$Vt=Vto+\gamma\ Vd \tag{8}$$

Where:

Jdss – saturation current;

Vto – threshold voltage of an ideal FET;

Vt – effective threshold voltage;

α, γ – parameters of the model

The above DC model has been further improved in [19, 10] to give accuracy similar to that of the Tajima model, but the number of parameters used is less and they are easily identified.

The gate–drain breakdown current was simulated by Materka & Kacprzak by the reversed diode equation in which the effect of the drain–source current was not included:

$$Jbgd(Vgd) = -Jbo\ [\exp(-\alpha_b\ Vgd)-1] \quad (9)$$

Where:

Jbo, α_b – model parameters.

D. Curtice Cubic Model [7] *(1985)*

In this model, a third order polynomial is used to relate Igs and Vgs. This gives a better fit to DC characteristics, than in the first Curtice model, although some non–physical behaviour may result for voltages below pinch–off [12]. Also, model parameter identification is difficult. The model uses a simplified form of the Equation (6) to simulate the breakdown current. The forward conduction current of the Dgs diode is approximated by the piece–wise linear function. Cgs capacitance depends on Vgs as described in Equation (2), but additionally, it increases linearly with voltage Vd. The model is quite suitable for most microwave applications, but its overall accuracy is sometimes insufficient. There could also be convergence problems during the simulation due to discontinuous first derivatives of some of the modelling functions.

E. Statz (Raytheon) model [8] *(1987)*

This model gives good accuracy for simulation of the channel current of a MESFET, but some effects are not taken into the account (for example, threshold voltage does not vary with the drain–source voltage). The salient features of the model are the symmetric expressions for Cgs and Cgd capacitances, valid for both normal and inverse operation of the device. However, the overall model is not fully symmetrical.

3.2.3 Limitations of Existing MESFET Circuit Models

Each MESFET model has some specific constraints and inaccuracies, but additionally, there are more general limitations related to the fact that equivalent circuit models are only convenient approximations of the much more complicated physical reality. In this section, the general limitations of the existing MESFET models will be discussed.

Quasi–static Assumption and Frequency Dispersion

All the MESFET models discussed above rely on the quasi–static assumption. This means, that the charge in the channel is uniquely defined from the instantaneous terminal voltages. This assumption is quite justified for the undepleted area beneath the gate, because in a typical MESFET the dielectric relaxation time in the undepleted channel is about 0.1 ps. Therefore, free charges can redistribute themselves almost instantaneously in comparison with the period of the usable microwave signal frequencies. This however, is not true for charges trapped by the surface states between the gate and the drain and at the interface between channel and the substrate. Changes in the surface and interface charges are relatively slow, having characteristic frequencies typically in the range 1–100KHz [26–28]. Therefore, the shape of the extension of the depletion region towards the drain and the shape of the interfacial depletion region do not follow the changes of microwave signals. The relationship between the total charge in the undepleted channel and the instantaneous voltages on the channel is different at microwave frequencies than at DC. The channel current in the device model shown in Figure 3.1a, and therefore the charge in the channel are determined by the current source Jds(Vd,Vg). As the quasi–static assumption is incorrect for the model, then the DC I–V characteristics cannot be used to describe Jds(Vd,Vg) at microwaves. Either the model must be modified to include elements simulating the effects of traps, or the modelling approach not relying on the quasi–static assumption must be used.

The above phenomena cause frequency dispersion of the parameters in the MESFET model. The most affected parameters are transconductance and drain–source (output) resistance of a MESFET. Values of these parameters at RF are much smaller than at DC, i.e. $g_{mrf}<g_{mo}$ and $R_{drf}<R_{do}$. The decrease in the transconductance is attributed to the surface states [27,28], while the decrease in the drain–source resistance is due to the traps in the channel–substrate interface [26].

It is relatively simple to account for dispersion in the output resistance by decoupling the DC and AC parts of the model with the Rf, Cf branch, as shown in Figure 3.1c. Unfortunately, there is no easy way to modify frequency dependence of the transconductance. Therefore, the approach to modelling of MESFETs depends on whether dispersion is present or not and on the kind of dispersion. This determines not only the modelling procedures but also accuracy and limitations of the models. The following three cases can be considered:

(1) Negligible dispersion in both g_{mrf} and Rdrf.

The quasi–static assumption can be used and DC and AC models can simply be superimposed. As the DC characteristics can be simulated very accurately, then the models can also be very accurate – theoretically from DC, through RF, and up to the microwave frequencies.

(2) Negligible dispersion in g_{mrf} but strong dispersion in Rdrf.

This is probably the most typical case for modern self-aligned short gate MESFETs. As mentioned before, the Rf, Cf branch is used to decouple the DC and AC models, and the modified quasi-static modelling approach can still be applied [19,20]. The total output conductance at microwave frequencies will be approximately equal to the parallel connection of Rf and Rdo and for a constant value of Rf, its variation with bias will be prescribed by the variation of Rdo at DC. However, the Rdrf dispersion depends on the bias and so, there is an error in the model prediction of the output resistance for larger ranges of bias conditions.

Additional error arises as a result of single pole representation of charging and discharging of the traps. In reality, several time constants, and not one time constant should be used to represent these effects. Therefore, the model is not very accurate at RF, even if the Rf x Cf time constant corresponds to the dominant characteristic frequency of the traps. In most applications, however this error is not very important, and the time constant Rf x Cf can be chosen quite arbitrarily, as long as it is much greater than the largest period of the signal of interest. Smaller values of Rf x Cf are preferable in time-domain simulators, where the large spread of time constants can cause prolonged computing times.

(3) Strong dispersion in both g_{mrf} and Rdrf

In this case, the quasi-static modelling approach cannot be used. The Jds(Vd,Vg) current source must be chosen to simulate pulsed or RF rather than DC characteristics, which means, that the model will not predict correctly the bias conditions of the device [19,20]. Nevertheless, the accuracy of the large-signal microwave analysis is much better in this case, than for models based on the DC characteristics [29-31].

Thermal Effects

Influence of temperature on the device operation is not simulated properly in the existing non-linear MESFET models. Existing parameter extraction methods do not take into the account the device temperature during the measurements, either. These errors are particularly important in power devices, hence there are great difficulties with accurate simulation of those devices.

Model Symmetry

None of the models discussed here are fully symmetrical, and therefore they cannot predict accurately device operation in the inverse mode. The inverse mode is more often seen in digital circuits, but it may be also present in highly non–linear microwave oscillators during part of a cycle. Proper implementation of the model on a computer should take care of such a possibility. Asymmetric models should be extended into the inverse bias conditions, even if the expressions are not accurate there. In this case, the model would not fail completely and a simulator would at least be able to predict, that the device entered the inverse mode.

3.3 An Improved Non–linear MESFET Model

The MESFET model proposed below combines the best features of the models discussed in the previous Section and also includes some new features. The complexity of the model is kept to a minimum without sacrificing its accuracy in microwave applications. It is assumed that a device operates in the normal mode, i.e. Vds>0, because not all of the expressions are symmetrical.

The model structure is as shown in Figure 3.1. Firstly, the dependence of each non–linear element on bias conditions will be discussed and then, the key requirements for the modelling functions will be identified. From those requirements, an optimum form of the modelling equations will be proposed.

3.3.1 DC Characteristics

DC characteristics of a MESFET are described by the non–linear current source Jds(Vd,Vg) in Figure 3.1a. Jds(Vd,Vg) is the main non–linear element in the MESFET model which simulates the control of the channel current by voltages applied to the device terminals. It predicts various DC features, such as bias points, pinch–off and saturation. Transconductance g_m associated with this source (see Figure 3.1.b) determines the gain of the device and the drain–source conductance Gds – the driving properties. This means, that both the function Jds(Vd,Vg) and its derivatives must be in good agreement with the real device behaviour. Typical DC characteristics for the 0.5 μm x 300 μm epitaxial MESFET are shown in Figure 3.2.

Analysis of DC characteristics of various types of MESFETs (i.e. epitaxial or ion–implanted; with large negative, small negative or positive threshold voltages; with submicron or longer gates) shows a large variety of different types of behaviour. Main features associated with differences in the form (or "shape") of the characteristics have been identified in [10]. Models which have sufficient flexibility to modify the "shape" of the characteristics can be used for all types of MESFETs and technologies. The requirements for such an accurate and general model are specified below [10]:

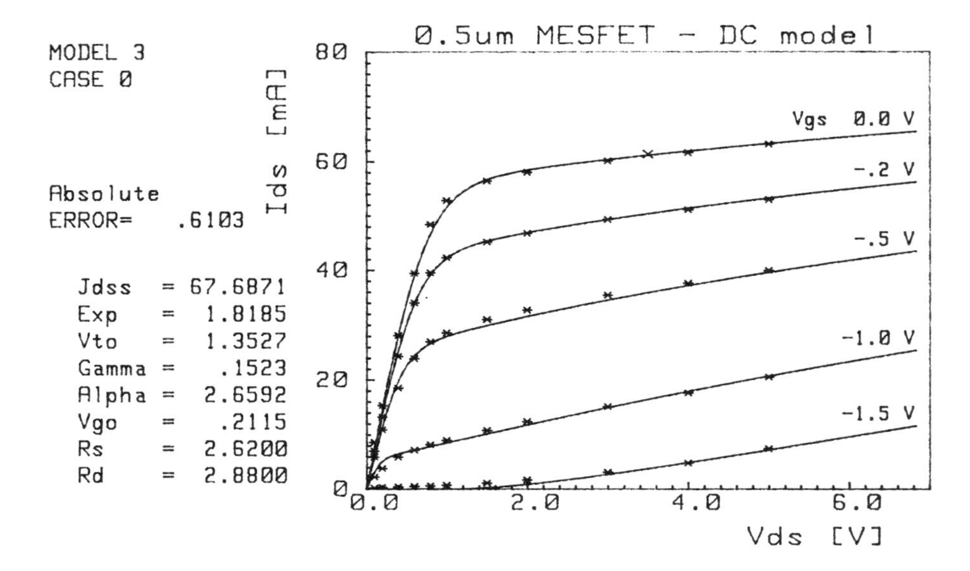

(a) Measured and calculated I–V characteristics

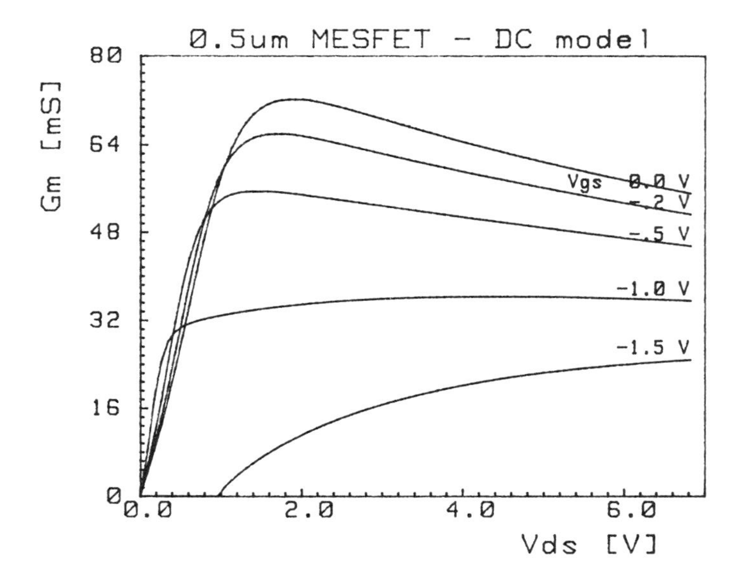

(b) Calculated transconductance g_{mo}

Figure 3.2 DC characteristics of the 0.5 μm x 300 μm low-noise MESFET. Device was modelled using Equations (10–11) with Go=0.

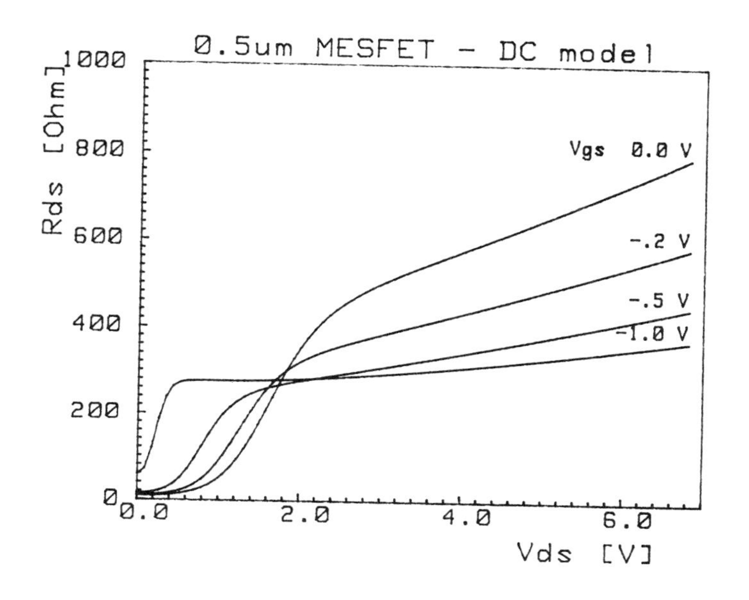

(c) Calculated output resistance Rdo

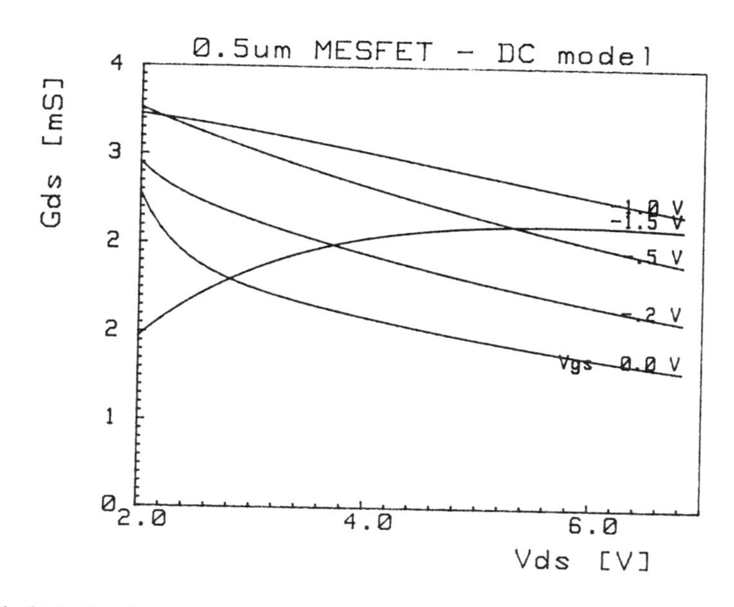

(d) Calculated output conductance Gds in the saturation region.
Figure 3.2 DC characteristics of the 0.5 μm x 300 μm MESFET (Contd)

(1) Flexible dependence of Jds on Vg.

The dependence of Jds on Vg should have a parameter allowing for adjustment of its character. The fixed square–law dependence of Jds on Vg used in Equations (1) and (7) is often significantly different from the behaviour of an actual device [8,9,11] and gives a large error in the calculated transconductance. Variation of this relationship for various devices is mainly attributed to differences in the channel doping profiles.

(2) Control over conductance dependence.

After removing the effects of series resistances, the output conductance of a MESFET in saturation is mainly attributed to three phenomena:

(a) Channel length modulation - gives finite positive conductance, which increases with the channel current. More pronounced for longer gate devices;

(b) Leakage current due to electrons being injected into the substrate – this effect also gives a positive conductance, but dominates for small current values. More pronounced for shorter gate devices;

(c) High–field dipole domain formed near the drain - decreases the output conductance and dominates for large channel currents. More pronounced for shorter gate devices.

The relative contributions to the conductance from the channel length modulation, leakage current and dipole domain may be very different for different devices. Therefore, good control over both the value and the rate of change of Gds with respect to voltage Vg is required. Two features related to this property and often not properly simulated in existing MESFET models are specified below.

(3) Conductance can decrease with Jds.

Channel conductance Gds in the saturation region should be allowed to decrease with the increasing Vg voltage. This effect can be observed in most short–gate devices. The positive contribution from the leakage current dominates near the pinch–off, while the negative contribution of the dipole domain can be observed for higher currents. Typical DC characteristics of such a device are shown in Figure 3.2. It should be noted, that in some devices completely different variations of Gds with Vg are observed than those described above, and the output conductance can be seen to increase with Vg.

(4) Conductance can be negative.

For large channel currents (in particularly for longer gate devices), the high–field dipole domain formed near the drain may produce a negative slope on the DC characteristics (the so called Gunn effect). The model should be able to simulate this negative conductance. An example of the

DC characteristics of the device with the Gunn effect are shown in Figure 3.3.

(5) Pinch–off dependent on Vds.

The effective pinch–off voltage should be dependent on Vds. This effect is due to the leakage current through the substrate, particularly in devices with non–abrupt doping profiles [4].

(6) Saturation dependent on Vgs.

Control over the change of the saturation voltage with the gate–source voltage is required [9]. Devices with a large negative threshold voltage, –Vt, exhibit an "early" saturation, which is apparently independent of the gate–source voltage Vgs. On the other hand, devices with small threshold voltage saturate at the drain–source voltage $Vds \approx Vgs + Vt$ [9].

Empirical functions used in various MESFET models to describe the Jds(Vd,Vg) current source are compared in Table 3.1. In addition to the models discussed previously, a variant of the Materka & Kacprzak model proposed by Jastrzebski [19,10], and a recent model by Larson [9] are included in the Table. For each model, the number of parameters used in the Jds(Vd,Vg) function is given to indicate its complexity. Greater complexity usually results in greater versatility of the model, but also increases difficulty in identification of its parameters.

Also specified in Table 3.1 is the form of the function used for the simulation of the saturation of the channel current. The three different types of functions used in the relevant models are:

- hyperbolic tangent function (TANH),
- exponential function of a 3'rd order polynomial (EXP3),
- polynomial of a 3'rd order (POLY3).

All models were analysed from the point of view of their ability to simulate the six required properties of the DC characteristics. Results are displayed in Table 3.1. A plus sign "+" in this table means, that the model possesses the required property, while the minus sign "–", that it does not. The plus in parentheses "(+)" shows, that the model only partially fulfills the requirement. An asterisk "*"indicates, that in some cases the model may fail to predict the real behaviour of the device.

It can be seen from Table 3.1, that the DC models proposed by Tajima [2,3] and Jastrzebski [19,10] seem to predict correctly almost all the important device trends. Tajima's model is fairly complex and a majority of its parameters have no physical interpretation. The model proposed by Jastrzebski, however, is much simpler and its parameters can be interpreted on the DC characteristics. This model has the following form:

Table 3.1 : Properties of some DC MESFET Models

Name Reference	Curtice 1 [1]	Tajima 1,2 [2],[3]	Materka [4,5]	Curtice 2 [7]	Statz [8]	Larson [9]	Jastrzebski [19],[10]
No. of parameters	4	8	4	6	5	5	6,7
Saturation function	TANH	EXP3	TANH	TANH	POLY3	POLY3	TANH
Flexible dependence of JDS on Vg	–	+	–	+	+	+	+
Control over conductance dependence	(+)*	+,+*	–	–	(+)*	(+)*	+
Conductance can decrease with Jds	–	+	+	+	–	–	+
Conductance can be negative	–		–	–	–	–	+
Pinch-off dependent on Vds	–	+	+	+*	–	*	+
Saturation dependent on Vgs	–	–,+	+	–	–	+	+

Where:

+	the model has the required property
(+)	the model possesses only part of the required features
–	the model does not have the required property
*	the property may be wrongly simulated for some devices or there can be non-physical behaviour of the model

$$Jds(Vd,Vg) = \begin{cases} Jdss\ V^{\epsilon}\ [\tanh(\alpha Vd/Vgg+Vt)) + G\ Vd] & \text{for } Vgg \geq -Vt \\ 0 & \text{for } Vgg < -Vt \end{cases} \tag{10}$$

$$V = (1+Vgg/Vt)\ (1+Vgo/Vto) \tag{11a}$$

$$Vgg(t) = Vg(t-\tau)-Vgo \tag{11b}$$

$$Vt = Vto + Vgo + \gamma *Vd \tag{11c}$$

$$G = \frac{Go}{Jdss(1+Vgo/Vto)^{\epsilon}} \tag{11d}$$

Parameters of the above functions have the following interpretation (values for an "ideal" device are given in parentheses):

Jdss – saturation current – can be used as a scaling factor;

ϵ – exponent of the relationship between current and Vg, (ϵ =2);

α – parameter controlling saturation voltage Vsat. Approximately $Vsat \approx (Vg+Vto)/(\alpha/3-\gamma)$, ($\alpha$=3);

Vto – threshold voltage of an ideal FET, (*Vto=Vp–Vbi*, where Vp is a pinch–off voltage);

γ – coefficient of the threshold voltage dependence on Vd, (γ=0);

Vgo – If $\gamma>0$ and *Go*=0, then Vgo is equal to the Vg voltage, at which the Jds(Vd,Vgo) saturated characteristic has zero slope. For *Vg<Vgo* the slope is positive, while for *Vg>Vgo* the slope is negative, *(Vgo=Vbi)*.

Go – Additional component of the conductance. Go is equal to the conductance in the saturation at Vg=Vgo, *(Go =0)*.

Comparison of the above model with the original formula used by Materka & Kacprzak (Equation (7)) shows, that three additional parameters were introduced: ϵ, Vgo, and Go. The new term (1+*Vgo/Vto*) in Equation (11a) is only a normalising factor, which preserves the physical interpretation of the voltage Vgo and simplifies identification of model parameters. Additional conductance control is introduced by the term *G* x *Vd*, which for most devices can be omitted at DC, but is useful for modelling the I–V characteristics in presence of frequency dispersion.

An example of the quality of the modelling using Equations (10–11) is shown in Figures 3.2 and 3.3. The model gives an excellent agreement between measured and calculated DC characteristics both for low–noise (Figure 3.2a) and power (Figure 3.3a) devices. The modelled behaviour of the transconductance and the drain–source conductance (Figures 3.2 and 3.3), demonstrates all the features observed experimentally [11].

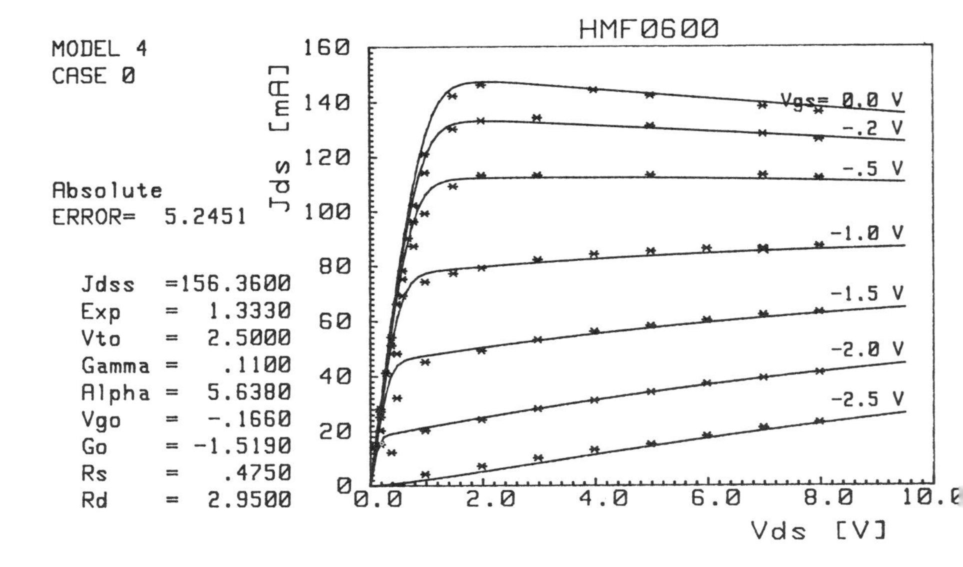

(a) Measured and calculated I–V characteristics ;

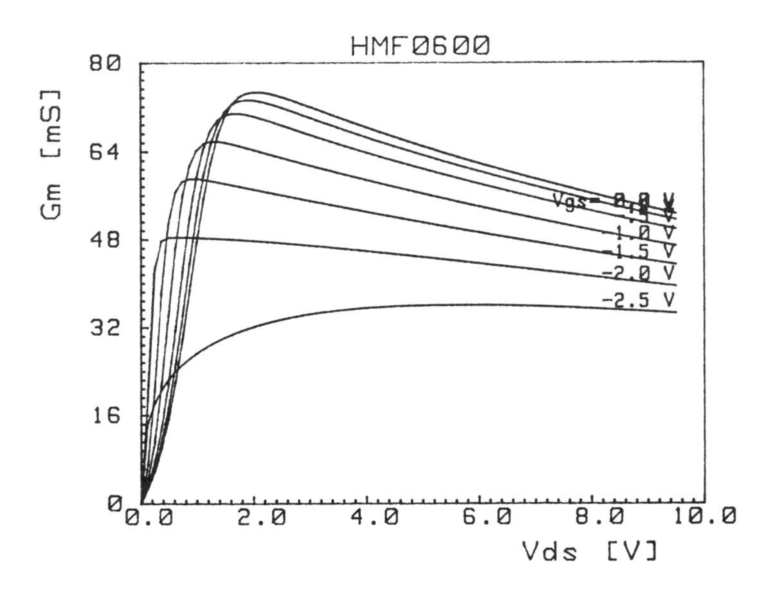

(b) Calculated transconductance g_{mo} ;

Figure 3.3 DC characteristics of the 600 μm power MESFET with the Gunn effect. Device was modelled using Equations (10–11)

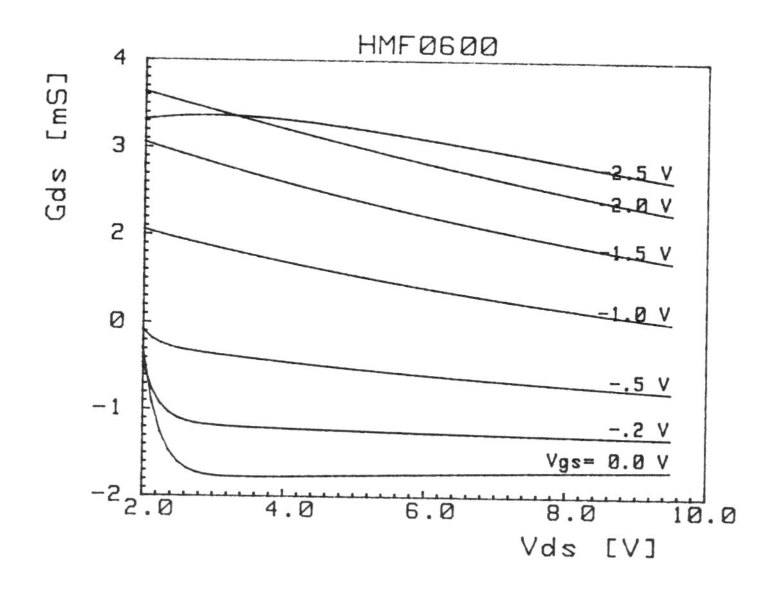

(c) Calculated output conductance Gds in the saturation region
Figure 3.3 DC characteristics of the 600 μm power MESFET with the Gunn effect (Contd)

3.3.2 Gate–source and Gate–drain Capacitances (Cgs, Cgd)

In a device operating in the active region, the gate–source capacitance Cgs describes changes of the charge in the depletion layer with respect to the voltage changes between gate and source. This is a lumped approximation of the essentially distributed capacitance of the gate. The gate–drain capacitance represents another part of the charge on this distributed capacitance. This artificially divides up the single gate charge and associates a portion of it with each capacitor. The main problem with modelling of Cgs and Cgd capacitances is, that they are non–physical [14]. Although, there have been a number of physically based models derived for the entire gate charge (see for example [33]), they are incompatible with the lumped approximation and cannot predict the division of the total capacitance between Cgs and Cgd. The only solution within the constraints of the existing lumped MESFET models seems to be the use of empirical formulae. Some of the most comprehensive, and therefore very complicated expressions, were proposed by Statz et al. [8]. They provide a smooth transition between all modes of MESFET operation including the inverse mode. For the normal active mode, a simple diode-like dependence given by Equation (2) is still used to describe variation of the depletion layer capacitance with the gate-source bias. All other models discussed in Section 3.2.2 also use the same equation.

Equation (2) does not describe correctly the value of the capacitance beyond the pinch-off voltage and gives infinite or complex values for voltages equal to or larger than the built-in voltage Vbi. For junction voltages beyond pinch-off, the depletion layer extends to the substrate and the associated capacitance falls to zero. Only small fringing capacitances remain. Conversely, when the voltage on the junction reaches the built-in voltage, the depletion layer disappears altogether. An infinite capacitance would simulate this fact quite well, but unfortunately, most of the non-linear computer programs (like for example SPICE) require a value of the charge for a given voltage and are not able to accept such an element (1). Therefore, for practical implementation in a computer program, Equation (2) has to be extended below the pinch-off and above the built-in voltage. The formulae proposed below are simplifications of the Statz capacitance model [8]. As symmetry is not required, then the change of the charge between Cgs and Cgd capacitances accompanying transition between normal and inverse modes is not simulated. In many models [1-3,5,7] the gate-drain capacitance Cgd is set to a constant value corresponding to the fringing capacitance between electrodes. In this model, however, the same equations are used for both capacitances.

$$C(V) = Cgf + \begin{cases} Cg(V) & \text{for } V<Vbi-dVbi \\ Cg(Vbi-dVbi) & \text{for } V>=Vbi-dVbi \end{cases} \tag{12a}$$

$$Cg(V) = \frac{Cg}{\sqrt{(1-Ve/Vbi)}} \quad \frac{(1+E)}{2} \tag{12b}$$

$$E = \frac{V+Vto}{\sqrt{(V+Vto)^2 + (dVto)^2}} \tag{12c}$$

$$Ve = \frac{1}{2}\left[V-Vto+\sqrt{(V+Vto)^2 + (dVto)^2}\,\right] \tag{12d}$$

Where:

V - voltage on the capacitance (either Vg or Vgd);

Cgf - gate fringing capacitance (can be different for Cgs and Cgd);

(1) One of the exceptions is CAD program ANAMIC developed at the University of Kent, where non-linear capacitor can be described as a voltage source dependent on charge, i.e. V=f(Q). An infinite capacitance can therefore be simulated as V=const(Q). Capacitances in MESFET models implemented in ANAMIC use this feature with good results.

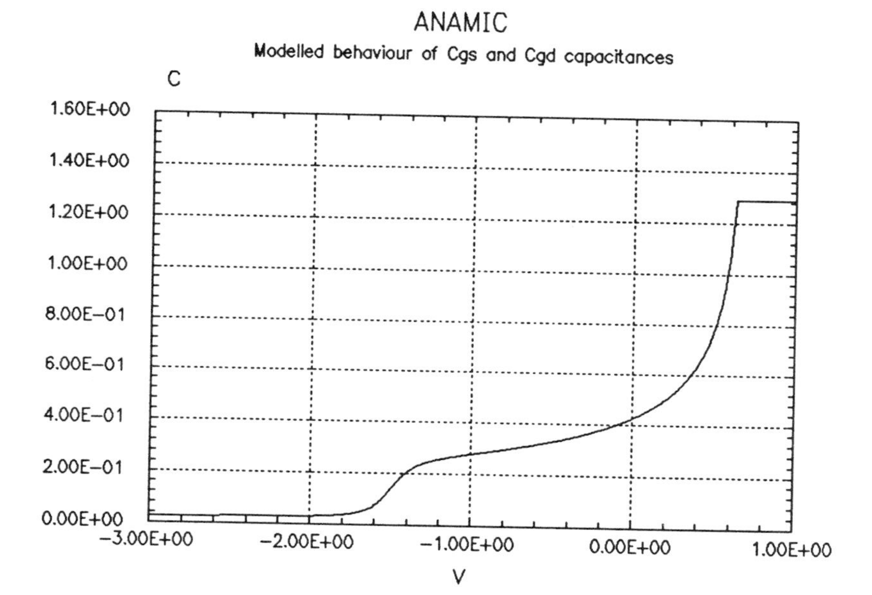

Figure 3.4 Simulated variation of the gate–channel capacitance using Equation (12). Parameter values are: Cgo = 0.4pF, Vbi = 0.7V, Cgf = 0.025pF, Vto = 1.5V, d = 0.1.

Cgo – depletion capacitance at zero bias;

Cg(V) – non–linear depletion capacitance;

Vbi – built–in voltage of the gate–channel Schottky barrier junction;

Vto – threshold voltage (parameter of the DC model);

d – a small positive number defining relative range of voltages over which fast change of the capacitance value takes place (can be set to 0.1).

The non–linear dependence of a capacitance described by Equation (12) is shown in Figure 3.4. When the voltage, V, on the capacitor is negative and well below the threshold voltage, –Vto, then the switching function $E \approx -1$ and $C(V) \approx Cgf = const(V)$. When the voltage, V, is well above the threshold voltage, –Vto, then $E \approx 1$ and the effective voltage $Ve \approx V$. Equation (12b) reduces then to the diode–like form (2). When the voltage, V, is just below or above Vbi, then the capacitance is set to a constant value $C = Cgf + Cgo/\sqrt{d}$.

3.3.3 Schottky Barrier Diodes (Dgs, Dgd)

These diodes have a distributed character, similar to the capacitances Cgs and Cgd, and cannot be accurately modelled by the lumped approximation. They represent the resistive component of the current through the Schottky barrier between the gate and channel. Under typical bias conditions and at small-signal levels both diodes are reverse biased and do not influence the device performance. For large signals, the gate–source part of the Schottky barrier may be forward biased. The main function of the Dgs component is to simulate that forward current. A typical diode Equation (3) is usually used to describe the gate–source current.

The Dgd diode is symmetrical with Dgs, but its main function is to model the breakdown current between gate and drain. In most MESFET models either a piece–wise Equation (6) or a diode–like Equation (9) is used to describe the breakdown current.

A simple new form of the diode equation is presented here, which simulates both forward and reverse breakdown currents. The effective breakdown voltage is dependent on the drain–source current Jds, and in this way the effect of the space charge dipole formed at the drain is taken into account. The same form of the function is used to model the current in both diodes, but the parameter b is set to zero for the Dgs diode.

$$Jd(V,Jds) = Jo\ [\exp(V/VT)-1]\ [1+\exp(-V/V_B\ )]\ 0.5 \qquad (13a)$$

$$V_B = V_{Bo} + b\ Jds^2 \qquad (13b)$$

$$VT = nkT/q = n\ 0.026\ V \text{ for } T=300°\ K \qquad (13c)$$

Where:

n – ideality factor;

Jo – zero current of the diode;

V_B – effective breakdown voltage coefficient. If for the voltage $-V_{br}$ the breakdown current is $-I_{br}$, then V_B can be approximated as:

$$V_B = \frac{-V_{br}}{ln(-2I_{br}/Jo} \quad ; \qquad (14)$$

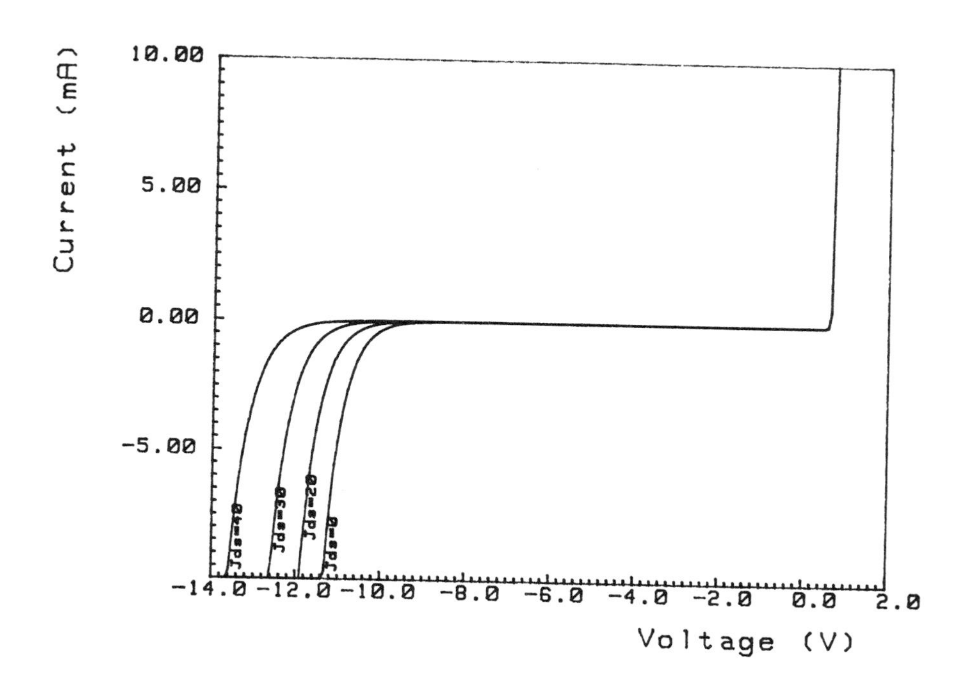

Figure 3.5 Simulated diode current using Equation (13). Parameter values are: *Jo* = 10^{-11} *mA*, *VT* = *0.026V*, V_B = *0.4V*, *b* = 5 x 10^{-5}*V* mA^{-2}, *Jds* = [0,20,30,40] *mA*.

V_{Bo} – breakdown voltage coefficient when Jds=0;

b – coefficient relating breakdown voltage to the channel current.

A plot of the simulated characteristics of the Dgd diode in a MESFET model is shown in Figure 3.5. Both the breakdown current and the forward conduction current can be clearly seen. The effective breakdown voltage is modified by the drain–source current Jds. For larger Jds currents, the breakdown begins at more negative voltages.

3.3.4 Undepleted Channel Resistance Ri

The Ri resistance is a charging resistance of the Cgs capacitance. In some models [1,8,9] this resistance is not included at all, while in models [5] and [7], it has a constant value. In the Tajima model [2,3], Ri is dependent upon Vg and is inversely proportional to Cgs, so that the charging time constant

Ri x Cgs remains the same. In the model presented here, the expression used in [6] is adopted, which for the special case of r=0.5 is equivalent to the Tajima's formula.

$$Ri(Vg) = Rio\ |1 - Vg/Vr|^r \qquad (15)$$

Where:

Rio – value of the resistance at Vg=0;

Vr – voltage at which Ri becomes zero (theoretically Vr=Vbi);

r – exponent of the value between 0.5 and 2 (theoretically r=0.5)

3.4 Non–linear Modelling From Measured S–parameters

The need for modelling of MESFETs in the design stage and sometimes even during the production of microwave circuits can be attributed to two main factors.

Firstly, in typical applications, the high requirements set by the microwave industry push the operation of MESFETs up to the limits of their physical capabilities. MESFETs rarely are used below one tenth of their current gain cut–off frequencies f_T. Microwave circuit designs are usually required to extract maximum performance from the device and so, it is extremely difficult, often impossible, to make them insensitive to changes in device parameters. In the CAD of such circuits, relatively small errors in the device model may cause large errors in the simulation results (2). Therefore, very accurate models are required.

Secondly, considerable variability is observed between devices of the same type. Manufacturer's published data or models included in libraries of the CAD programs rarely describe a particular device accurately enough. Although, there are variations among devices produced on the same wafer, the differences are much smaller than between devices from different wafers or produced at different times. Therefore, an optimum way to obtain a reasonable accuracy is to derive a model for each batch of MESFETs. Preferably, several devices from the same batch should be characterised to assess tolerances. The model which has parameter values closest to average should be selected. It is not a good practice to use average values of model parameters, in particularly if spreads of values are quite large. In this case, the intrinsic consistency of the model could be broken up, and such properties like maximum available gain and stability factor might be wrongly predicted.

(2) For the same reasons, it is difficult to achieve good production yields of circuits designed for performance without tweaking and tuning.

In case of non-linear microwave CAD, derivation of user's own models is additionally necessitated by an almost complete absence of the relevant information in the manufacturer's data. This fact is often attributed to the lack of good, standard non-linear MESFET models, but difficulties in extracting these models are probably as important.

Identification of non-linear circuit models of microwave devices in general, and of MESFETs in particular, is a very complex and difficult problem. The main reason is, that model elements cannot be directly measured at microwave frequencies. They cannot be accurately calculated from physics based analytical expressions either, because they depend on such unpredictable factors like variations in processing cleanliness, surface condition, doping profile tail, and defects in the semiconductor. Although some of the model parameters can be determined from DC or low frequency measurements, their values at microwave frequencies may be different due to the frequency dispersion phenomenon [26-28]. Therefore, a majority of model parameters have to be extracted from the small-signal scattering parameters measured over broad range of microwave frequencies.

3.4.1 Optimisation of MESFET Models – Problems and Difficulties

In a typical parameter extraction approach, an optimisation method is used, which varies model elements until the calculated S-parameters fit the measurements. There are well known difficulties with MESFET model extraction from measured S-parameters [17,18]. Firstly, the results are often inconsistent, i.e. they depend upon the starting point and a particular procedure used. Secondly, they may be not unique, i.e. several almost equally good solutions may exist with completely different element values. Finally, at the point of the best fit, values of some model elements may be non-physical.

Some of these problems can be explained by the fact, that models do not simulate adequately enough all the important aspects of the device behaviour. The MESFET model shown in Figure 3.1 does not correspond directly to the physical structure of the device (see Ladbrook [35, Chapter 6]), but it is a convenient, simplified approximation. Consequently, there is no single set of model parameters which would give a perfect match between measured and calculated terminal properties over wide frequency range.

The intrinsic inadequacy of the model adds up to the much more severe consequences of the S-parameter measurement and de-embedding errors. Random errors can usually be eliminated in the optimisation process. However, systematic calibration and de-embedding errors cannot be removed and contribute to further substantial increase in the inconsistencies of the results. For larger de-embedding errors the model may not be able to follow the measured behaviour at all or some model parameters may become non–physical. For example, if the bond wire inductances were over–stripped, then the extracted values of series inductances might be negative.

Additional difficulties with parameter extraction are related to the general properties of optimisation methods.

(a) Computer time and uncertainties increase substantially with the number of optimised variables. Current optimisation procedures cannot efficiently handle circuit optimisation problems with more than ten variables. However, the basic linear MESFET model has 13 parameters (Figure 3.1b), while the non–linear model has even more than that.

(b) Some model elements have little influence on the circuit response. They are difficult to identify, because any values of these elements (within certain range) give almost the same S–parameter fit.

(c) Elements and parameters of the model are not always independent. Such elements , like for example Rg and Ri in Figure 3.1, cannot be easily distinguished from each other when looking at the device terminals [18]. If optimised at the same time, they create a singularity in the optimisation problem. In this case, the optimisation converges very slowly and the whole family of solutions may exist, which may include non–physical values of parameters.

3.4.2 Method Based on Simultaneous Optimisation at Many Bias Points

Even for a single linear model, some elaborate methods and complicated optimisation procedures have to be employed in order to increase reliability of the extracted model parameters [17,18]. In case of the non–linear model, the complications are multiplied, because several linearised models over a range of bias conditions need to be considered. In a typical approach, each linearised model is optimised separately. Analysis performed in [20] shows, that this is equivalent to the constrained hierarchical two–level optimisation with a very large number of unknowns and functions. In concept, the obtained values of linear model elements should be the same at all bias points, while the non–linear elements should vary with bias according to their physical behaviour. In practice, however, nonuniqueness and inconsistencies of the parameter extraction at individual bias points, coupled with measurement errors, make this procedure very unreliable and extremely difficult. Probably because of these difficulties, the recommended parameter extraction procedures for most of the models reviewed in Section 3.2.2 rely on the optimisation to only one set of S–parameters. This, obviously, does not give sufficient information to identify all the parameters of the non–linear modelling functions.

The improved non–linear modelling method, first proposed and developed by Jastrzebski(3) [19,20] and also recently published by Bandler et al. [21] resolves the parameter extraction problems discussed above very efficiently. The basic idea behind the method is to optimise the whole non–linear model simultaneously at all measured bias points. The non–linear elements are described by functions which constrain their variations with the bias, while the linear elements are kept the same at all bias points. This reduces the degrees of freedom during the optimisation and at the same time, introduces perturbations into the model. The probability of finding a global and meaningful minimum of the error function is increased. Consistency and uniqueness of the model parameters are also better, because the non–linear elements are forced to obey physical relationships, and therefore some wrong solutions to the optimisation problem at individual bias points are eliminated. The effects of random measurement and extraction errors between the bias points are smoothed out, in a similar fashion to that in the curve fitting process. One of the important benefits of this method is, that it can be applied to extract parameters of the non–linear MESFET models with frequency dispersion [19,20]. As it was pointed out in Section 3.2.3, for devices with dispersion in both transconductance and output conductance, the Jds(Vd,Vg) current source (see Figure 3.1) should simulate pulsed or RF rather than the DC I–V characteristics. This method allows the dispersive I–V characteristics to be found directly from S–parameter measurements. In this case, the non–linear function Jds(Vd,Vg) is optimised to the measured S–parameters over large range of bias points to match the required transconductance and output conductance values at microwave frequencies [19,20].

3.4.3 Computer Algorithm

The modelling method described above requires a dedicated computer procedure, because existing general microwave CAD programs do not have all the necessary features. The method has been implemented in the computer program SOPTIM, developed at the University of Kent, which has been extensively tested on a large number of examples, and is currently used in several industrial laboratories and universities. Main aspects of

(3) This method has been implemented in the computer program SOPTIM for non–linear modelling of microwave devices, developed at the University of Kent in 1984. It was first used to model MESFETs for the monolithic VCO described in [6].

this computer algorithm, important from the point of view of the parameter extraction procedure discussed in Section 3.5, are presented below.

Although, the program extracts non-linear MESFET models, the large-signal analysis methods are not required. The linear AC analysis is used to calculate S-parameters of the model at each of the measured bias points. Measured and calculated S-parameters are compared and the difference between them is minimised. The error function during the minimisation represents combined errors of the small-signal models at all bias points.

The internal loop of the algorithm, which is performed for each bias point, consists of the following steps:

1. Solve the DC circuit and calculate the internal controlling voltages Vgo, Vdo, and Vgdo.
2. Calculate values of non-linear elements from the given functions and create a linearised model.
3. Perform AC analysis of the linearised model at the same frequencies, at which S-parameters were measured.
4. Calculate differences between simulated and measured S-parameters and add a contribution to the error function from the analysed bias point.

A. Solving for DC Bias Point

The DC analysis of a MESFET can be simplified, if typical bias conditions are assumed, i.e. if the Schottky barrier junction is reversed biased and there is no breakdown. DC model of a MESFET for this case is shown in Figure 3.6. Intrinsic controlling voltages Vgo, Vdo, and Vgdo can be found by solving numerically the following system of non-linear algebraic equations:

$$Vgo = Vgso - Jds(Vdo, Vgo)\ Rs \tag{16a}$$

$$Vdo = Vdso - Jds(Vdo, Vgo)\ (Rs + Rd) \tag{16b}$$

$$Vgdo = Vgo - Vdo \tag{16c}$$

Because of the possible presence in the I–V characteristics of the region with negative differential conductance (Gunn effect), general numerical methods may experience convergence problems in solving Equations (16).

An equivalent formulation of the DC problem, which results in a single non-linear equation with one unknown variable, the channel current Jdso, has the form:

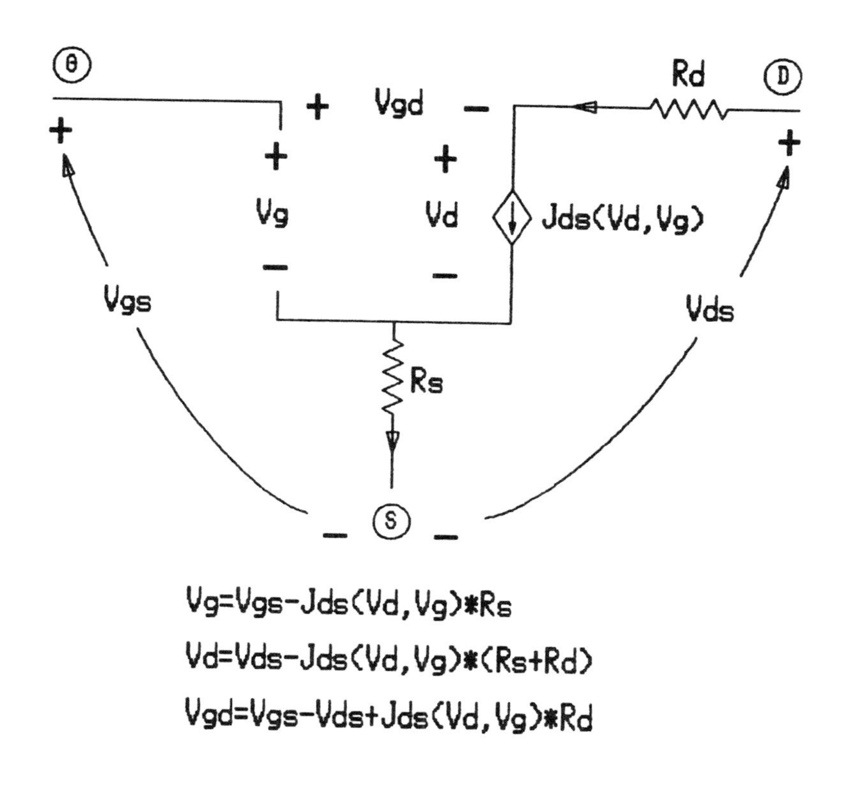

Figure 3.6 Simplified DC MESFET model.

$$Jdso = Jds(Vdso - Jdso\ (Rs+Rd),\ Vgso - Jdso\ Rs) \tag{17}$$

In this case, the solution can be found with a variety of well convergent methods (for example *regula falsi*), and then the controlling voltages can be calculated by a simple substitution for Jds in (16).

B. Non-linear Functions and Their Parameters

Values of non-linear elements are calculated for a given bias point from non-linear functions. Because there is a large variety of possible optimisation tasks, then the function structure must be sufficiently general. The form shown below fulfills most requirements.

$$X = f_X\ (V,P,E) \tag{18}$$

Where:

X – Non-linear element value;

$f_X(.)$ – Function describing element X;

V – Parameters, which have a different value at each bias point (so called vector or point–dependent parameters). For example, bias voltages Vgso and Vdso are of this type;

P – Parameters, which have the same value at all bias points (ordinary parameters). An example is the built–in voltage Vbi;

E – Element values as parameters (element parameters).

As an example, a function to calculate value of the capacitance Cgs from Equation (12) can be defined as follows:

$$Cgs = f_{CGS}\ [V_1 = Vgo,\ P_1 = Cgo,\ P_2 = Vbi,\ P_3 = Cgf,\ P_4 = d] \tag{19}$$

In this case, V_1 is equal to the internal controlling voltage Vgo, and four ordinary parameters $P_1 - P_4$ specify coefficients of the Equation (12). A separate computer subroutine to calculate value of the function f_{CGS} for given values of parameters must be attached to the program. In practice, there is a number of built–in subroutines for various models in the library. If these are insufficient, then a user can write his own subroutines and link them with the program.

A function can be arbitrarily complicated. For example, it can perform numerical solution of the DC Equations (16–17). The appropriate format for this case is shown below.

$$Jds = f_{JDS}\ [V_4 = Vgso,\ V_5 = Vdso,\ E_1 = Rs,\ E_2 = Rd,\ V_1 = Vgo,\ V_2 = Vdo,\ V_3 = Vgdo\ ] \tag{20}$$

Where:

- vector parameters V_4 and V_5 are the given terminal voltages Vgso and Vdso, respectively;
- element parameters E_1 and E_2 are series resistances Rs and Rd;
- calculated by the subroutine internal controlling voltages Vgo, Vdo, and Vgdo are stored in the vector parameters V_1, V_2, and V_3, respectively.

In the last example, the function assigned values to the vector parameters, which then can be used by other functions (see, for example V_1 in f_{JDS} and f_{CGS} above). This feature can be employed to describe elements, for which the unknown functional dependence is to be identified from the variations of their values with bias voltages. For example, the drain–source resistance Rds can be described by the identity function:

$$Rds \equiv V_6 \tag{21}$$

If the optimisation is performed in terms of the vector parameter V_6, then separate values of Rds at each bias point will be found by the program.

C. Optimisation Variables

In general, all parameters of the MESFET model can be optimised at the same time, including values of circuit elements and coefficients of non–linear functions. In case of linear elements, their values are adjusted simultaneously in all linearised models and are forced the same at all bias points. Non–linear elements have different values for different biases and they can be optimised in two different ways:

(a) Separately at each bias point, which gives a set of values, from which bias dependence of the element can be determined (see, for example, the discussion of the function (21) above);

(b) As a non–linear function with adjustable coefficients, which are fitted to match measured S–parameters at all bias points.

Whenever general forms of the equations describing non–linear elements are known, method (b) is used, and the optimised variables are coefficients of these equations. Equations constrain the way non–linear element values can change, which gives good convergence and uniqueness of the solution. If the equations are not known or there is a doubt about their ability to simulate a particular device behaviour, then method (a) should be applied.

D. Error Function

The error function for this parameter extraction problem is a measure of the difference between calculated and measured terminal properties of the device at all bias points and at all measured frequencies. Convergence and robustness of the optimisation algorithm is, to a large extend, determined by the form of the error function used. To allow for flexibility and easy implementation of various optimisation schemes, the error function has the following general form:

$$Error(P,E) = \left[\sum_{k=1}^{K} \sum_{f=F_{min}}^{F_{max}} \sum_{i,j=1}^{2} (W^{kf} W_{ij} \left| Sc(P,E)_{ij}^{kf} - Sm_{ij}^{kf} \right|)^p \right]^{\frac{1}{p}} \tag{22}$$

Where:

P,E – parameters of non–linear functions and values of linear elements, respectively;

$\sum_{k=1}^{K}$ – sum over K bias points ;

$\sum_{f=F_{min}}^{F_{max}}$ – sum over the range of frequencies from F_{min} to F_{max};

$Sc(P,E)_{ij}^{kf}$ – calculated S_{ij} parameter at the bias point k and frequency f;

Sm_{ij}^{kf} – measured S_{ij} parameter at the bias point k and frequency f;

W^{kf} – weighting factor for the measurement at the bias point k and frequency f – the same for all S–parameters;

W_{ij} – weighting factor for S_{ij} – the same at all frequencies and bias points;

p – type of the norm; if p=1, then the l_1 norm; if p=2, then the least–squares norm is used.

All parameters of the error function can be adjusted by the user. In particular, a set of the optimised bias points and a range of active frequencies can be selected and changed during the optimisation. The weighting factors can be also modified from their default values.

Initially, all W^{kf} are set to one, which corresponds to an equal weight of all measurements. Default values of weighting factors W_{ij} are set by the program in such a way, that contributions to the error function of different S–parameters are made equal. W_{ij} can be calculated from the inverse of the norm $|| Sm_{ij} ||$, which gives an error function normalised to one:

$$W_{if} = \frac{0.25}{|| Smij ||} \tag{23a}$$

$$|| Sm_{ij} || = \left[\sum_{k=1}^{K} \sum_{f=F_{min}}^{F_{max}} \sum_{i,j=1}^{2} | Sm_{iJ}^{kf} |^{p} \right]^{\frac{1}{p}} \tag{23b}$$

E. Optimisation Methods

Optimisation methods which utilise both function values and their first derivatives are in most applications faster than those relying on the function values alone. However, the situation is not so clear for the parameter extraction problem, which suffers from ill–conditioning and multiple solutions, and some of the modelling functions may have discontinuous first derivatives. Robustness rather than speed is here more important and from that point of view, methods *not* based on first derivatives are better. Additionally, first derivatives of user–defined non–linear functions are, in general, not available and would have to be calculated numerically. For all these reasons, the optimisation methods chosen for the SOPTIM program require only function evaluations.

Two different stages can be identified during the process of optimisation of a complicated multidimensional error function, such as (22). In the first stage, while the solution is still far from the minimum, the best suited is a direct search optimisation method. Because of the possibility of local minimums, the method should have a certain element of randomness and must be very robust. The non–linear *simplex method* due to Nelder and Mead [36, Chapter 10] with random choice of the initial simplex points has proved to be very reliable in this application.

In the proximity of the minimum, the direct search methods converge relatively slowly. Therefore, in the final stage of the optimisation, a switch to the more efficient subroutine, such as for example *conjugate directions method* due to Powell [36, Chapter 10], should be made.

3.5 Parameter Extraction Procedure For Non–linear MESFET Models

If extrinsic parasitics are excluded, then the MESFET model in Figure 3,1c has 9 linear elements. For the model given in Section 3.3 and assuming a symmetric MESFET structure (i.e. Cgs(V)=Cgd(V), Dgs=Dgd), there are 19 unknown coefficients in the Equations (10)–(15) describing non–linear elements. This gives a total of 28 parameters of the non–linear MESFET model to be identified. The modelling procedure for such a large number of parameters needs to be very carefully designed, even if the improved optimisation method described in the previous section is used. In order to obtain reliable and repeatable results, the problem must be partitioned into smaller and simpler tasks, which can be performed sequentially.

The algorithm of the modelling approach developed and tested at the University of Kent [19,20] is specified in Figure 3.7. (4). The first part of the procedure (steps 1–5) is common to all cases and gives values of series resistances and inductances, DC model of the Schottky–barrier junction, and approximate values of all other model parameters. If external parasitics are not stripped off from the measurements, then they are determined in step 4, together with series inductances.

The second part of the procedure (steps 6–8) depends on type of the available I–V measurements. The preferable route is to use pulsed

(4) The complete parameter extraction algorithm for non–linear MESFET models has been implemented in a computer program FETMEX, recently developed at the University of Kent.

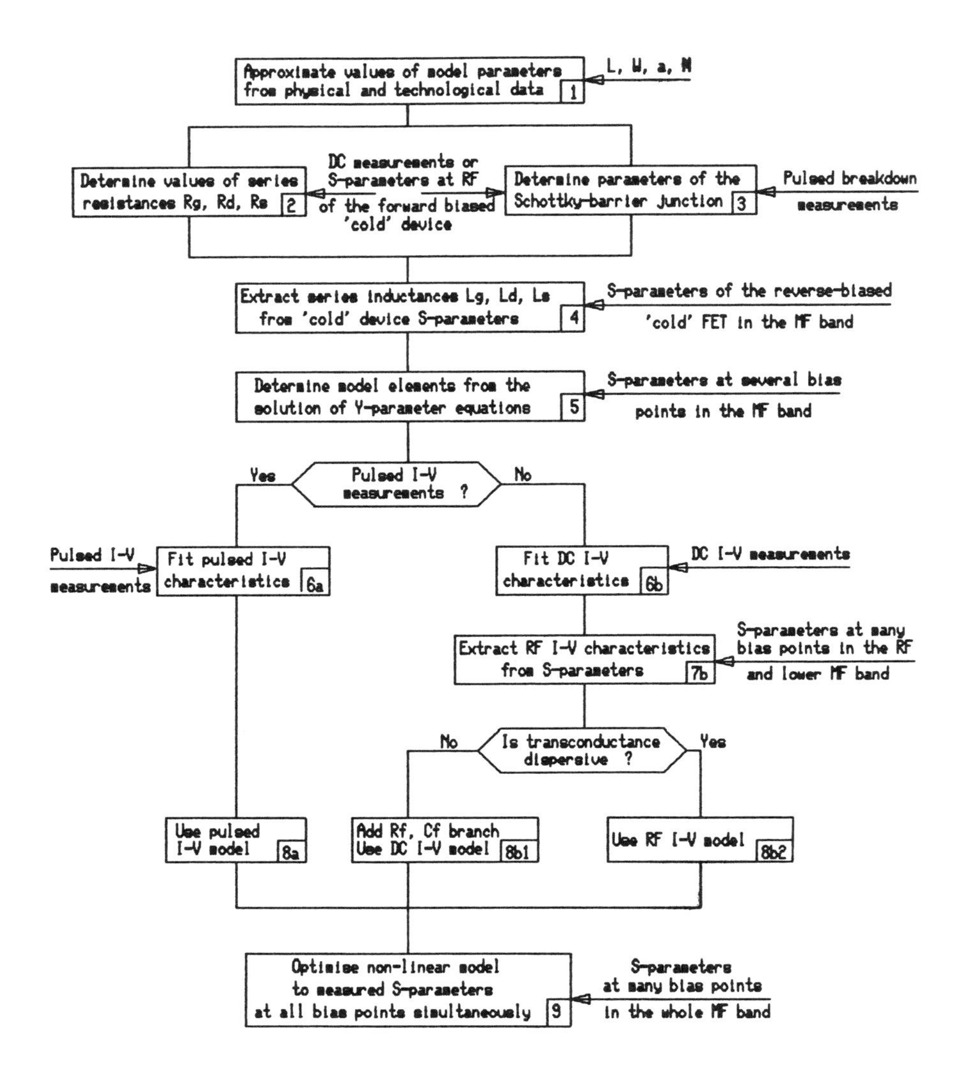

Figure 3.7 Parameter extraction procedure for non-linear MESFET models

(instead of DC)I–V characteristics [30,31]. As stated in [30], the best method for characterising a FET in the whole plane $Ids = f(Vds, Vgs)$, is to bias the transistor at $Vgso = 0$ and $Vdso = 0$ and measure it by synchronous short pulse voltages of variable amplitudes applied to the gate and the drain. If pulses are shorter than the recombination rates of the traps (about 20 μs), then the trap occupancy does not change and their effects (i.e. dispersion) are removed from the measurements. Additionally, for pulse widths not exceeding about 200 ns, the temperature effects are also removed [30], which gives improved accuracy of the I–V model.

The modelling is greatly simplified with pulsed I–V characteristics, because the quasi–static approach can always be applied, regardless of whether a device exhibits frequency dispersion or not. (5) Unfortunately, sufficiently fast pulsed measurement systems are rather complicated [30,31] and are not commercially available. An alternative I–V characterisation, which also removes the effects of dispersion is based on the large–signal RF measurements [29]. However, the required hardware and software are even more complicated than in the pulsed systems.

If only DC I–V characteristics are available, then a dispersive modelling approach proposed by Jastrzebski [19,20] can be used. First, a DC I–V model is derived by curve fitting (step 6b). (6) Next, the RF I–V model is obtained by optimisation of the Jds current source to S–parameters measured at many bias points in the RF band and lower end of the microwave band (step 7b, frequency range 45MHz–2GHz). Small–signal transconductance calculated from the RF I–V characteristics is then compared with the transconductance from the DC characteristics. If there is a substantial difference, then the device is strongly dispersive and the RF I–V model is used to simulate Jds current source. Otherwise, the DC I–V model can be used with the Rf, Cf branch added to account for a dispersion in the drain–source conductance.

In the final stage of the model extraction (step 9), accurate values of parameters describing intrinsic passive elements Cgs, Cgd, Ri, and Cds and the transit time τ are found from a family of small–signal S–parameter measurements using the optimisation algorithm described in Section 3.4. If necessary, all other parameters may also be adjusted. S–parameters should cover the whole range of operating conditions of the device, in both saturation and resistive regions. The frequency range should start from the smallest possible value (for example 45 MHz) and extend to the maximum value, at which the measurement and de–embedding errors are still acceptable. With the current hardware, the maximum frequency is typically between 18 and 26.5 GHz.

Details of some of the steps of the parameter extraction procedure are discussed below.

3.5.1 Approximation Of Model Parameters From Physical And Technological Data

The problem of calculating parameters of the MESFET model from physical and technological data is extensively treated by Ladbrook [35,

(5) Limitations of the models of dispersive devices derived in this way are discussed in Section 3.2.3.

(6) Computer program MESFET DC developed at the University of Kent can be used to identify an I–V model of a MESFET very efficiently, with the aid of interactive graphics and optimisation.

Chapter 6]. The equations given below are only first-order approximations, used as a starting point for the modelling procedure. These expressions are only valid for submicron GaAs MESFETs with aluminium gates, and at a room temperature.

$$Vp = Vto + Vbi = 7.23\ N_D\ a^2\ (V) \tag{24a}$$

$$Vbi = 0.706 + 0.06\ log(N_D)\ (V) \tag{24b}$$

$$Jdss = 0.224W\ N_D\ a \left[1 - \sqrt{\frac{Vbi + 0.234\ L}{V_p}}\right] (mA) \tag{24c}$$

$$Jo = 3.4\ x\ 10^{-8}\ L\ W\ (A) \tag{24d}$$

$$Cgo = 3.33\ x\ 10^{-4}\ (L + 2a)\ W\ \sqrt{N_D}\ (pF) \tag{24e}$$

$$Cgf = 1.7\ x\ 10^{-4}\ x\ W\ (pF) \tag{24f}$$

$$\tau = 7.4\ L^{1.56}\ (ps) \tag{24g}$$

Where:

N_D – channel doping density in $10^{16} cm^{-3}$;

L – channel length in μm;

W – channel width in μm;

a – thickness of the channel layer in μm.

3.5.2 'Cold' FET Measurements

When the drain–source voltage of the FET is set to zero, then the device becomes passive, i.e. 'cold'. Under such conditions, operation of the MESFET is relatively simple to describe analytically. Usually, a distributed gate structure is considered to increase accuracy of the model. There are several techniques for extraction of extrinsic series elements based on 'cold' measurements [22–24]. In this sub-section a unified approach will be presented, which is derived from the lumped equivalent circuit of the passive MESFET shown in Figure 3.8.

This equivalent circuit, contains the same external series elements and fringing capacitances as the model in Figure 3.1. Note, however, that the fringing capacitances in this model are separated from the main gate capacitance. Other elements are different in both models. (7) Element Cg in Figure 3.8 represents a total capacitance of the depletion layer, and Dg

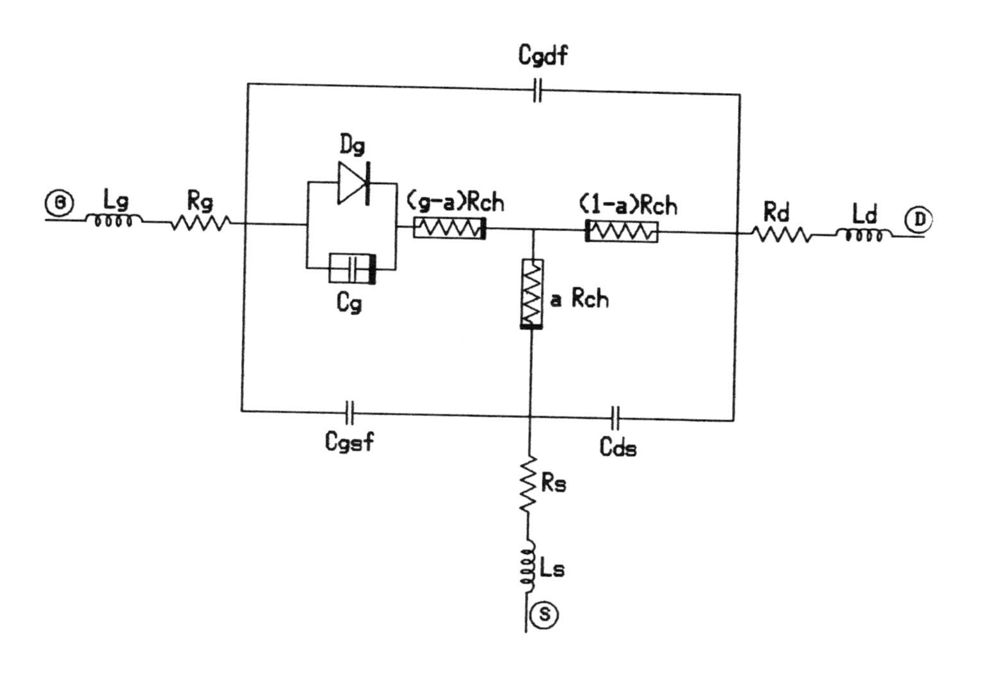

Figure 3.8 Equivalent circuit of a 'cold' MESFET (Vds = 0)

simulates a Schottky–barrier diode between gate and channel. Rch is a distributed non–linear channel resistance, which in this lumped representation is split into three parts: aRch, gRch, and (1–a)Rch. Coefficients a and g depend on the value of the gate current and on whether small– or large–signal case is considered. Formulae for a and g were derived in [24] and were also quoted in [33]. When the voltage drop across the aRch resistance is much smaller than the thermal voltage $VT = qkT/q$, i.e. for small gate currents, then it can be assumed that $a = 1/2$ and $g = 1/3$ [24]. In this case, the circuit in Figure 3.8 is the same for the common–source and the common–drain configurations. The maximum gate current for which coefficients a and g can be treated as constants can be calculated from the approximate expression:

(7) This is one of the examples of the limitations of the generally acceptable MESFET model in Figure 3.1 – it cannot be used to predict accurately 'cold' device measurements.

$$Igmax \leq 0.2 \quad \frac{kT/q}{0.5\ Rch} \approx 10^{-5}\ \frac{a\ W}{L} \quad N_D^{0.82} \quad (A) \tag{25}$$

Where: W, a, and L are in μm, and N_D is in $10^{16}cm^{-3}$.

The passive MESFET equivalent circuit can be further simplified for particular ranges of frequencies and gate bias voltages.

A. Determination Of Series Resistances And Junction Parameters From Rf S–parameter Measurements.

The method presented here is related to the 'end' resistance measurement technique developed by Lee et al. [24] for the DC case, and it is an extension of the work by Vogel [25], who proposed the usage of RF S–parameter measurements to extract series resistances. The main benefit of this modified method is, that the only data required are the low–frequency (below 500 MHz) S–parameters for the forward biased MESFET. The device can be mounted in the same fixture as used for other Network Analyzer measurements, and no reversing between common–source and common–drain configurations is needed. An additional advantage over the method based on DC measurements is that it takes into account dispersion effects in Rs and Rd series resistances [27].

The method will be explained for the case of the device in the common source configuration. Small–signal S–parameters are measured at one or more RF frequencies at various values of the forward DC gate current and the gate–source voltage. S–parameters are converted to Z–parameters, which in this case have a very simple form, because reactive components can be omitted.

$$Z_{RF} = \begin{bmatrix} Rg+Rj+Rs+gRch & Rs+aRch \\ Rs+aRch & Rs+Rd+Rch \end{bmatrix} \tag{26}$$

In the above Equation, Rj represents the differential resistance of the forward biased gate junction Dg. It can be derived from the diode Equation (3):

$$Rj = \frac{n \times VT}{Ig} = \frac{n\ k\ T}{q\ Ig} \tag{27}$$

For the uniform doping profile, the channel resistance Rch is a function of the gate bias

$$Rch = Rcho \times \eta \tag{28a}$$

$$\eta = \frac{1}{1-\sqrt{\dfrac{-Vg + Vbi}{V_p}}} \tag{28b}$$

For simplicity, it will be assumed that condition (25) is fulfilled. Otherwise, expressions for a and g given in [24] would have to be used (denoted there as α' and α_g, respectively). In general, a and g are functions of $(Ig\,Rch/VT)$ and can only be found by iterations. The passive MESFET model is reciprocal, i.e. $Z_{21} = Z_{12}$ and the Z matrix (26) has only three independent components. However, there are five unknown elements: Rg, Rd, Rs, Rj, and Rch. Therefore, some additional equations or measurements are needed. If the bias dependencies (27)–(28) are utilised, then the series resistances can be extracted as follows.

(i) Z–parameters are linearly dependent on Rch and consequently, are also linear functions of η given by Equation (28b). Providing Vbi and Vp are accurate, then the plots of $Z_{21}(\eta)$ and $Z_{22}(\eta)$ should follow straight lines.

$$Rs = Z_{21}(\eta \longrightarrow 0) \tag{29a}$$

Similarly

$$Rs+Rd = Z_{22}(\eta \longrightarrow 0) \tag{29b}$$

and Rd can be obtained. In practice, however, Vp must be adjusted until the straight lines are obtained. Also, the internal voltage Vg is initially not known and the external voltage Vgs must be used instead. Therefore, the second iteration is required, with voltage Vg properly calculated.

(ii) By eliminating Rch from the equations for Z_{11} and Z_{21}, the following function $G(1/Ig)$ can be defined:

$$G(\frac{1}{Ig}) = Rg + \frac{n \text{ x } VT}{Ig} = Z_{11} - \frac{g}{a} Z_{21} - (1-\frac{g}{a})Rs \tag{30a}$$

For small Ig currents, Equation (33a) becomes

$$G(\frac{1}{Ig}) = Rg + \frac{n \text{ x } VT}{Ig} = Z_{11} - \frac{2}{3} Z_{21} - \frac{1}{3} Rs \tag{30b}$$

Plot of $G(1/Ig)$ should be a straight line and its intercept with the ordinate axis gives Rg

$$Rg = G(\frac{1}{Ig} \rightarrow 0) \qquad (31)$$

(iii) If the temperature is known, then the ideality factor n of the diode can be found from the slope of the function G

$$n = \frac{dG}{d(1/Ig)} \; \frac{q}{kT} \qquad (32)$$

Zero current Jo of the diode can be calculated from the measured bias points (Ig, Vgs)

$$Jo = Ig \; \exp(\frac{-Vg}{n \; \mathrm{x} \; VT}) \qquad (33a)$$

$$Vg = Vgs - Ig \; [Rg + \frac{g}{a} Z_{21} + (1 - \frac{g}{a})Rs \,] \qquad (33b)$$

A more accurate value of Jo can be obtained by averaging the results from several bias points.

The above calculations (i)–(iii) should be repeated starting from Equation (29), with values of the internal voltage Vg evaluated from the Equation (33b). One iteration is usually sufficient.

If S–parameters were measured for a number of RF frequencies, then an average of the results should be taken.

B. Extraction Of Series Resistances From Dc Measurements

The same technique as described above can be also used if Z–parameters (26) are obtained from the slopes of the DC I–V curves measured for the forward biased gate: $Vgs = fgs(Ig,Id)$ and $Vds = fds(Ig,Id)$. In this case, the Ig current should change from 0 to Igmax given by Equation (25), while the Id current should be set to zero and a few values much smaller than Igmax $(Id \ll Igmax)$.

Another approach is to use the modified Fukui method [22], which requires three sets of DC measurements in different bias configurations:

(1) The forward biased I–V characteristics of the gate–source junction with the drain terminal open.

(2) Voltage and current of the forward biased gate–drain junction with the source terminal floating .

(3) Voltage and current of the device with drain and source connected together.

From the I–V characteristics, parameters of the gate–channel junction are obtained [22] and the sum of resistances ($Rg + Rs + gRch$) is evaluated at a certain value of the gate voltage Vg. If measurements in points 2 and 3 are performed at the same value of Vg, then the voltage drop on the diode Dg is also the same for all three cases. Assuming, that the voltage drop on the channel does not change either (which is not very accurate !), then a set of three equations corresponding to the three measurements can be derived from the model in Figure 3.8. The equations can be easily solved for Rs, Rd, and the sum of Rg+gRch. Rg can then be separated from Rch using for example the method based on Equations (28a,b).

C. Determination Of Extrinsic Series Inductances

The method, which will be briefly described below was originally proposed by Diamand and Laviron [23]. The required measurements are S–parameters of the reversed biased 'cold' MESFET in the microwave frequency range (MW), between about 2 and 18 GHz [17]. By assuming that the channel opening is above 20% and therefore, the channel resistance Rch and the capacitive reactance 1/ωCg are much smaller than the reactances of the fringing capacitances, a simplified Z–parameter matrix of the circuit in Figure 3.8 is [23]

$$Z_{MW} = \begin{bmatrix} Rg+Rs+g'Rch & Rs+a'Rch \\ Rs+a'Rch & Rs+Rd+Rch \end{bmatrix} + j\begin{bmatrix} \omega(Lg+Ls)-(\omega(Cg+Cgsf+Cgdf))^{-1} & \omega Ls \\ \omega Ls & \omega(Ls+Ld) \end{bmatrix} \quad (34)$$

Where, g' and a' are modified coefficients given by

$$g' = g\frac{Cg+(Cds+Cgdf)/g}{Cg+Cgsf+Cgdf} \quad (35a)$$

$$a' = a\frac{Cg+Cgdf/a}{Cg+Cgsf+Cgdf} \quad (35b)$$

Inductances Ls and (Ls+Ld) can be directly calculated from the slopes of the plots of the imaginary parts of the matrix Z_{MW} : $X21(\omega)$ and $X22(\omega)$. Inductance Lg can be found by the curve fitting of the function $f(\omega) = (Lg+Ls)-1/\omega C$ to the measured $X_{11}(\omega)$. In order to increase accuracy of the method, several gate biases may be used.

3.5.3 Extraction Of The Mesfet Circuit Model From Y-parameters [8]

If extrinsic series elements Lg, Rg, Ld, Rd, Ls, and Rs are stripped off, then the remaining intrinsic part of the linear MESFET model in Figure 3.1 has a simple Π-type structure. Y-parameters of such a structure are directly related to circuit elements, leading to the following set of equations:

$$Ri = Re\{\frac{1}{y_{11}+y_{12}}\} \tag{37a}$$

$$Cgs = -\left[\omega \ \mathrm{Im}\{\frac{1}{y_{11}+y_{12}}\}\right]^{-1} \tag{37b}$$

$$Cgd = -\frac{1}{\omega} \ \mathrm{Im}\{y_{12}\} \tag{37c}$$

$$Rds = \frac{1}{\mathrm{Re}\{y_{11}+y_{12}\}} \tag{37d}$$

$$Cds = \frac{1}{\omega} \ \mathrm{Im}\{y_{11}+y_{12}\} \tag{37f}$$

$$g_{mo} \ \exp(-j\omega\tau) = (y_{21}-y_{12}) \ (1+j\omega Cgs \ Ri) \tag{37g}$$

Model elements can be directly calculated from these equations at each given frequency value and bias condition. Providing that S-parameter measurement and de-embedding errors are small, and the extrinsic series elements are accurately known, then the extracted intrinsic elements should be constant with frequency. However, this is not usually the case, and it is impossible to fit the model with this direct method over wider frequency range. Typically, averaging results from the lower microwave frequencies between 2 and 7 GHz gives a reasonable accuracy for most model elements, but Ri and, in some cases, τ. After extracting linear models at all measured bias points, the modelling functions can be provisionally fitted to the non-linear element values.

3.5.4 Final Optimisation Of Non-linear Model

The intrinsic MESFET model identified in Section 3.5.3 usually does not fit well to the measured S-parameters over the entire range of microwave frequencies. Therefore, a final optimisation of the non-linear model is required. This should not be very difficult, because good approximations of all model parameters are available at this stage. The optimisation is

(8) Further details about this technique will be presented in the paper: A.K.Jastrzebski, A.Davies, "Direct extraction of non-linear MESFET model from Y-parameters", currently in preparation.

performed simultaneously at all bias points in the following order where values of linear elements and parameters of functions describing non–linear elements are adjusted:

1. Cds, Cgs, Cgd (and possibly Rf, Cf) over entire frequency range.
2. Ri and τ in the upper half microwave frequency band.
3. All, or only selected model parameters with the increased weighting factors at the frequencies, bias points and S–parameters, at which the errors are still not acceptable. In order to identify those elements, which are mainly responsible for the particular errors, the table of optimisation steps given by Kondoh [18] can be used.

3.5.5 Example Of Non–linear Modelling Of Dispersive Mesfet

The parameter extraction procedure described above will be demonstrated on the example of the monolithic epitaxial MESFET, which was already modelled in [19, 20] using a different algorithm. (9) It should be pointed out that the new results are practically the same as those reported earlier, but much less time and effort was spent on obtaining them.

The device was mounted on a carrier, and parasitic elements corresponding to the gate and drain pads (Cpgs, Cpds) and the source pedestal (Cps, Lps) were not stripped off from the measurements. Therefore, these parasitics were included in the equivalent circuit of the linearised model shown in Figure 3.9.

The following modelling steps, numbered identically as in Figure 3.7, were performed:

Step 1.

The average values of the physical parameters were specified by the manufacturer as: gate length $L = 0.5\ \mu m$, gate width $W = 300\ \mu m$, channel layer thickness $a = 0.135\ \mu m$, and doping density $N_D = 16.5 \times 10^{16}\ cm^{-3}$. Substituting these values into Equations (24a–g) gave the following approximate values of model parameters: $Vp = 2.17V$, $Vbi = 0.779V$, $Jdss = 53.5mA$, $Jo = 5.1\mu A$, $Cgo = 0.31pF$, $Cgf = 0.051pF$, $\tau = 2.51ps$. Additionally, values of parasitic elements were calculated from their physical dimensions.

Steps 2 and 3.

Series resistances and the gate–channel diode model were extracted from the low–frequency S–parameters of the 'cold' device, as described in

(9) Plessey Research (Caswell) sponsored the initial modelling exercise for that device and provided some measurements and other required data.

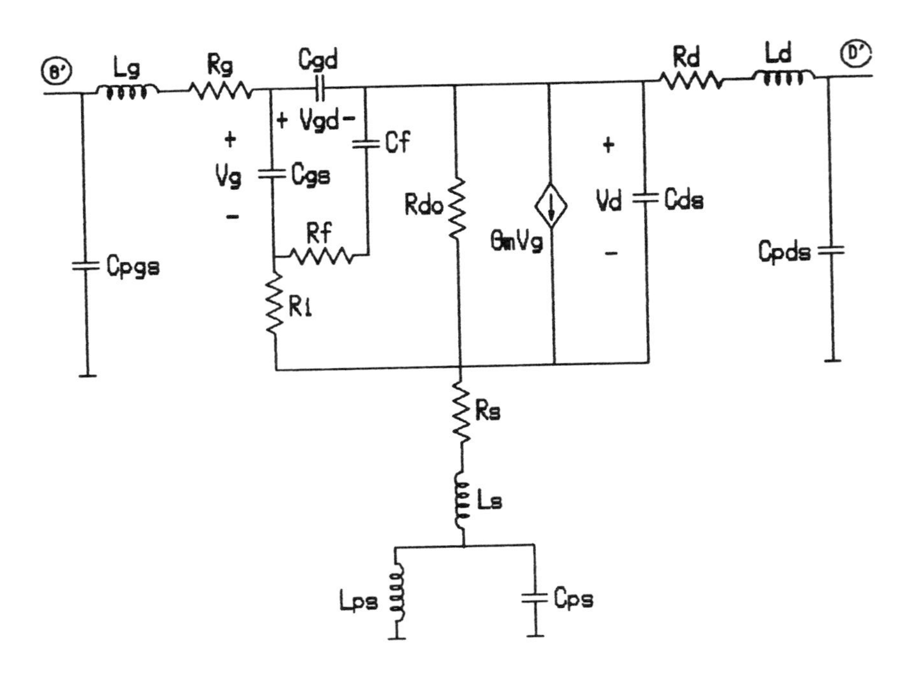

Figure 3.9 Linearised circuit model of the 0.5 μm monolithic MESFET including parasitics

Section 3.5.2A. No pulsed measurements facility was available, so the diode breakdown model was approximated from Equation (14) using manufacturer's data.

Step 4.

S–parameters of the passive MESFET biased at the point Vgs=0V, Vds=0V were measured over frequency range 2 to 17 GHz. At first, parasitics were stripped–off from the measurements, and the method described in Section 3.5.2C was used to find initial values of series inductances in the lower half of the microwave frequency band (i.e. 2–10GHz). Next, the equivalent circuit of the 'cold' MESFET in Figure 3.8 was expanded to include parasitics, and the whole circuit was optimised to the unstripped S–parameters in the entire frequency range. In this way, both series inductances and parasitic elements were obtained.

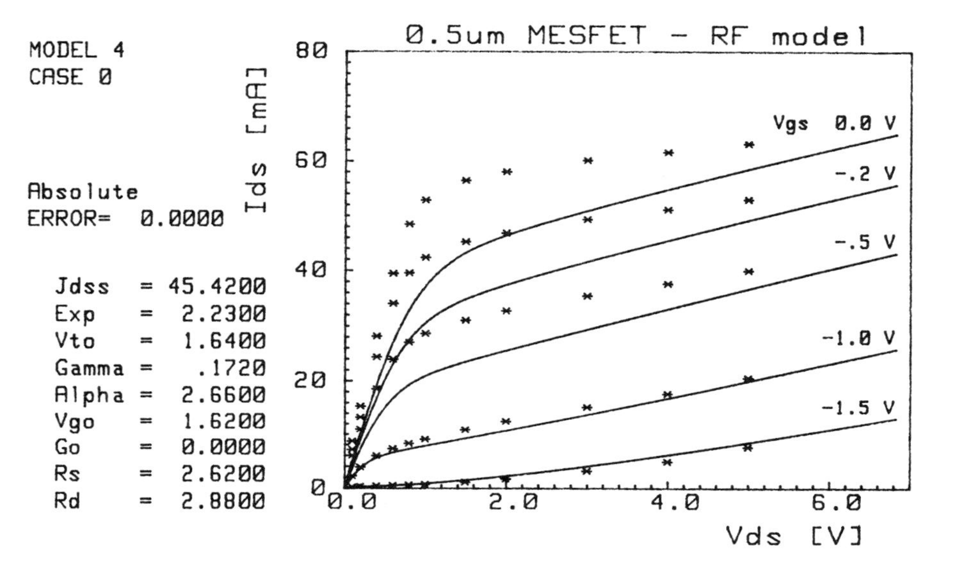

(a) Calculated RF I–V characteristics and the measured DC points

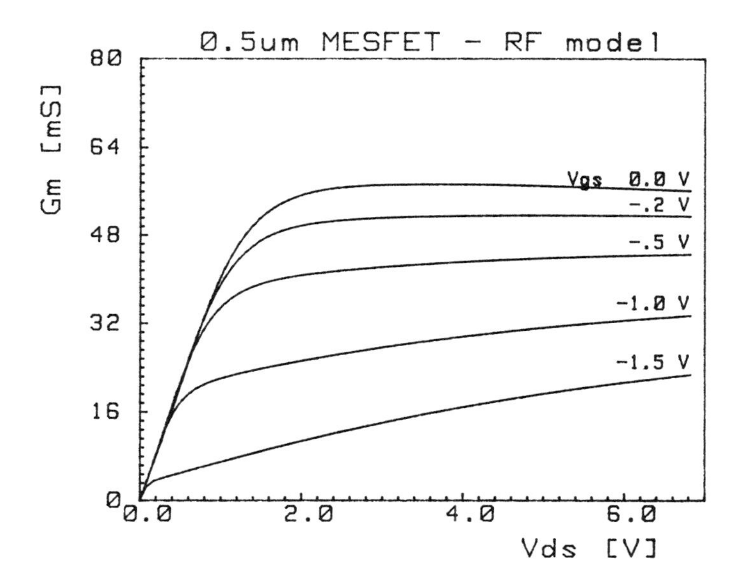

(b) Calculated transconductance *gmo*

Figure 3.10 RF characteristics of the 0.5 μm x 300 μm low–noise MESFET

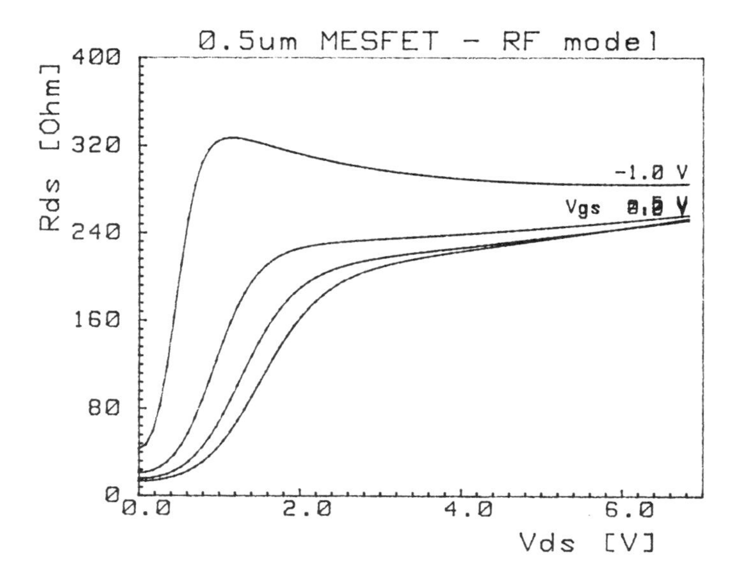

(c) Calculated output resistance Rdo

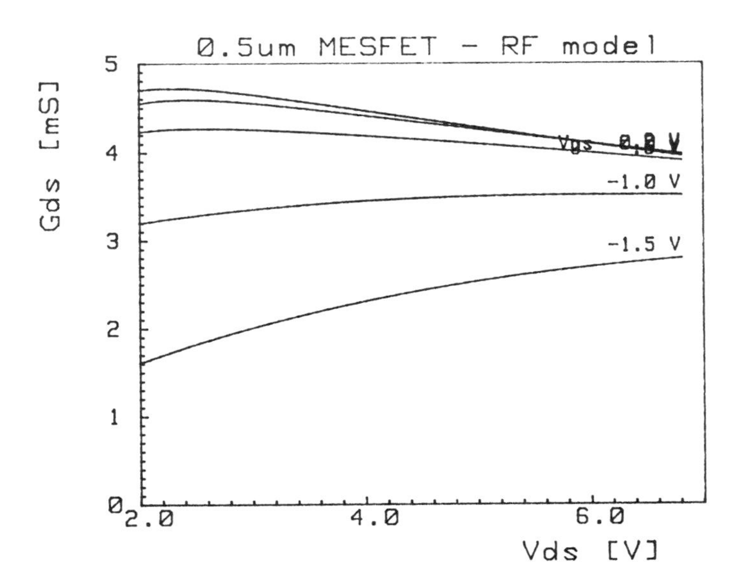

(d) Calculated output conductance Gds in the saturation region

Figure 3.10 RF characteristics of the low-noise MESFET (Contd)

Step 5.

S–parameters of the active MESFET were measured at six bias points in the frequency range 0.1 to 17GHz. The measurements were stripped to the terminals of the intrinsic MESFET and converted to Y–parameters. Approximate values of the intrinsic model elements were then calculated from Equations (37a–g). Two sets of results were recorded at each bias point: one – for the frequency 1GHz, and the second – calculated as an average value over the frequency range. Values of Cds and τ were calculated as averages from all bias points. Functions describing non–linear elements Cgs, Cgd, and Ri (Equations (12) and (15)), were fitted to the second set of results.

No pulsed measurements were available, and therefore, path 'b' in the modelling algorithm in Figure 3.7 was followed.

Step 6b.

The DC model based on Equations (10)–(11) was fitted to the measured I–V characteristics. The results are shown in Figure 3.2, indicating an excellent agreement between measured and calculated characteristics.

Step 7b.

The RF I–V model of the device was derived by simultaneous optimisation to the measured S–parameters of a family of linearised models (Figure 3.9) at all bias points and for frequencies below 1 GHz. Initial parameters of the non–linear controlled source Jds, were assigned from the DC model, derived in Step 6b. Other model elements were set to the values obtained from Step 5 for the frequency 1GHz. Non–linear elements were treated as vector parameters (see Section 3.4.3B), with different values at each bias points. All parameters of the source Jds, except α, were optimised (see Equation (10)). Other circuit elements were kept constant during the optimisation. The derived RF I–V characteristics are shown in Figure 3.10a. Measured DC points are also included in Figure 3.10a in order to allow for a comparison between the two models.

Step 8b2.

Calculated plots of RF transconductance and drain–source resistance and conductance are shown in Figure 3.10b–d. There was a large difference

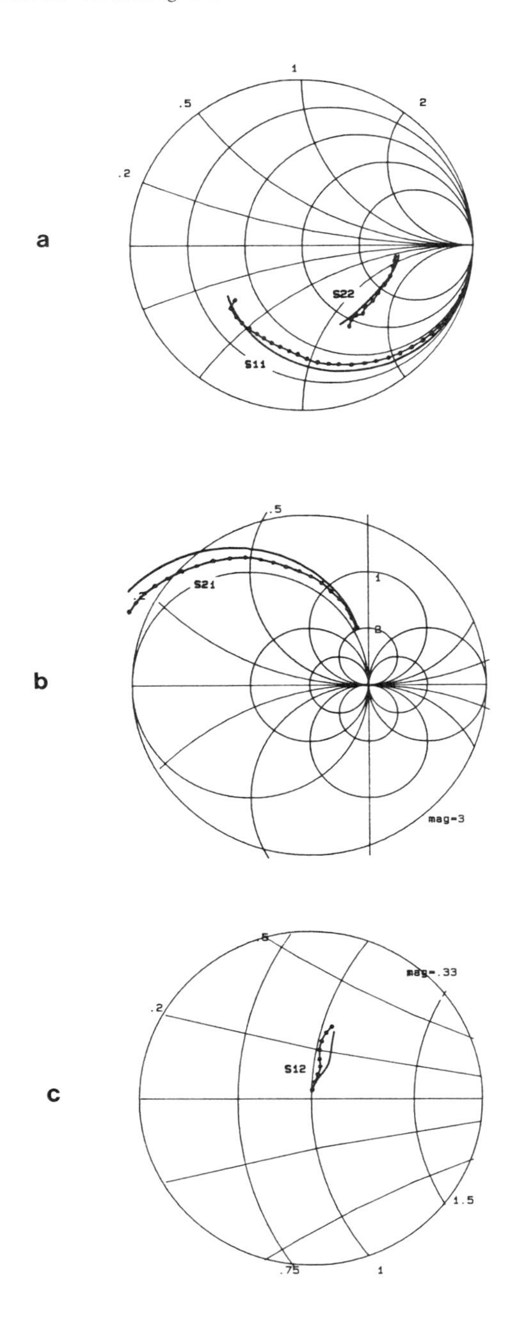

Figure 3.11 Measured and calculated S–parameters for the MESFET at Idss near the knee point (Vgs=0V, Vds=2V) (a) S_{11} and S_{22}, (b) S_{21}, (c) S_{12}

between these RF parameters and their DC equivalences shown in Figure 3.2b–d. This indicated, that the device was strongly dispersive and the RF I–V model was used.

Step 9.

At this stage, a relatively good non–linear model was already available. Therefore, the final optimisation was performed with the Powell's method (see Section 3.4.2E). First, elements Cds, Cgs, and Cgd were optimised in the whole frequency range from 0.1 to 17 GHz. Next, Ri and τ were optimised in the upper MF band from 10 to 17 GHz. It was noticed, that the S22 parameter had rather irregular behaviour, probably due to large measurement errors, and it was difficult to fit. Therefore, an additional branch Rf, Cf was included in the model (see Figure 3.9) and the circuit was re–optimised.

Figures 3.11 and 3.12 contain comparisons between the measured and calculated S–parameters for the two bias points near the ends of a hypothetical load–line of a large–signal amplifier: (i) at Idss near the knee point (Vgs=0V, Vds=2V), and (ii) at the low noise (LN) point (Vgs=–1.41V, Vds=6V). Results for the other bias points show a similarly good agreement between the model and the measurements.

3.6 Conclusions And Future Developments

The primary objective of this Chapter was to propose some practical solutions to the all too well known problems and difficulties experienced by microwave engineers when trying to use or derive non–linear MESFET models. Various specific and also, more general limitations of the MESFET models currently used in the CAD programs were analysed. One of the more surprising results was, that the most popular of the models were the least accurate. A second generation of MESFET models is needed which will have a structure more closely related to the device physics, and which will be capable of simulating such aspects of the device behaviour like dispersion, thermal effects and inverse operation. However, before these big steps can be made, some small improvements to the existing models are still possible. The model proposed in Section 3.3 is such an attempt, in which it is felt, a sensible compromise between complexity and accuracy in microwave applications has been achieved.

An algorithmic approach to non–linear model parameter extraction from measured S–parameters, presented in Section 3.5, is a substantial improvement over the existing methods. Although, it can be used with a graph paper, calculator, and any of the optimising linear CAD programs, its main benefit is the possibility of automating the modelling task. This

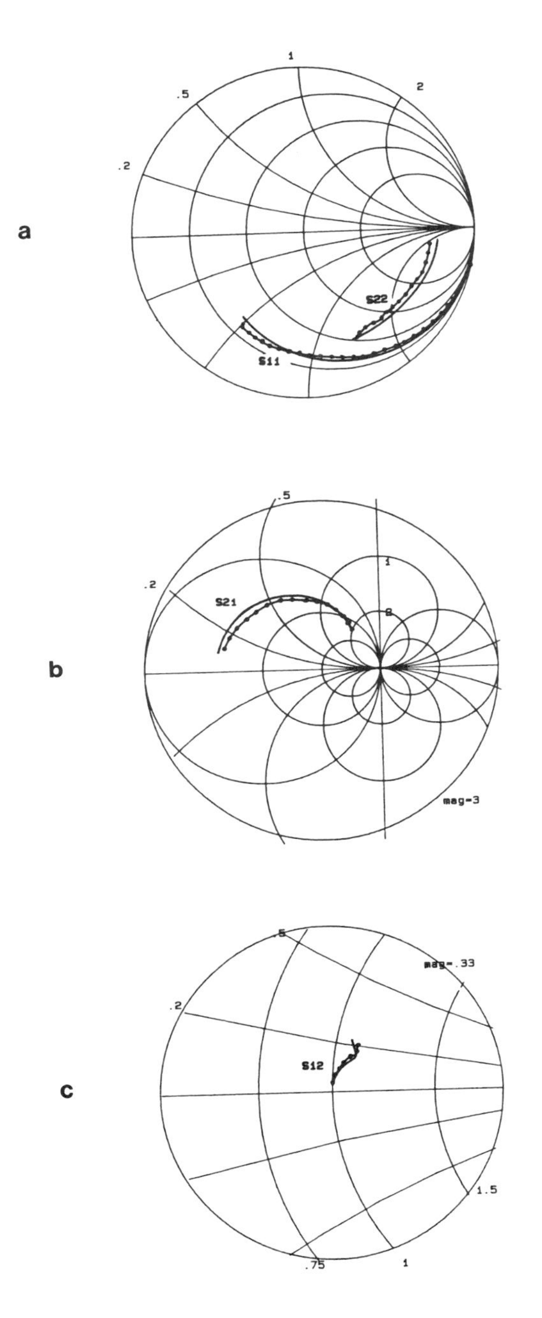

Figure 3.12 **Measured and calculated S–parameters at the low noise point (Vgs=–1.41V, Vds=6V) (a) S_{11} and S_{22}, (b) S_{21}, (c) S_{12}**

however, requires new and dedicated software tools, not yet commercially available in one package.

The problem often not sufficiently appreciated is the influence of the measurement errors on the accuracy of the derived models, which is detrimental and difficult to assess. This aspect was discussed in general terms in Section 3.4.1, but its solution lies in the better quality of the de-embedding methods and the use of integrated device measurement and parameter extraction systems (10).

An interesting approach to MESFET modelling has been recently proposed by Ghione et al. [16]. In this case, a physics based numerical model was used to generate I–V characteristics and sets of S–parameters for the intrinsic device. Those results were then used to derive MESFET circuit models. The major advantage of that method is, that physical simulation offers much more information than the measurements can provide, since the values of electrical variables inside the device become available. This makes identification of circuit models much easier. Additionally, more accurate, though more complex structures of the equivalent circuit models can be adopted without the problems encountered with parameter extraction.

Another important step in the MESFET and HEMT modelling has been recently reported by Ladbrook [35], who developed accurate and fast methods of relating elements of the equivalent circuit to physical parameters of the device and vice versa. These so called forward and reverse modelling techniques are invaluable tools in the yield–oriented design approach, so critical in the case of MMICs. An important benefit of utilising technological relationships in the MESFET modelling lays in the fact, that the correlations among the equivalent circuit elements can be used to identify a unique solution for the parameter extraction problem.

3.7 References

Non–linear MESFET Circuit Models

1 W.R.Curtice, "A MESFET Model for Use in the Design of GaAs Integrated Circuits", IEEE Trans. MTT, vol.MTT–28, pp.448–456, May 1980

2 Y.Tajima, B.Wrona, K.Mishima, "GaAs FET Large–Signal Model and its Application to Circuit Designs", IEEE Trans. ED, vol.ED–28, pp.171–175, Feb. 1981

3 Y.Tajima, P.D.Miller, "Design of Broad–Band Power GaAs FET Amplifiers", IEEE Trans.MTT, vol.MTT– 32, pp.261–267, March 1984

---

(10) Detailed analysis of this problem will be presented by O.P.Leisten, A.Davies, R.J.Collier in the paper "On the performance of various de–embedding algorithms on the modelling of devices at micro-wave frequencies", ARMS Conference, University of Kent, Sept. 1989

4 T.Kacprzak, A.Materka, "Compact dc Model of GaAs FET's for Large-Signal Computer Calculation", IEEE J. Solid-State Circuits, vol.SC-18, pp.211-213, April 1983

5 A.Materka, T.Kacprzak, "Computer Calculation of Large-Signal GaAs FET Amplifier Characteristics", IEEE MTT, vol.MTT-33, pp.129-135, Feb.1985

6 M.I.Sobhy, A.K.Jastrzebski, R.S.Pengelly, J.Jenkins, J.Swift, "The Design of Microwave Monolithic Voltage Controlled Oscillators", 15 European Microwave Conference, Paris, pp.925-930, 1985

7 W.R.Curtice, M.Ettenberg, "A Nonlinear GaAs FET Model for Use in the Design of Output Circuits for Power Amplifiers", IEEE Trans. MTT, vol.MTT-33, pp.1383-1394, Dec.1985

8 H.Statz et al., "GaAs FET Device and Circuit Simulation in SPICE", IEEE Trans. ED, vol.ED-34, pp.160-169, Feb.1987

9 L.E.Larson, "An Improved GaAs MESFET Equivalent Circuit Model for Analog Integrated Circuits Applications", IEEE J. Solid-State Circuits, vol.SC-22, pp.567-574, Aug. 1987

10 A.K.Jastrzebski, "Non-linear MESFET Models", IEE Colloquium on Large Signal Device Models and Parameter Extraction for Circuit Simulation, London, April 1988

11 N.Scheinberg et al., "An Accurate MESFET Model for Linear and Microwave Circuit Design", IEEE J. Solid-State Circuits, vol.SC-24, pp.532-539, April 1989

12 W.R.Curtice, "GaAs MESFET Modeling and Nonlinear CAD", IEEE trans. MTT, vol.MTT-36, pp.220-230, Feb. 1988

Physical MESFET models

13 C.M.Snowden, "Computer-aided design of MMICs based on physical device models", IEE Proc., vol.133, Pt.H., No.5, Oct. 1986

14 R.H.Johnson, B.W.Johnson, J.R.Biard, "A Unified Physical DC and AC MESFET Model for Circuit Simulation and Device Modelling", IEEE Trans. ED, vol.ED-34, pp.1995-2001, Sept. 1987

15 M.A.Khatibzadeh, R.J.Trew, "A Large-Signal, Analytic Model for the GaAs MESFET", IEEE Trans. MTT, vol.MTT-36, pp.231-238, Feb. 1988

16 G.Ghione, C.U.Naldi, F.Filicori, "Physical Modelling of GaAs MESFET's in an Integrated CAD Environment: From Device Technology to Microwave Performance", IEEE Trans. MTT, vol.MTT-37, pp.457-468, March 1989

MESFET model parameter extraction

17 W.R.Curtice, R.L.Camisa, "Self-Consistent GaAs FET Models for Amplifier Design and Device Diagnostics", IEEE Trans. MTT, vol.MTT-32, pp.1573-1578, Dec.1984

18 H.Kondoh, "An Accurate FET Modelling from Measured S-Parameters", IEEE MTT-S Digest, pp.377-380, 1986

19 A.K.Jastrzebski, "Non-linear MESFET Modelling", 17 European Microwave Conference, Rome, pp.599-604, 1987

20 A.K.Jastrzebski, "Parameter Extraction for Non-linear MESFET Models", IEE Colloquium on Large Signal Device Models and Parameter Extraction for Circuit Simulation, London, April 1988

21 J.W.Bandler et al., "Integrated Model Parameter Extraction Using Large-Scale Optimization Concepts", IEEE Trans. MTT, vol.MTT-36, pp.1629-1638, Dec. 1988

Determination of extrinsic series elements

22 H.Fukui, "Determination of the Basic Device Parameters of a GaAs MESFET", Bell Syst. Tech. J., vol.58, No.3, pp.771-797, March 1979

23 F.Diamand, M.Laviron, "Measurement of the Extrinsic Series Elements of a Microwave MESFET Under Zero Current Conditions", 12 European Microwave Conference, Helsinki, pp.451-456, 1982

24 K.W.Lee et al., "Source, Drain, and Gate Series Resistances and Electron Saturation Velocity in Ion-Implanted GaAs FET's", IEEE Trans. ED, vol.ED-32, pp.987-992, May 1985

25 R.Vogel, "Determination of the MESFET Resistive Parameters Using RF-wafer Probing", 17 European Microwave Conference, Rome, pp.616-621, 1987

Dispersion in MESFETs

26 C.Camacho-Penalosa, C.S.Aitchison, "Modelling Frequency Dependence of Output Impedance of a Microwave MESFET at Low Frequencies", Electronics Letters, vol.21, No.12, pp.528-529, June 1985

27 J.Graffeuil, Z.Hadjoub, J.P.Fortea, M.Pouysequr, "Analysis of Capacitance and Transconductance Frequency Dispersions in MESFETs for Surface Characterization", Solid-State Electronics, vol.29, No.10, pp.1087-1097, 1986

28 P.H.Ladbrooke, S.R.Blight, "Low-field Low-frequency Dispersion of Transconductance in GaAs MESFETs" IEEE Trans. ED, vol.ED-35, pp.257-267, March 1988

Pulsed and RF MESFET modelling

29 M.A.Smith et al., "RF Nonlinear Device Characterization Yields Improved Modelling Accuracy", IEEE MTT-S Digest, pp.381-384, 1986

30 C.Guo et al., "Optimum Design of Nonlinear FET Amplifiers", IEEE Trans. MTT, vol.MTT-35, pp.1348-1354, Dec. 1987

31 M.Paggi, P.H.Williams, J.M.Borrego, "Nonlinear GaAs MESFET Modelling Using Pulsed Gate Measurements", IEEE Trans. MTT, vol.MTT-36, pp.1593-1597, Dec. 1988

Books

32 R.S.Pengelly, "Microwave Field-Effect Transistors - Theory, Design and Applications" (2nd ed.), Research Studies Press, 1986

33 M.Shur, "GaAs Devices and Circuits", Plenum Press, 1987

34 R.Soares (ed.), "GaAs MESFET Circuit Design", Artech House, 1988

35 P.H.Ladbrooke, "MMIC Design: GaAs FETs and HEMTs", Artech House, 1989

36 W.H.Press et al., "Numerical Recipes, The Art of Scientific Computing", Cambridge University Press, 1986

Chapter 4

CAD Tools for Gallium Arsenide Design

S. J. Newett, D. R. S. Boyd and K. Steptoe

4.1 Introduction

4.1.1 Background

In common with users of other integrated circuit technologies, designers in Gallium Arsenide (GaAs) now depend heavily on computer aided design (CAD) tools. These are used both to develop and refine the circuit design and then to produce a physical realisation of the circuit in the form of mask patterns used to fabricate the device. CAD tools are also used to reduce measurement data and produce accurate component models for use in circuit simulation.

Historically these tools have come from two backgrounds. The tools originally used to design hybrid microwave circuits have evolved and developed to now offer a comprehensive range of design aids to the monolithic microwave integrated circuit (MMIC) designer. On the other hand, the world of silicon integrated circuit (SIC) design has generated a wide spectrum of CAD tools spanning all stages in the design of electronic systems from conception to implementation. Many of these tools are also applicable to GaAs design.

MMIC CAD tools tend to cater for the design of circuits of modest complexity where the difficulty of the analysis problem dominates. Here accuracy of modeling is of paramount importance and the approximation of lumped components made at lower frequencies is no longer valid. As designs become more ambitious and frequencies of operation increase, the need to get more involved with the physics of the active devices and to model the passive components and interconnections and their interactions with each other, has become inescapable.

The main thrust with SIC CAD tools is to handle potentially very large circuits while preserving a sufficient degree of accuracy at each stage of the design process. Complexity is managed through abstraction of detail as the design is analysed at higher and higher levels. These techniques are only valid while the abstractions are valid and this limits the frequency range within which these tools can be applied. SIC CAD tools exist for the design of both analogue and digital circuits.

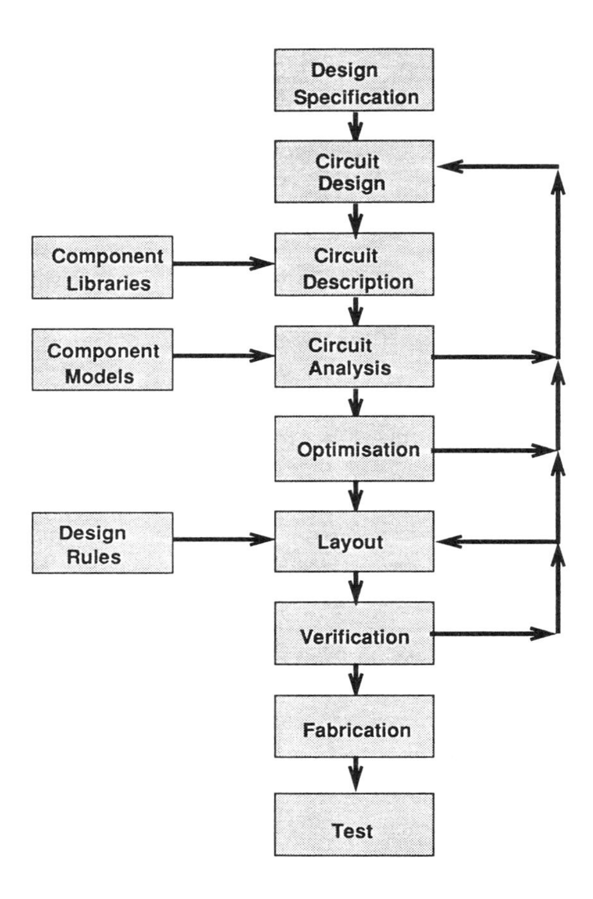

Figure 4.1 Typical design development route

It is unlikely that a single CAD supplier will meet all the designer's needs and so it will be necessary to assemble a set of tools from several sources. This will involve the transfer of design data between CAD tools during the design process. In the past, different data formats have made this difficult but the emergence of standards in this area is easing the problem. It is an interesting fact that many of the tools now on the market started as university research projects. This process continues today with several active research groups in the UK and Europe pursuing this area. It will be important to establish routes for commercial exploitation of the software developed.

4.1.2 Design Cycle

Before studying in detail some of the specific CAD tools available today, it is worthwhile to take a general look at the overall design process to get a feel for the steps involved and their relationship.

Figure 4.1 shows in outline a typical design development route. The major intellectual contribution of the designer is in turning the initial design specification into a potentially realisable circuit design. This design must then be captured into the chosen CAD system either in graphical or text form. The resulting circuit description will probably contain references to components from a pre-defined library available in the target process technology. This circuit description is frequently used as the basis for an analysis of the design by the process known as simulation. Here input signals are applied to a model of the circuit in the computer and the predicted outputs are observed to see if the design behaves as required by the specification. The simulator uses models for each of the circuit components which predict their behaviour, either from first principles or on the basis of measurements made on actual devices. For MMICs, the interactions between components which take place in the physical circuit layout must also be taken into account during the earliest stages of circuit design. The design is modified until the specification is met. It may then be desirable to further improve the circuit design by optimisation where circuit design parameters are varied to see if further improvements in performance can be achieved. This may lead to the conclusion that further design refinement is necessary.

When the circuit design meets all requirements as predicted by simulation, physical design of the circuit can begin. This involves the construction and placement of all the mask shapes necessary to implement the circuit. These shapes must conform to the process design rules which are specified by the fabrication house to ensure that the circuit can be manufactured with acceptable yield. Because the layout process is frequently manual and therefore error prone, a verification step closely follows most layout work to check that the design rules have been obeyed. A further check analyses the layout to ensure that its electrical performance still meets the specification. Failure at this step may involve a substantial redesign but this is still preferable to a costly and lengthy fabrication exercise only to find that the circuit is unacceptable. The final steps are then fabrication of the circuit and testing to select good circuits and characterise their performance.

4.2 Microwave Design Methods and Models

The most extensive use of Gallium Arsenide ICs to date has been in the microwave field (in MMICs). Before looking at specific programs and

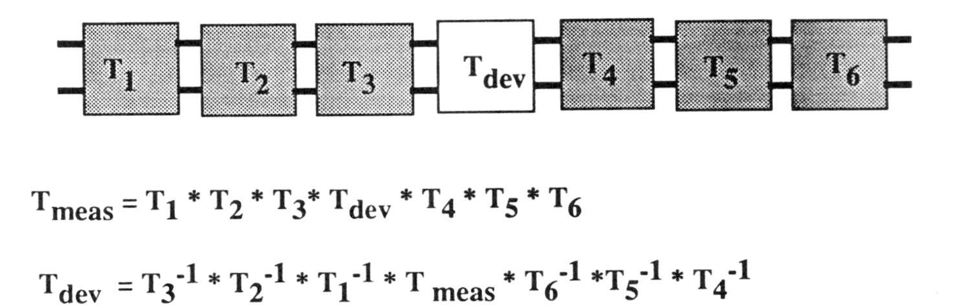

T_1
T_2
T_3
T_{dev}
T_4
T_5
T_6
$T_{meas} = T_1 * T_2 * T_3 * T_{dev} * T_4 * T_5 * T_6$
$T_{dev} = T_3^{-1} * T_2^{-1} * T_1^{-1} * T_{meas} * T_6^{-1} * T_5^{-1} * T_4^{-1}$

Figure 4.2 Illustration of de-embedding

their techniques, it is worthwhile looking at some of the important pre-requisites. In particular, models are required for both active and passive devices; their extraction from measured data requires a de-embedding procedure, and an optimiser is usually used to fit the equivalent circuit to this data.

4.2.1 De-Embedding

When a test device is measured it has to be mounted in some sort of jig or package. The model required by the designer, however, is that of the intrinsic device. The process by which the device characteristics can be extracted is called de-embedding.

The components between the connection planes of the network analyser can be represented by a cascade of T matrices with the device in the middle as illustrated in Figure 4.2. The T matrix in this instance is called the "transfer scattering matrix" which like an ABCD matrix is cascadable. The individual T matrices describe the components used to represent the discontinuities of the coax to microstrip transitions, the microstrip launchers and the bond wires. (These often have to be derived by electrically modelling of the construction of the jig used). By pre and post multiplying by the inverse T matrices, the matrix for the device can be extracted. Microwave designers use s-parameters rather than other parameters such as y, h or z since the latter rely on the ability to produce short and open circuits which are not easily producible at high frequencies. The s-parameters relate incident and reflected power at the ports of a circuit within a system of a defined characteristic impedance. They are, however, not cascadable.

$$\begin{vmatrix} b_1 \\ b_2 \end{vmatrix} = \begin{vmatrix} S_{11} & S_{12} \\ S_{21} & S_{22} \end{vmatrix} \begin{vmatrix} a_1 \\ a_2 \end{vmatrix}$$

By rearranging these equations such that the variables at port 1 are the dependent variables and those at port 2 are the independent ones, the T matrix is defined [1] which is cascadable.

$$\begin{vmatrix} b_1 \\ a_1 \end{vmatrix} = \begin{vmatrix} T_{11} & T_{12} \\ T_{21} & T_{22} \end{vmatrix} \begin{vmatrix} a_2 \\ b_2 \end{vmatrix}$$

$$T_{11} = \frac{(-S_{11}S_{22} + S_{12}S_{21})}{S_{21}} \qquad T_{12} = \frac{S_{11}}{S_{21}}$$

$$T_{21} = \frac{-S_{22}}{S_{21}} \qquad T_{22} = \frac{1}{S_{21}}$$

$$S_{11} = \frac{T_{12}}{T_{22}} \qquad S_{12} = T_{11} - \frac{T_{12}T_{21}}{T_{22}}$$

$$S_{21} = \frac{1}{T_{22}} \qquad S_{22} = \frac{-T_{21}}{T_{22}}$$

4.2.2 Optimisation

The use of optimisers is frequently encountered in circuit design. Their function is to modify the component values in a circuit to bring its response closer to a set of goals [2]. The most common examples of their use are in adjusting the components in a circuit (e.g. a filter) to meet a specification and in adjusting the components of an equivalent circuit model to fit some measured data.

The two important aspects of optimisers are:
(1) The error formulation used and
(2) The search algorithm used.

At various steps in the optimisation process, an error function is calculated which gives a measure of how close the circuit is to meeting the desired specification. The search algorithm attempts to find a set of component values which give a lower error value until, hopefully, the

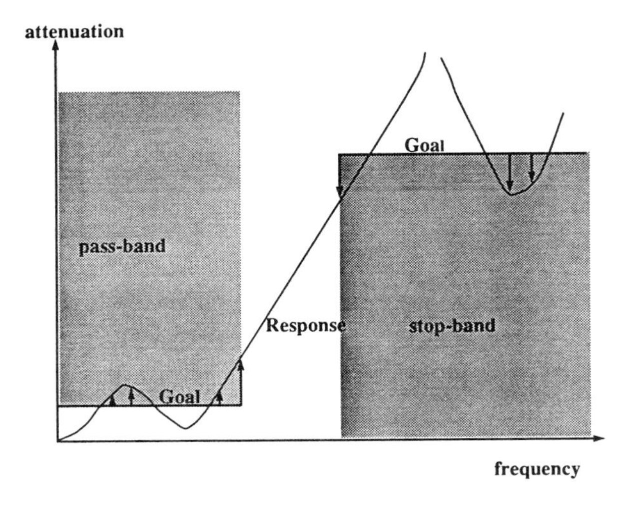

Figure 4.3 Calculation of the error value in the least squares formulation

specification is met. The process will also stop if no further reduction in error value can be obtained.

There are many ways in which the error function can be formulated. An example, illustrated in Figure 4.3, is known as the least–squares method. At a number of frequencies where the specification is violated, the magnitude difference is included in a sum to generate an overall error value.

$$\text{error} = \Sigma \frac{W_i\,(\text{response}-\text{goal})^2}{N}$$

where N is the number of individual frequencies at which the specification is not met. Notice that the individual errors add together positively and that the individual weightings W_i can be applied by the user to give emphasis to certain requirements over others. It is also usually possible to define the error calculation to be made up from several responses of the circuit (e.g. Gain and Noise).

The optimiser must start from a set of initial component values given to it by the user. It must find a set of component values which gives a zero error value (when the specification is met). Figure 4.4 shows a simple case where only two variables are involved. Contours of constant error value are shown. In this case a global minimum is easily found. However, it is possible that individual local minima, saddle points or long ravines might

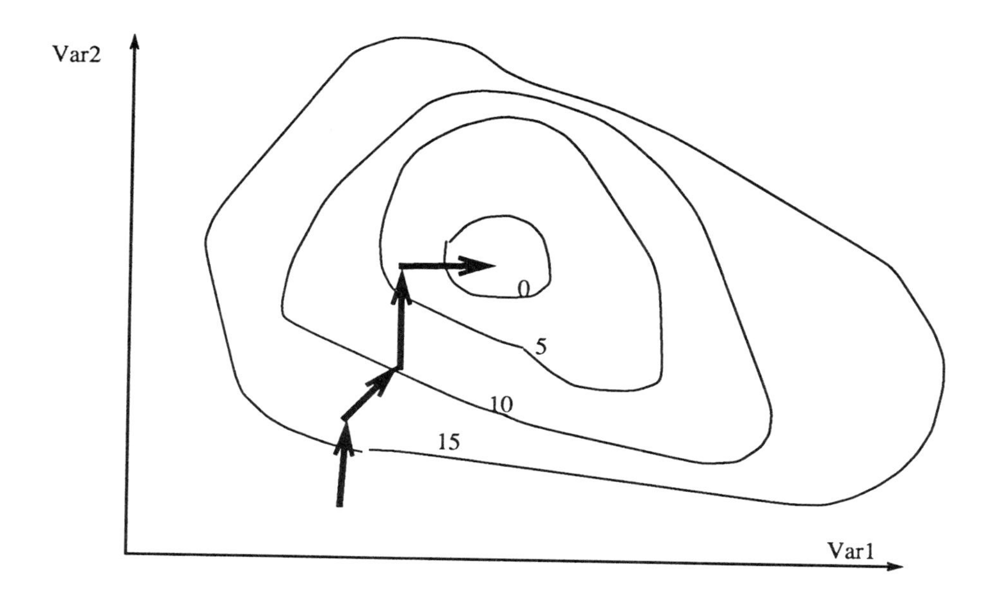

Figure 4.4 Possible path taken by an optimiser

exist which may cause the optimiser difficulty. The various different algorithms available can cope with such geographical features with varying degrees of success. As an illustration of error function formulations and search methods, some common ones are discussed below.

(i) Some Other Error Function Formulations

The minimax error function formulation uses the maximum difference which exists between the response and goal values as its error value. Fits obtained using this error function therefore tend towards an equi–ripple type of approximation.

The least–pth formulation uses a formula similar to that of the least squares algorithm, except that the individual errors are raised to a power of p. The value of p is raised in a sequence 2–4–8–16 during the optimisation process. As the value of p increases, the formulation looks more like the minimax formulation since the largest differences are accentuated.

(ii) Some Search Methods

In the random search algorithm, the error value is first calculated with a set of initial component values, the component values are perturbed in a

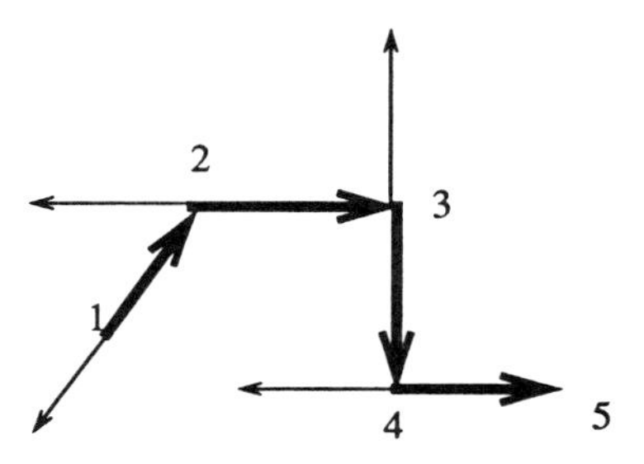

Figure 4.5 Illustration of the Random Search process

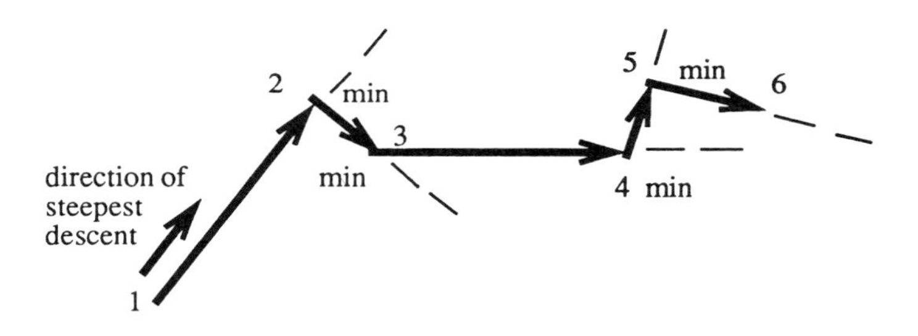

Figure 4.6 Illustration of the Gradient Search process

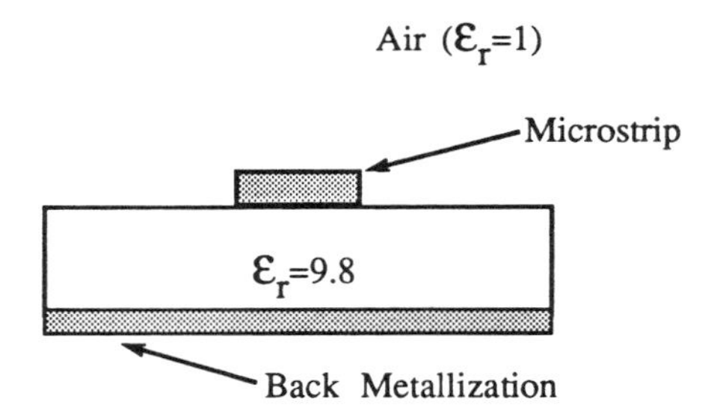

Figure 4.7 Construction of microstrip lines on an alumina substrate for MICs

random manner and an error value calculated for the new point and for the opposite point with respect to the starting point. The point with the minimum value is used as the starting point for the next evaluation. This process is illustrated in Figure 4.5.

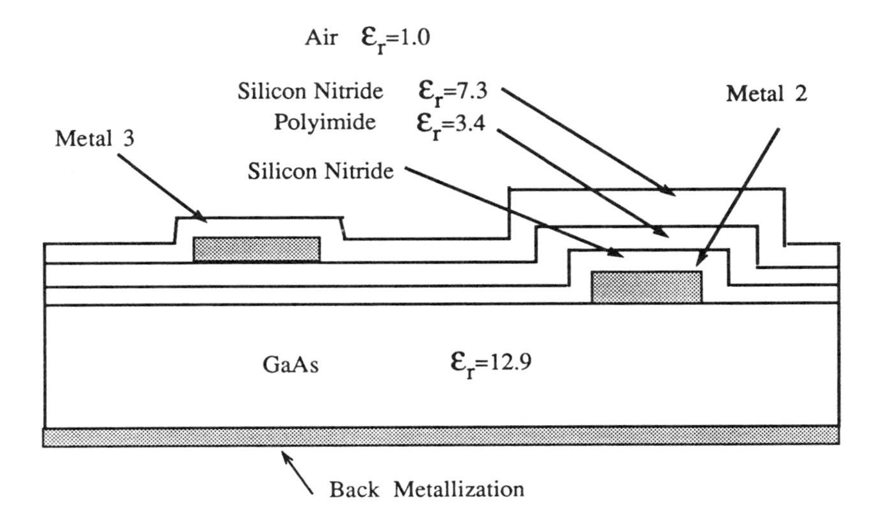

Figure 4.8 Multiple dielecric layers surrounding metallizations on a GaAs MMIC

In the Gradient search method, the direction of steepest descent is determined. Component values are moved in that direction until the minimum error point along that line is found. Then the process is repeated as illustrated in Figure 4.6.

Optimisation can be carried out using the TOUCHSTONE program (discussed later). For example, the components of an equivalent circuit for a spiral inductor can be fitted to measure data. TOUCHSTONE can be run within the LIBRA harmonic balance simulator, and provides a menu of different optimisation processes.

4.2.3 Passive Components

(i) Microstrip Lines

The components built in a Microwave Integrated Circuit (MIC) are obviously of a Microstrip nature, i.e. they are constructed by metallizations on top of a substrate of one dielectric constant, with air above and a back metallization layer beneath as shown in Figure 4.7.

The construction of a GaAs MMIC is similar, but the metallizations are built between a number of passivation layers having different dielectric constants as shown in Figure 4.8. The effective value of ϵr (ϵr_{eff}) is of course frequency dependent (exhibits dispersion) and depends on the strip width. It usually has a value of between 6 and 8.

It is important to realise that the component sizes on an MMIC can become significant in comparison to the wavelength. When a component's size reaches about 1/20th of the wavelength, it is no longer adequate to model it as a lumped component and it becomes necessary to regard it as a distributed one.

Table 4.1 illustrates the wavelengths involved for an effective dielectric constant of 7.

$$\frac{C}{\sqrt{\epsilon_r}} = f\,\lambda$$

where c = speed of light in air ($3 \times 10^8 ms^{-1}$)

Table 4.1 : Wavelength/Frequency Relationships

Frequency f	Wavelength λ	λ/20
1 GHz 12 GHz	113 mm 9.5 mm	5.6 mm 0.47 mm

(ii) Inductors

Figure 4.9 shows the basic equivalent circuit for a spiral inductor. It contains the primary inductance value L, the resistance of the metallization R, shunt capacitances to the backplane C_1 and C_2 and a bypass capacitor to account for coupling between the windings and between the windings and the underpass C_{12}.

The component values can be calculated using formulae which have been derived from the Grover formulae [3] previously used for MICs. For instance the primary inductance for a spiral inductor can be calculated using the following formulae:

$$L = 0.0008\, N^2 S[\mathrm{In}(S/b) + 0.726 + 0.178\,(b/S) + b^2/(8S^2) - (G_1 + H_1)/N]\ \mathrm{nH}$$

Where:

N = Number of turns

S = Mean side of spiral = I+(N−1/2)p

I = Inside dimension of spiral

p = Pitch

b = pN

$G_1 = \ln((w+3)/P)$

$H_1 = H - (0.04/N)[0.23 + 0.38N + \ln(N/2)]$

$H_1 = H - (0.04/N)[0.23 + 0.38N + \ln(N/2)]$
H is a function of N obtained from a graph
w is the track width

Values for the other components in the circuit are obtained from simpler equations. This equivalent circuit however is adequate only up to about 80% of the S_{21} resonant frequency. When higher frequencies (and harmonics) are to be considered, a more complex equivalent circuit model is required. Figure 4.10 shows the extended model used by the LINMIC+ program.

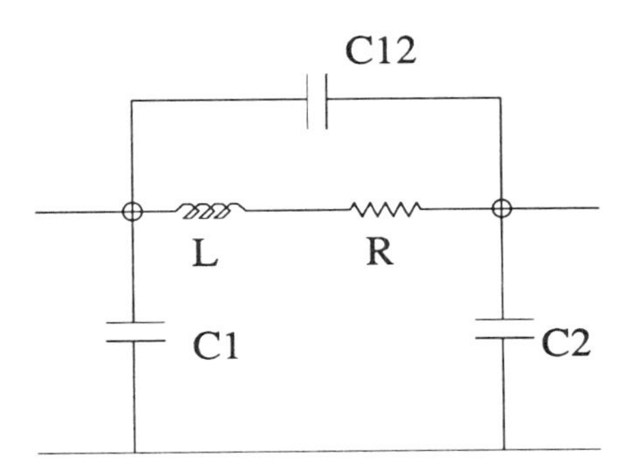

Figure 4.9 Basic equivalent circuit for spiral inductors

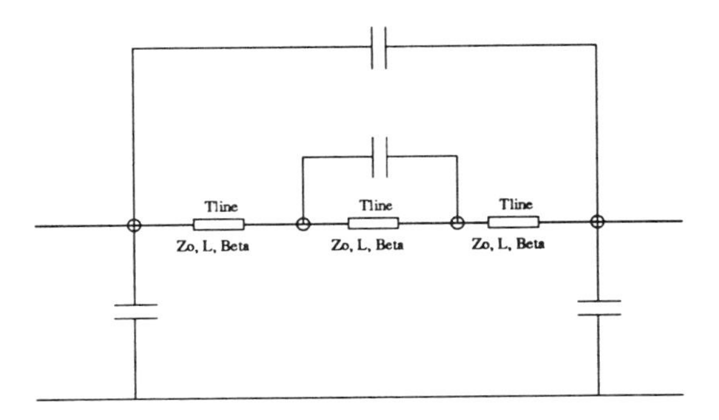

Figure 4.10 Extended model of a spiral inductor used by LINMIC+. Reproduced with permission of Jansen Microwave

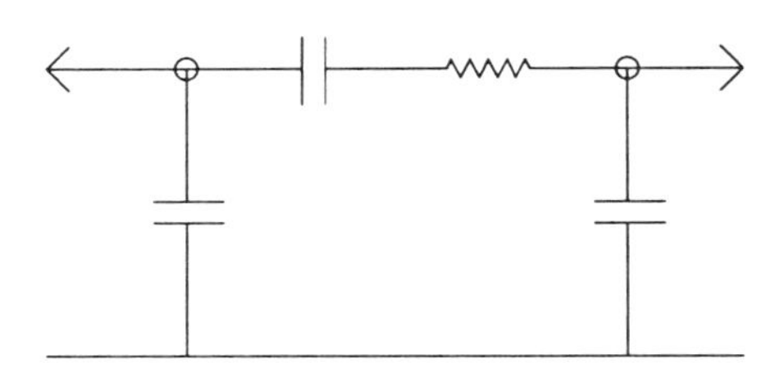

Figure 4.11 An equivalent circuit model for overlay capacitors

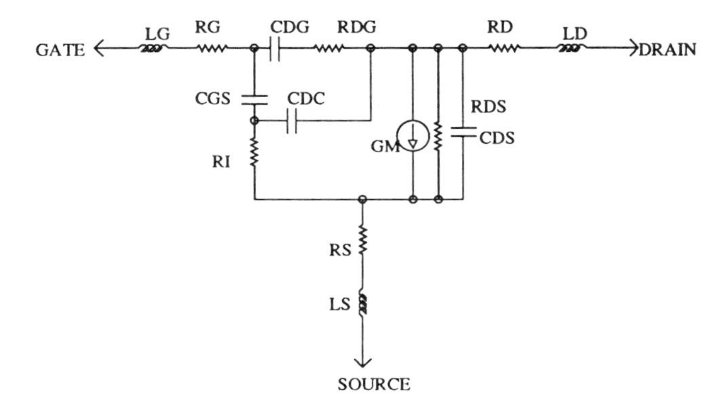

Figure 4.12 Small-signal linear model for a GaAs MESFET

(ii) Capacitors

Figure 4.11 shows the equivalent circuit used for overlay capacitors. It includes metallization resistances and capacitance to the backplane.

4.2.4 Linear Transistor Model and Device Scaling

From a set of measured s-parameters, a small signal model can be fitted using an optimiser. A standard model is shown in Figure 4.12, where all of the components have their origin in the physical structure of the device. This model can be fitted separately to data obtained for the device at different bias points. Typically, a foundry might give data for a MESFET at 20%, 50% and 100%I_{DSS}.

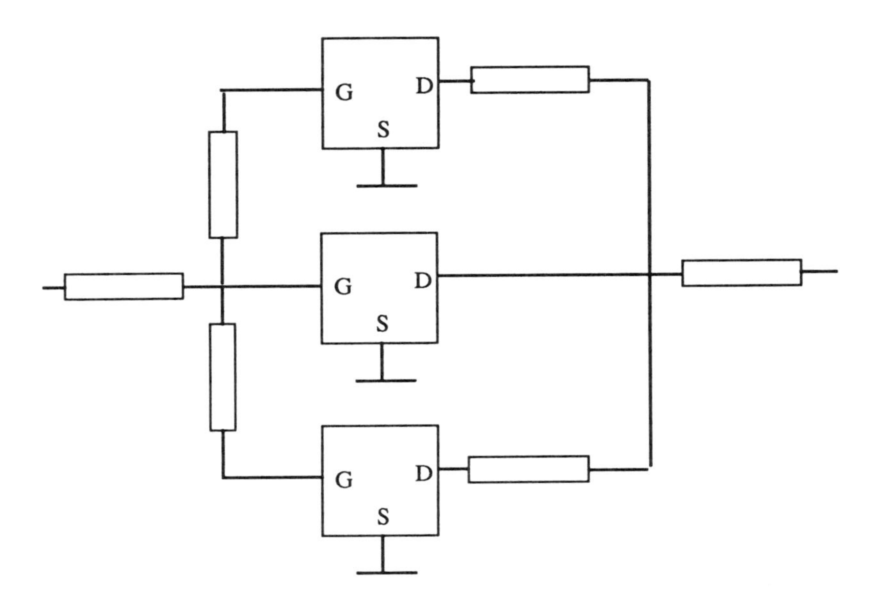

Figure 4.13 Parallel connection of sub-cells

One way in which device scaling can be achieved is by connecting a number of these equivalent circuits in parallel as shown in Figure 4.13. Each transistor represents a 100 μm element. Three such transistors are fitted to the data for a 300 μm transistor. Different sized transistors can then be made using the 100 μm transistor as a building block. The transmission lines have been included on the inputs and outputs to account for delays between the sub-cells due to interconnect metallization. An alternative method of device scaling is to fit a set of polynomials which describe the component variation with gate width to data obtained from measuring a number of different sized devices.

4.2.5 Non-linear Transistor Models

When non-linear circuits are to be analysed, different models are used. Most common of these are the Curtice Level I [4] and Level II [5] models. They are used in digital GaAs circuits and in microwave applications.

Figure 4.14 shows the equivalent circuit used in the Curtice model, in which the drain current is described by equations dependent on V_{gs} and V_{ds}.

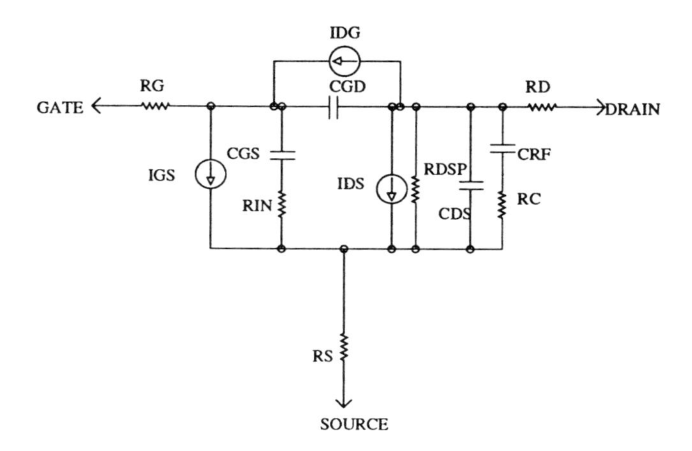

Figure 4.14 The Curtice equivalent circuit for a GaAs MESFET

Level I model

$$i_{ds} = \beta\,(v_{gs}-v_{to})^2\,(1+\lambda\,v_{ds})\tanh\,(\alpha v_{ds})$$

Level II model

$$i_{ds} = (A_0 + A_1\,v_1 + A_2\,v_1^2 + A_3\,v_1^3)\,\tanh\,\gamma v_{ds}$$

where $v_1 = v_{gs}\,(1 + \beta(v_{dso} - v_{ds}))$

Both of these models use a tanh function to describe the shape of the drain characteristic curves (which feature early saturation). The Level I model uses a square law relationship to describe the spacing between the i_{ds} curves for each gate voltage, whereas the Level II model uses a cubic polynomial. The R–C network R_c C_{rf} models the effect known as dispersion (seen in GaAs MESFETs) in which the output conductance increases to about 3 times its DC value for frequencies above about 100 kHz.

The equivalent circuits used for MESFETs in microwave design usually contain between 4 and 8 non–linear elements [6] as shown in Figure 4.15, where the non–linear elements are highlighted by enclosing rectangles. The Materka model [7] is a common one, which has 6 non–linear elements.

When a switch equivalent circuit is required, the one shown in Figure 4.16 can be used [8]. Here some of the component values are changed when the switch is OFF from what they are when the switch is ON.

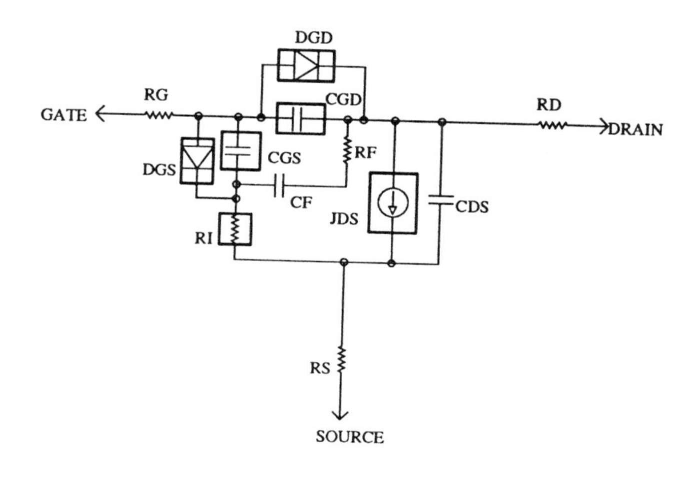

Figure 4.15 A General non-linear equivalent circuit

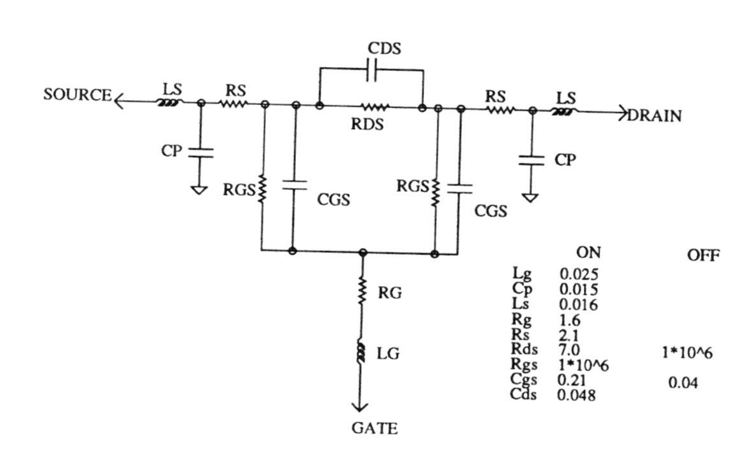

Figure 4.16 A typical equivalent circuit for the MESFET switch. Reproduced with permission of Plessey III/V

4.3 Microwave Design Tools and Techniques

The existence of design tools for MICs has enabled the rapid development of MMICs, and developments in this field are still adding to the designer's tool–kit. However, in this section besides mentioning these tools, a heavy emphasis has been placed on those tools which go beyond the basic needs of MIC designers to an environment in which complex mathematical modelling of the devices and layout is required. This is not intended to belittle the importance of the MIC design tools, which have a role in their own right, but to show that MMIC design has generated requirements of its own which have been the driving force behind these programs.

4.3.1 Small–Signal AC (Frequency Domain) Analysis

There are three main small–signal frequency domain programs which are most common in the microwave field: TOUCHSTONE from EESof, SUPER–COMPACT from Compact Software and MDS from Hewlett Packard. We have written here about the features of TOUCHSTONE, because this is the tool with which we have most experience. The other tools may differ in detail, but generally provide the same sort of functions. TOUCHSTONE, written by EESof, is marketed in the UK by Coss–Mic Ltd. It runs on an IBM–PC with maths co–processor, Sun, Apollo or Micro–Vax workstations. The facilities provided include sweeping a range of frequencies, plotting gain and noise circles, calculating simultaneous conjugate–match conditions and optimising circuits. TOUCHSTONE has 7 different optimisers available, plus a Monte–Carlo yield analysis capability. It has an easy to use menu. Circuits are described in an input file, using the library of components. These include: resistors, capacitors, inductors, voltage and current sources, transmission lines, stripline, microstrip, and co–planar waveguide components. Active devices can be described either by making up an equivalent–circuit, or by including in the circuit file a reference to a file containing the measured s–parameters. The primitives used by TOUCHSTONE have been developed for MICs, but can be useful in MMIC design. Within TOUCHSTONE, there is a choice of optimisers, in terms of both the search method used and the error function formulation used. The three search methods are Random, Gradient, and Quasi–Newton. The three error function formulations are least squares, least pth and minimax. EESof's ACADEMY microwave design environment incorporates TOUCHSTONE, LIBRA (Harmonic Balance) an advanced version of MICAD (Layout) and Schematic Capture around a common database.

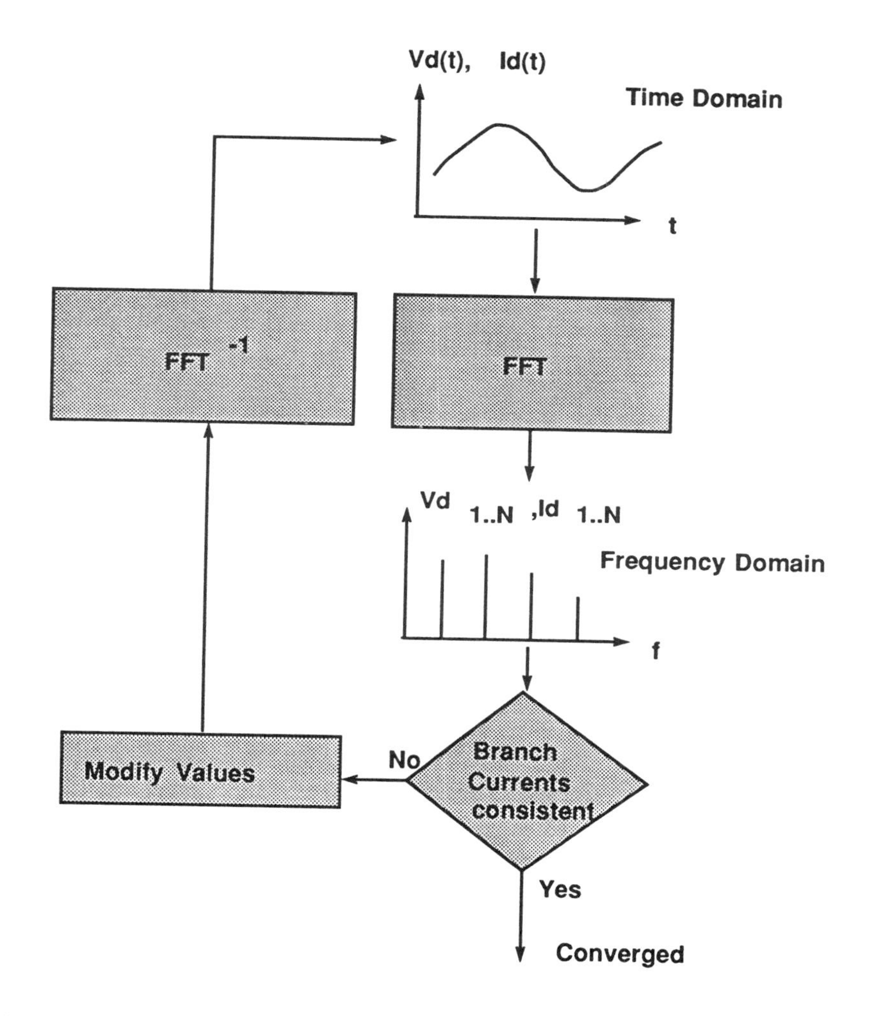

Figure 4.17 Illustration of the Harmonic Balance process

4.3.2 Harmonic Balance

Linear circuits (lumped and distributed) are easily simulated in the frequency domain, whereas non–linear circuits are more easily simulated in the time domain. It is often necessary to simulate circuits which contain non–linear elements (for instance large signal amplifiers, mixers and frequency doublers). In a time domain simulator, the simulations can take a long time to reach their steady state operating conditions. The

Harmonic Balance method of simulation allows non–linear circuits to be analysed efficiently and works well provided that they are not strongly non–linear. In this method, the circuit is split into linear and non–linear portions. The non–linear parts are analysed in the time domain, whereas the linear parts are simulated in the frequency domain. By using a Fast Fourier Transform (FFT) [9,10] and an inverse FFT, the currents in the circuit branches between these blocks are optimised until they are consistent with each other. The process is illustrated in Figure 4.17. There are two methods of making these circuits equate, namely, by optimisation or by the Kerr reflection algorithm [11,12].

The three main suppliers of microwave software also provide Harmonic Balance Simulators. LIBRA comes from EESof, its input file being compatible with TOUCHSTONE. It uses the Curtice GaAs MESFET models and has a user definable modelling capability. For example, a time domain waveform can be produced by LIBRA for a circuit, in which the magnitude of the input is increased until the waveform is distorted. On the screen, the harmonics can be seen at the output for any of those input levels. MICROWAVE HARMONICA comes from Compact Software. It is possible to include ones own models for non–linear devices in terms of state–equations. These are contained in a FORTRAN sub–routine. The program is then re–compiled with the new models.

4.3.3 Time Domain Analysis

(i) MICROWAVE SPICE

MICROWAVE SPICE is marketed by EESof. It has all the features of SPICE 2G6 with the addition of lossy TEM lines, microstrip lines, coupled microstrips, mutually coupled inductors and GaAs MESFET models. The input file is compatible with TOUCHSTONE. The Curtice Level I and Level II models are used for the GaAs MESFETs.

(ii) ANAMIC

ANAMIC was written at the University of Kent, and performs non–linear analysis in the time–domain, using the state–space method [13]. Although it was written initially for MIC design, it is sufficiently general to be applicable to many types of circuit design problems. It can simulate circuits containing both lumped and distributed components, including multi–coupled lines. In particular ANAMIC gives the user the flexibility to develop his own non–linear models for devices. Outputs from ANAMIC are in the form of tables or plots of time domain waveforms or

Fourier transforms of any voltage or current in the circuit. Another advantage of a time–domain simulator such as this is that it can handle highly non–linear circuits, which are not resolvable using a Harmonic Balance simulator. ANAMIC can be run in either batch or interactive mode. Along with the ANAMIC program are a number of supporting programs. TWOPORT calculates the large–signal s–parameters and large signal impedances, and aids in the design of matching circuits. MSTRIP performs analysis and synthesis of microstrip and strip–lines, (including discontinuities). There are two programs which are used to derive MESFET models. SOPTIM fits the non–linear model to measured s–parameters for several bias points simultaneously. MESFET DC optimises the DC characteristics of the device. ANAMIC runs on VAX (UNIX and VMS), SUN–3, ORION, ICL 2960 and PERQ computers.

4.3.4 Layout Editors – GAS STATION

Written by Barnard Micro–Systems, GAS STATION is marketed by BMS in the UK and by Compact Software in the US. GAS STATION is a stand–alone layout editor. It runs on an IBM–PC (which must have a Matrox graphics card fitted but without the need for memory enhancements), or on an HP–9000 series 300 or 800 computer. GAS STATION provides:

(1) Layout of semiconductor and super–conducting devices, ICs, MICs, MMICs, printed circuit boards and technical diagrams. Design rule checking is done on–line. Complex shapes can be generated, including circles, arcs, a general set of exponential functions and a sixth order polynomial shape definition. To obtain a constant impedance through a bend, the chamfering of transmission lines can be accomplished with a command implemented in software.

(2) Automatic layout from SUPER–COMPACT or TOUCHSTONE netlists. Netlists from either of these programs can be read in to generate a layout, and they can be generated from a GAS STATION layout.

(3) Bi–directional schematic capture of microwave circuits, in which netlists can be translated to and from schematic diagrams.

and

(4) Component compilation.

GAS STATION is mouse driven, with hierarchical and user definable menus and generates as input HPGL, GDSII, IGES and an EBL format.

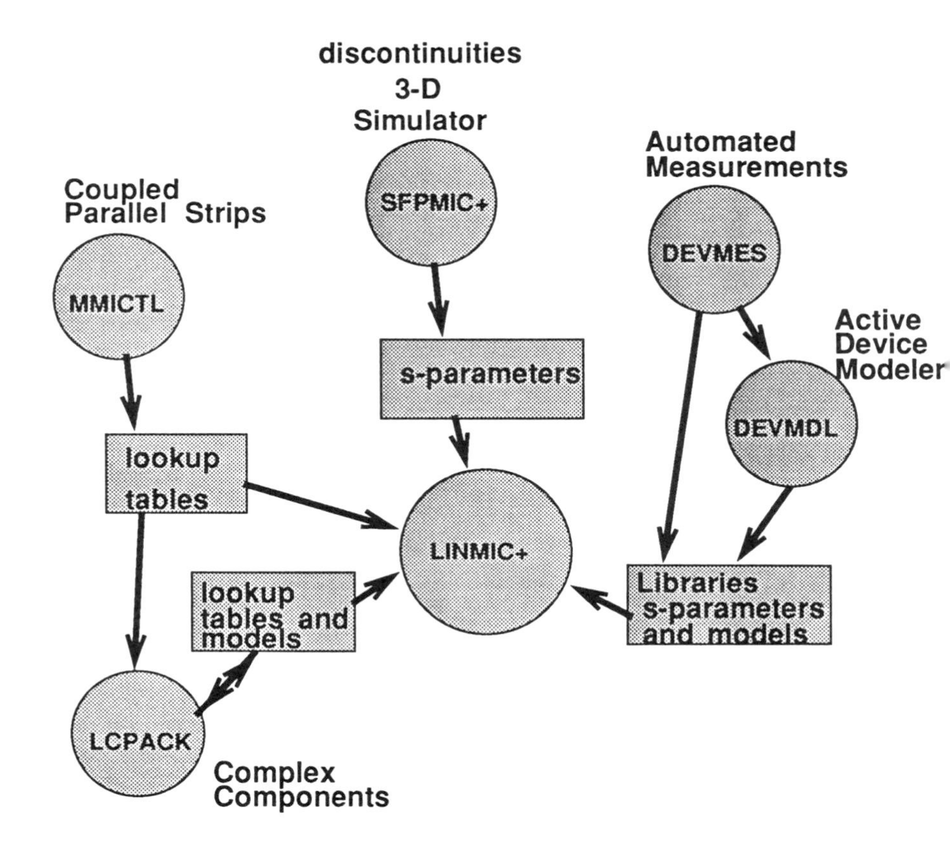

Figure 4.18 Relationship between the major programs in LINMIC+

4.3.5 Layout Analysis and LINMIC+

Some MMIC design can be done using those simulators (TOUCHSTONE and SUPER–COMPACT) which were originally developed for MIC design. However, the small size of components and inter–component spacings render their models inaccurate. Also, the construction of microstrip lines on an MMIC is much more complex than those on MICs; metallization now sits within a sandwich of layers of differing dielectric constants (GaAs, Polyimide, Silicon Nitride and Air). Using the basic simulators it is only possible to design circuits by approximating the construction to one of metal on a substrate of equivalent effective dielectric constant (ϵr_{eff}), making approximations to models and by ensuring that tracks are separated enough to prevent coupling. Clearly this does not make effective use of the substrate area. In order to produce competitive circuits, it is necessary to minimise the circuit size, and thereby increase the functionality that can be built onto a chip. Moreover, the cost involved in making prototypes makes this scenario unattractive. By using CAD which models the circuit at the layout level [14], it is possible to optimise the geometry of a layout and produce compact circuits that are "Right first time".

The LINMIC+ (version 2.1) suite of programs consists of the LINMIC+ simulator, plus 5 major sub-packages: MMICTL, LCPACK, DEVMDL, SFPMIC+ and DEVMES. These programs allow the design of MICs and MMICs, where inter-track coupling is analysed using various field-theory based portions. It was written by Jansen Microwave of West Germany. Figure 4.18 shows how the programs in the LINMIC+ suite interrelate.

LINMIC+ itself is a small-signal frequency domain simulator, like TOUCHSTONE or Super-Compact. It differs, however, in that distributed components can be modelled in 5 different ways:

(1) quasi-static analytical descriptions

(2) full-wave based analytical descriptions

(3) interpolation from various types of look-up tables

(4) s-parameters from the full-wave 3-D simulator and

(5) s-parameters generated by DEVMES (Measured) or files generated by other CAD packages.

The circuit is entered via a text file, which generates a layout in a graphics window. This feature gives reassuring feedback to the user that his circuit file is correct, and can be plotted to generate an artwork. LINMIC+ incorporates noise analysis and an optimiser, which uses the least-pth error function formulation.

The module MMICTL performs a high resolution full-wave analysis of coupled microstrip lines using the Spectral Operator Expansion Technique [15], which has evolved from the Enhanced Spectral Domain Method [16]. It covers extreme geometry ranges, and is valid up into high mm-wave frequencies. The substrate configuration can consist of up to 3 dielectric layers below and 3 above the metallisation. MMICTL can be used in two ways: where a regular array of microstrips is involved, MMICTL generates a 3 dimensional lookup table for coupled microstrip lines (regular strip width and spacing) and where more arbitrary arrays of parallel tracks are used, a 2-dimensional or a 3-dimensional table is generated with the variable parameters in the configuration being chosen by the user. Each run generates a look-up table relevant to a particular configuration i.e. a particular process and a particular number of coupled strips. It is numerically quite intensive and in extreme cases on

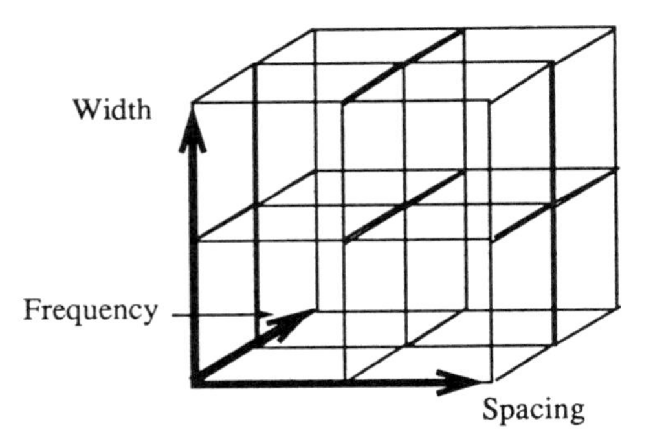

Figure 4.19 3-dimensional look-up table created by MMICTL

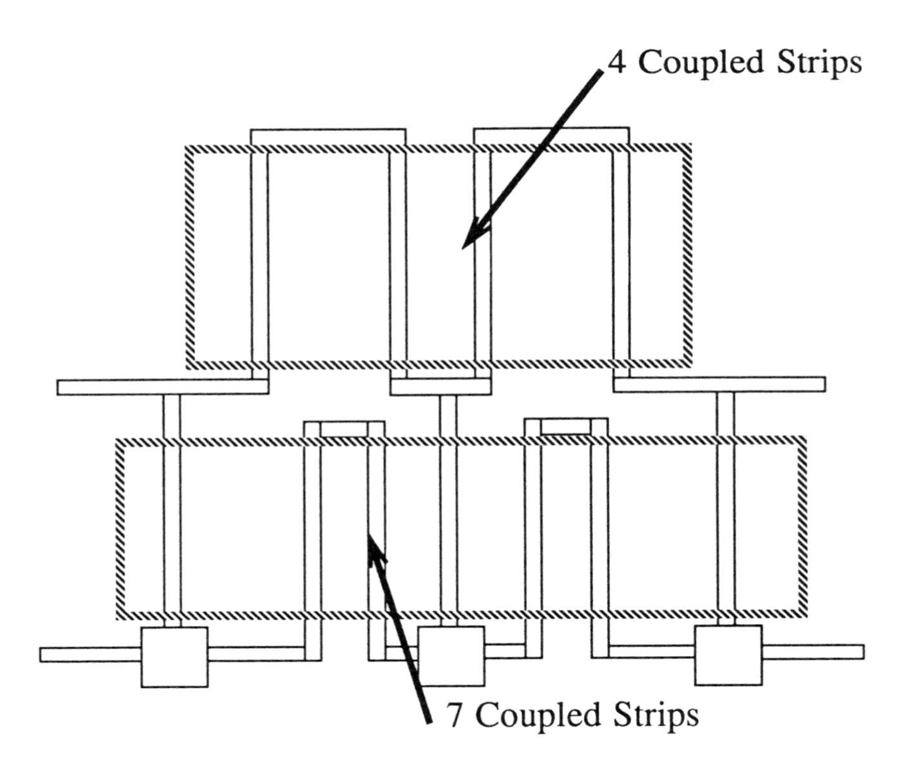

Figure 4.20 Partitioning of a travelling-wave amplifier

workstations may require an overnight run. For simple cases, look–up table generation typically takes a few minutes or tens of minutes.

However, it should be noted that tables can be defined in quite a general way, so that they only need to be generated once for repeated use. Once it has been run, LINMIC+ can interpolate between the results to simulate a given layout configuration. Typically the 3 dimensions of the look–up table are track width, track spacing and frequency. Solutions are obtained for the minimum, maximum and mid–point values along each axis giving 27 measurements (more can be specified by the user). Figure 4.19 illustrates the 3–dimensional look–up table. It makes use of symmetry to reduce the calculation time. More often though, a less regular structure requires analysis (when all the spacings between tracks are not equal). Figure 4.20 shows a section of a travelling–wave amplifier layout, and how it might be segmented by the user. Two sections require analysis in MMICTL: one of 4 coupled tracks and one of 7, before using LINMIC+ to analyse the circuit. Optional graphics with MMICTL allow the user to display current densities and transverse electric fields of strip configurations.

The module LCPACK analyses spiral inductors, spiral transformers, interdigitated capacitors and couplers. For instance, for spirals it uses the look–up tables generated by MMICTL and additional look–up tables for the corrections necessary (i.e. the corner and crossover capacitances). It generates lookup tables, in this case having 4 dimensions: number of turns, spacing between tracks, track width and centre span. LCPACK generates the s–parameters for these components and uses these directly in the simulator, or fits them to an equivalent circuit for repeated use. The user can go into an interactive mode and tune the construction of, say, a spiral inductor to give him the values which he will accept.

DEVMDL is the device modelling package. From files of s–parameters, an equivalent circuit model is fitted. The devices handled by DEVMDL are GaAs MESFETs (chip or packages, power FETs and internally matched FETs), Schottky Diodes and Bipolar Transistors. DEVMDL uses the same optimiser as LINMIC+.

SFPMIC+ is the full–wave 3–dimensional simulator. It models the fields associated with an individual discontinuity (e.g. bends, T junctions etc) or interacting group of them. It generates s–parameters which can be read by LINMIC+.

DEVMES, the automated measurement program simplifies the measurement of components. The hardware requirements are:

(1) An 8720 or 8510 Network Analyser with the Time Domain Option.

(2) A test jig, with microstrip launchers long enough to move the discontinuity of the coax to microstrip transition away from the device under test.

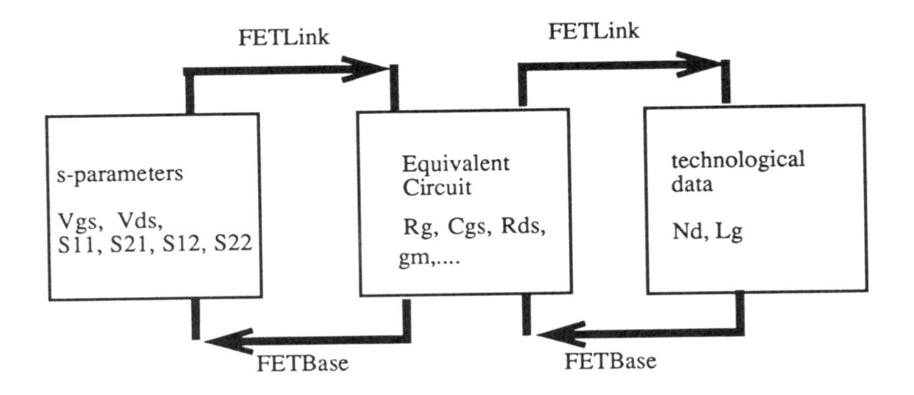

Figure 4.21 FETLink and FETBase

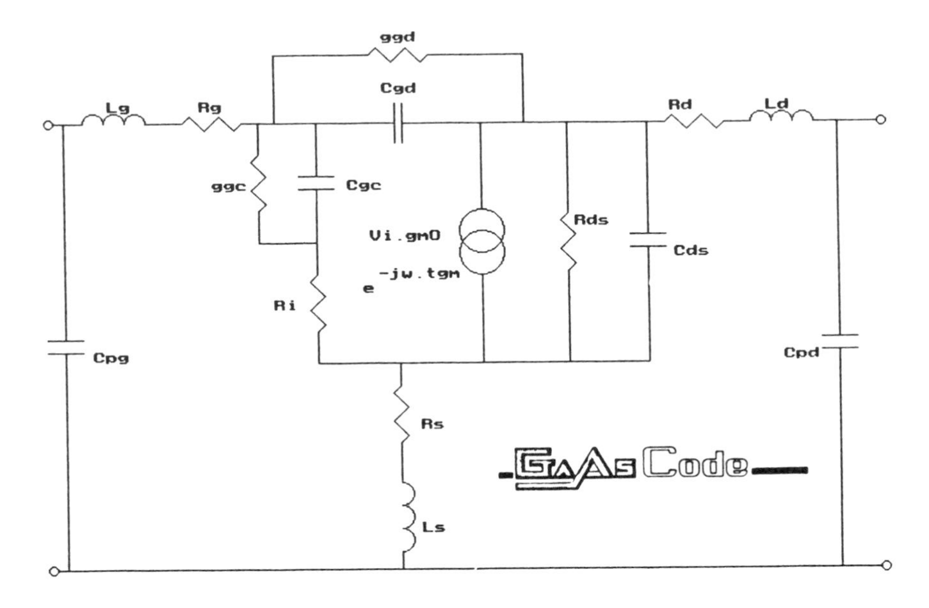

Figure 4.22 Equivalent circuit used by FETLink

The gating facility of the time domain option is used to capture the information required for de-embedding. The open ended microstrip line (no device inserted) is used as the calibration standard. Then the program can de-embed the devices measured.

LINMIC+ runs on VAX workstations and mainframes using GKS graphics, on HP 9000 series 300 and 800 workstations and on Apollo computers.

4.3.6 Device Modelling

It is evident that of all the components on an MMIC, it is the active device (the GaAs MESFET) which is the most critical. Its performance depends on the quality of the most difficult stages of processing, and it is drawn with the finest geometries on the chip (0.5 µm gate length compared to 10 to 150 µm geometries for microstrip lines). It is therefore easily degraded by deficiencies in the processing. One of the major setbacks which GaAs technology has had to endure has been that of poor yields. The situation has not been helped by the application of Monte-Carlo yield or performance spread analyses to the elements of the MESFET equivalent circuit. This has disregarded the fact that the component values are interrelated and depend on the fundamental device physics. Recently released software from GaAs Code Ltd allows the designer to understand more about his devices [17].

(i) GaAs Code

GaAs Code offer two programs, one called FETBase, the other FETLink. They work in opposite directions between s-parameters and technological data as shown in Figure 4.21. Both run at a reasonable interactive rate on an IBM-PC or equivalent with a maths co-processor. They work from compact device physical algorithms, and are relevant to GaAs MESFETs and HEMTs. These tools are valuable for foundry process monitoring, and in determining the relative merits of different foundries. Predictions of yield or spread in performance against processing parameters are obtainable.

(ii) FETLink 1.1

FETLink accepts measured s-parameters, and an equivalent circuit model (Figure 4.22) is fitted to each set. From the equivalent circuit model, technological data can be recovered; either channel doping density and gate length of MESFETs or gate length for HEMTs. FETLink requires one measured piece of data, the gate resistance of the MESFET, which is easily measured at DC. A set of s-parameter measurements consists of the s-parameters at a range of frequencies for one specific bias point. Bias information must be included in this file if technological data is to be

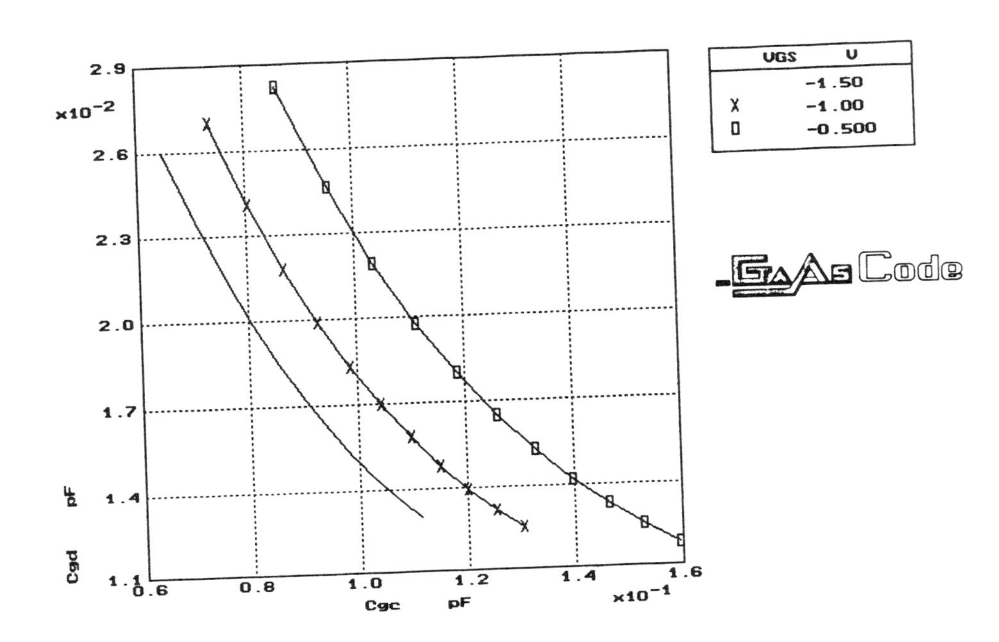

Figure 4.23 Graph showing the correlation between C_{gd} and C_{gc} over a range of V_{gs} and V_{ds}. (Reproduction with the permission of GaAs–Code Ltd Copyright Material)

recovered. The fitting routines work for the saturation region of the FET characteristic and run fast enough to keep up with a wafer prober as it steps across a wafer. An extended fitting routine now allows fitting to the FET characteristic below the knee. Data can be displayed in 3 different ways:

(a) Bar charts for statistical representation. For example, it is possible to show a histogram of metallurgical gate length as deduced by FETLink.

(b) Bias plots to show how elements vary with bias. It is also possible to show, for example, the bias variation of C_{gd} of a 0.5 μm Foundry FET, exhibiting the cross–over in the ordering of curves with V_{gs} that is characteristic of GaAs FETs and HEMTs. Such data can be obtained, together with all other equivalent circuit elements at 50 bias points in approximately 3 minutes using GaAsCode's equivalent circuit extraction program FETLink 1.1 running on a 20 MHz PC–AT. The C_{gd} cross–over forms a useful test of the validity of forward modelling routines or simulators; such a cross–over should always be predicted as a feature by any valid technology–based model or simulator.

(c) Scatter plots (Figure 4.23) which plot one element against another (these help in the detection of important correlations).

(iii) FETBase 1.1

FETBase works in the opposite direction to FETLink. From a menu of 6 common layout configurations (Figure 4.24), the dimensions are entered to describe the device (Figure 4.25).

From this the equivalent circuit values are generated, and from them the s–parameters of the device can be predicted for a given set of bias conditions. The DC and noise characteristics of the device (noise figure, pinchoff voltage, avalanche point etc) are also predicted. No measurement of physical devices is necessary.

(iv) Semiconductor Modeling

A more fundamental way of modelling semiconductor devices is to solve the semiconductor carrier transport equations on a 2–dimensional finite element grid [18]. Modelling at this level gives valuable insight into the operation of the device and avoids the restrictions imposed by equivalent circuits. Unfortunately, this approach is computationally intensive. An alternative approach uses a quasi–2D method of solution [19]. Rather than solving the semiconductor equations in 2 dimensions, an analytical solution defines the geometry of the depletion region and therefore the conducting channel. It is considerably faster than a full 2–D simulator since the semiconductor equations are solved in 1 dimension only. In the conducting channel the equipotential lines are assumed to be straight, also the transition between the conducting channel and the depletion region can be assumed to be abrupt (Figure 4.26). Using these approximations, a set of differential equations describing the charge transport in the channel can be written. In order to calculate the geometry factor, the relationship between the voltage along the channel and the depletion layer is required. A sloping, straight line approximation can be made for the equipotential lines in the depletion region. A solution based on Poisson's equation along the surface between x_r and x_{dep} finds the corresponding point x' for any point x along the channel, and the lines of constant electric field are assumed to be semi–circles of radius r centred at $x_r + x'$. The geometry of these sets of lines describes the depletion region edge. The simultaneous solution of Poisson's equation, the current continuity equation and the geometry factor in a 1 dimensional finite difference problem gives the solution.

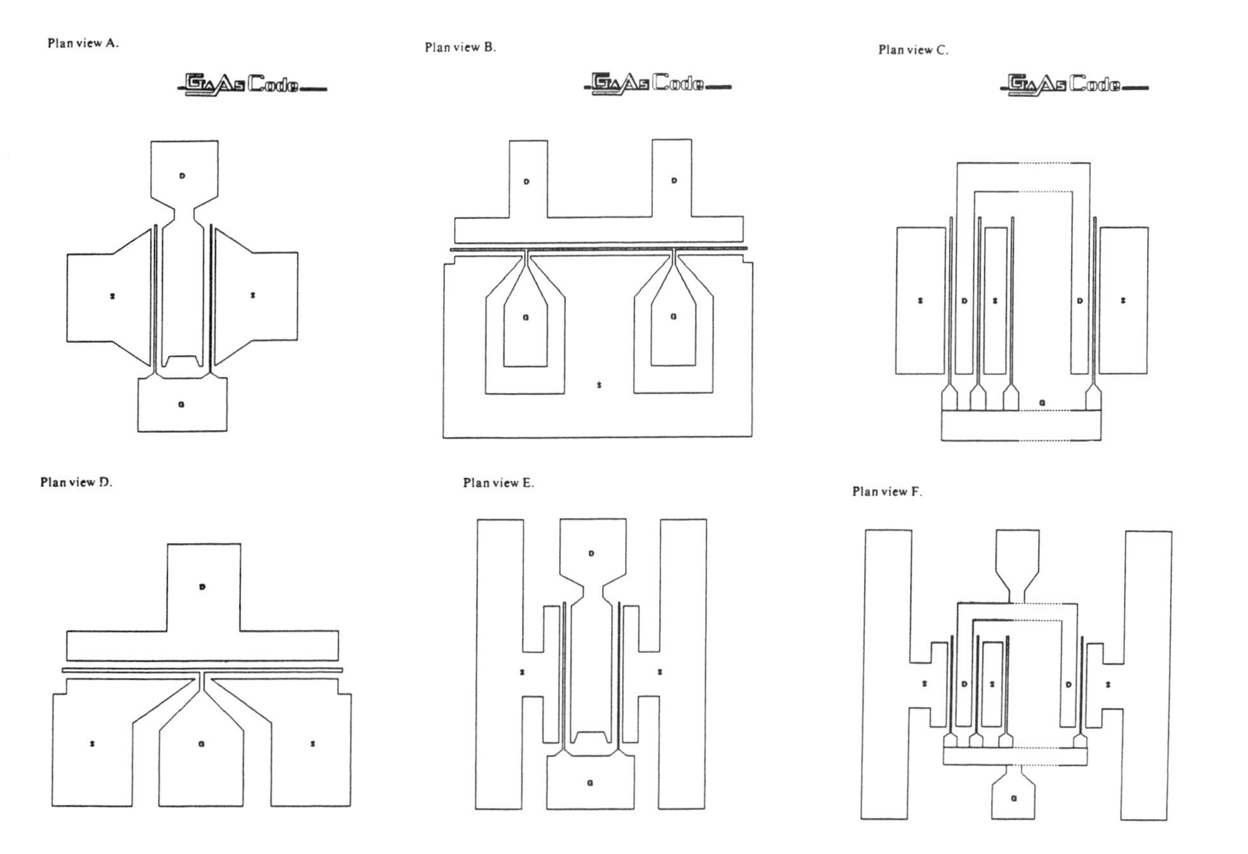

Figure 4.24 FET layout configurations available in FETBase. (Reproduction with the permission of GaAs-Code Ltd Copyright Material)

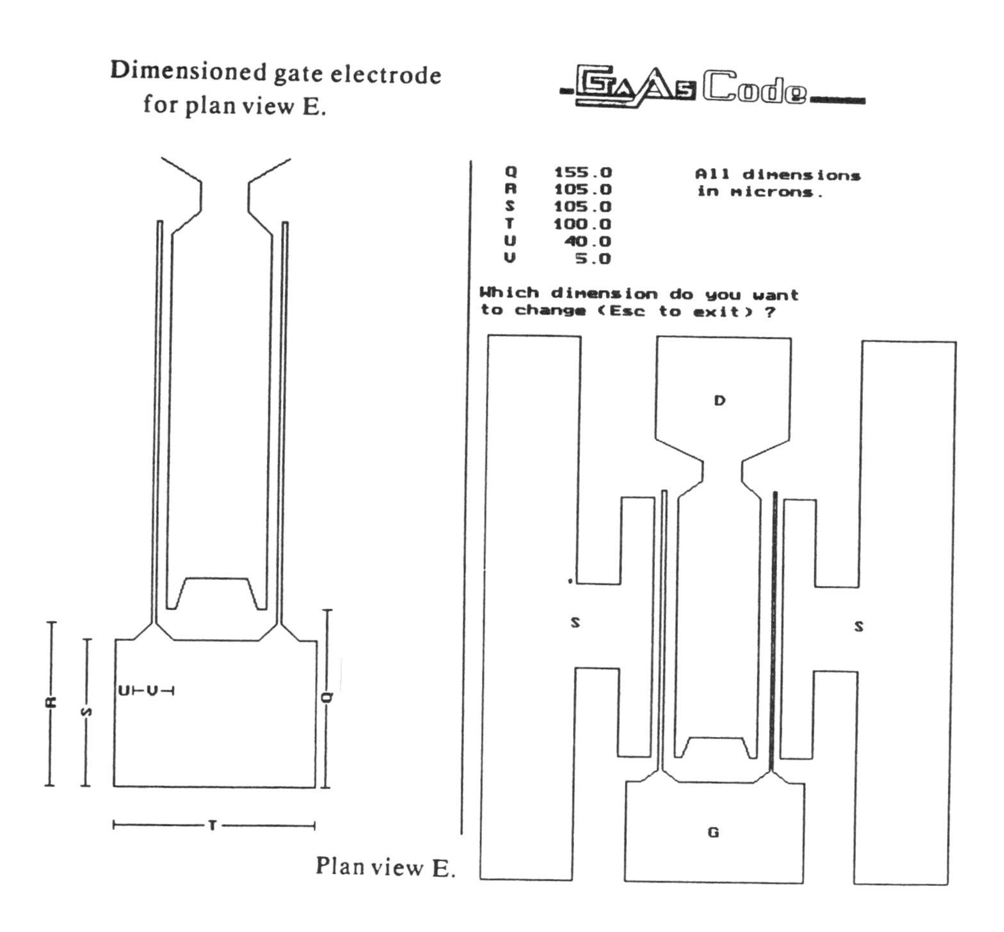

Figure 4.25 Dimensioned plan view of a GaAs MESFET for FETBase. (Reproduction with the permission of GaAs-Code Ltd Copyright Material)

The FETCAD program written at the University of Leeds uses a quasi–2D method and can produce DC characteristics and s–parameters rapidly on a PC [20]. Analyses have run 1000 times faster than full 2–dimensional simulations. Typically a 50 point DC characteristic can be calculated in 20 seconds on a PC–AT personal computer. Their model (Figure 4.27) allows recessed gates and non–uniform doping profiles. It accounts for surface–depletion effects, non–abrupt depletion region interface, substrate conduction, contact resistivity, three dimensional structure, avalanche breakdown and gate conduction. Its results have been applied successfully on MMIC designs. Research at the University of Leeds uses Transputers to produce full 2–D solutions. Some results obtained using it,

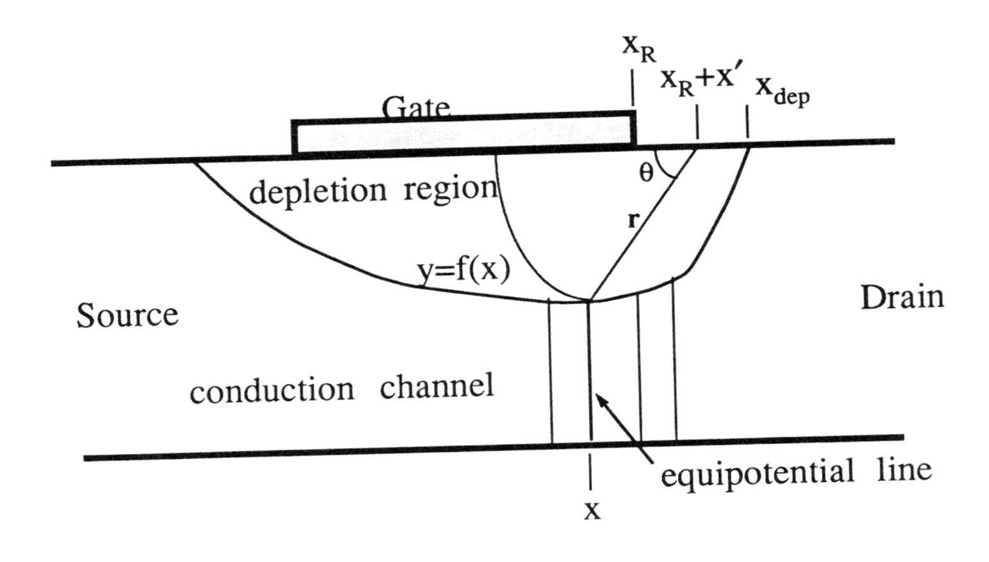

Figure 4.26 Depletion region boundary and equipotential lines

showing equipotential lines and electron concentration are given in Figure 4.28.

4.4 Analogue and Digital Design Tools

As mentioned in the introduction, CAD tools developed to assist the design of analogue and digital silicon ICs may be applied to the design of GaAs ICs. These tools span a wide range of functions from schematic input to mask layout and verification. This section reviews some of these functions and indicates the nature of the CAD packages which are available to assist the designer. Much of the authors' practical experience in this area has been with CAD packages available to the UK academic community as a result of the UK ECAD Initiative. Some of these are used in the examples which follow.

4.4.1 Schematic Input

Schematic input has for some time been the most common method of entering a circuit description into a CAD system. It typically involves using a graphic display and mouse to draw a circuit diagram on the screen.

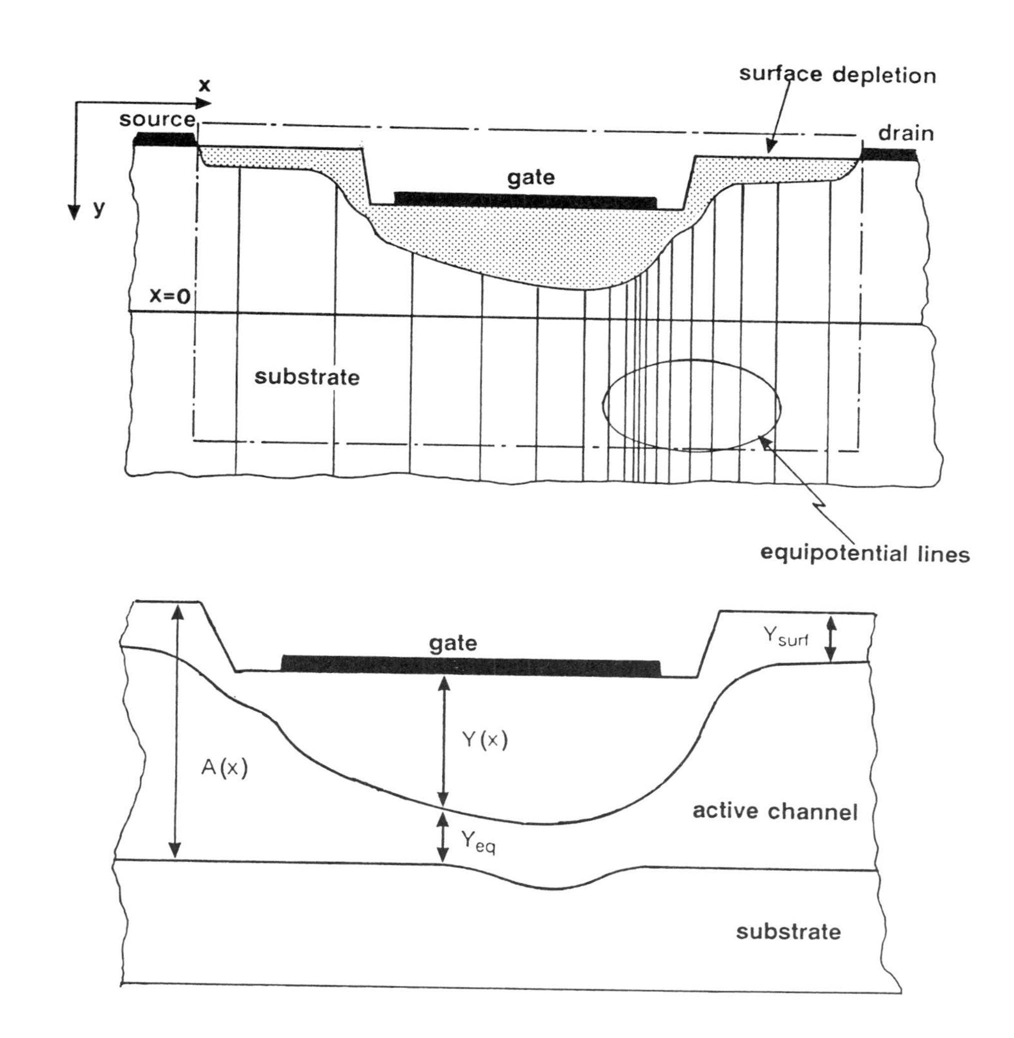

Figure 4.27 Physical model of a GaAs MESFET for the quasi-2D method (Reproduced with permission of the University of Leeds. Copyright material

Pre-defined components are selected from a library and placed at appropriate positions. The connection points on these components are then joined by lines which signify electrical connections. Most CAD systems allow the creation of a design hierarchy by letting the user represent a section of circuit by a single component symbol which is then used in other circuit diagrams. A very complex design can thereby be expressed in a compact form where the top-level schematic contains

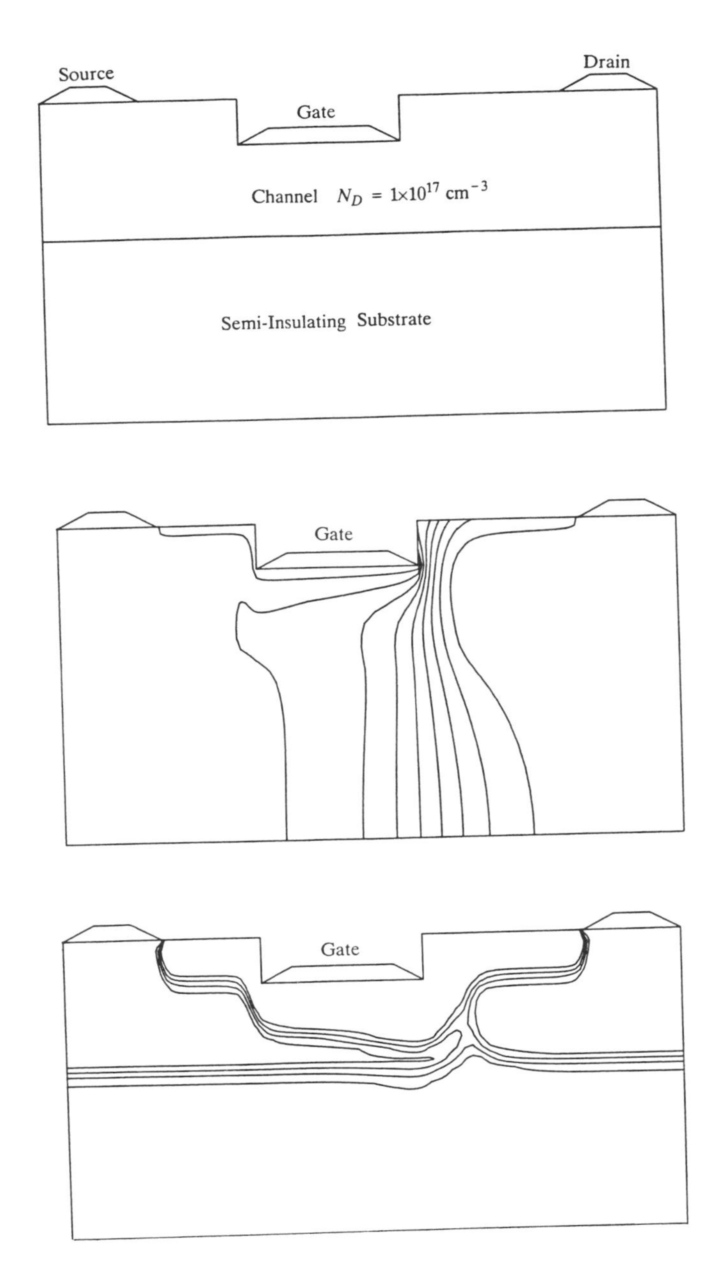

Figure 4.28 A 1 μm recessed gate MESFET using the full 2–D simulator (a) geometry (b) equipotential lines and (c) electron concentration (Reproduction with permission of the University of Leeds. Copyright Material)

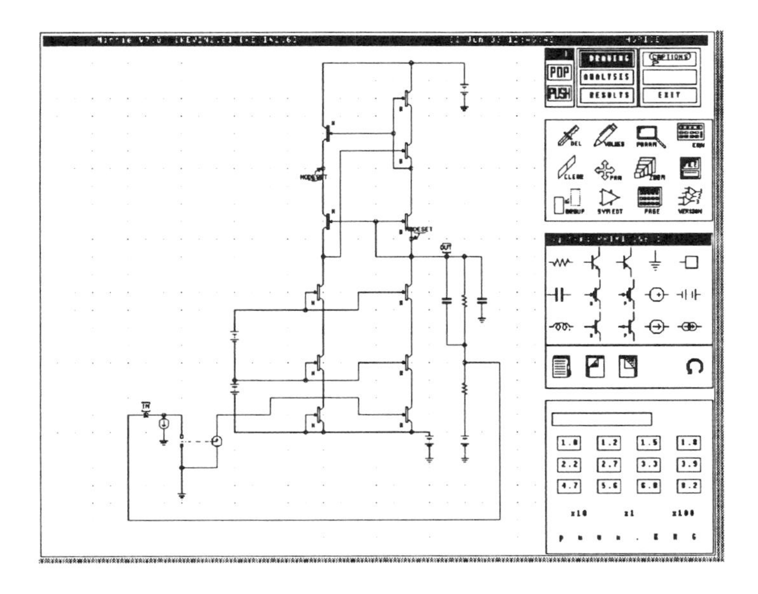

Figure 4.29 Screen display of MINNIE showing a double-cascade amplifier

component symbols which represent lower level circuit schematics and so on. Such schematic input systems are used for both analogue and digital SIC design and are increasingly being used to represent higher frequency designs to be implemented in GaAs.

The result of schematic input is a computer data base containing the components and all connectivity information about the user's circuit. This information may be used in a variety of ways. Initially, checks are usually applied to ensure that the circuit design conforms to some basic rules of construction. Following this the data may be translated into a form which can be understood by other CAD tools such as simulators. The principle is that this single data base is the primary circuit description from which all other forms are derived. Changes to the circuit must be made by modifying this data base, usually through the schematic input system.

For analogue circuit design, the basic components used in the schematic will be FETs, capacitors, resistors, inductors, etc. For digital circuits, the components will usually be logic gates such as AND, OR functions and more complex structures such as ADDERS, REGISTERS, etc. These in themselves are abstractions of complex analogue behaviour which results in an essentially digital behaviour characteristic.

As an example of analogue schematic input, Figure 4.29 shows the screen display of the MINNIE system from Interactive Solutions. This is typical of most CAD systems in having a drawing area and an on-screen menu to allow the user to control the program. The circuit display is a novel double-cascode amplifier the design of which is discussed in Chapter 10.

4.4.2 Analogue Simulation

The purpose of a simulation program is to show the designer how a circuit would behave without him having to actually build it. The first priority therefore must be accuracy of analysis or otherwise the designer will be misled. In time domain analogue simulation, circuits are analysed at a basic electrical level where voltage and current values are computed as a function of time throughout an electrical network. The circuit is described as an interconnection of active and passive devices such as FETs, diodes, resistors, capacitors, inductors, voltage and current sources, etc. The most popular circuit analysis program is SPICE which originated from the University of California at Berkeley and is now in widespread use throughout the electronics industry. Several commercial CAD vendors have improved the basic SPICE program to correct errors and improve its stability and accuracy. An example of this is HSPICE from Meta-Software.

One of the most important aspects of any analogue simulator is the accuracy of the device models which it uses. Extensive research has been carried out in this area and a variety of models are now available. A number of these particularly cater for the operation of active devices at the higher frequencies encountered in GaAs circuits, for example the Curtice models [4,5].

In performing a SPICE analysis, the first step is usually a DC analysis of the circuit to find the operating point. This is followed by either a transient or AC analysis to find the time-dependent or frequency-dependent behaviour of the circuit respectively. Models must have well-behaved numerical characteristics to allow the iterative process of time-stepping during transient analysis to proceed successfully. The process of numerical integration used in SPICE is slow but is still the most accurate method for analysing electrical circuit behaviour given good device models.

Most versions of SPICE now have some form of graphical post-processor to display the results of analysis in a form useful to the engineer. The MINNIE schematic system (by Interactive Solutions Ltd) is closely coupled with HSPICE to provide such a post-processing capability. Figure 4.30 shows the results of a transient analysis of the circuit shown in Figure 4.29. Both internal and external nodes of the circuit may be monitored.

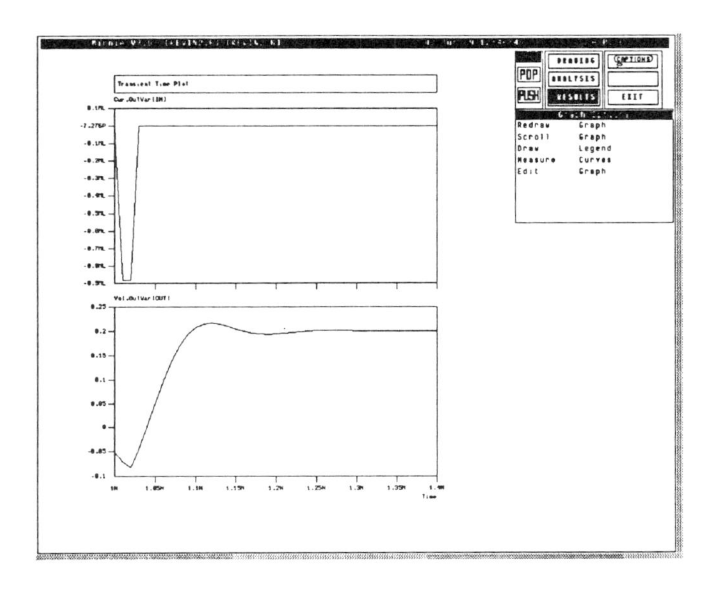

Figure 4.30 Screen display of MINNIE showing transient response of amplifier

A technique often used to model complex devices is to create an equivalent circuit or macromodel which has the desired terminal characteristics. The component values in this macromodel are determined by an iterative optimisation process which progressively adjusts them until the required behaviour is observed. HSPICE contains an optimiser which can be used for this purpose. Coupled with circuit analysis, sensitivity analysis allows the designer to investigate how sensitive some aspect of the behaviour of a circuit is to the variation of specific component values. This can be used both to adjust a circuit design to remove undesirable sensitivity to particular component values and to give some indication of the tolerances to be assigned to components to ensure the circuit yield is acceptable.

In an attempt to reduce the time taken to perform circuit analysis, alternative techniques have been developed which approximate some aspects of the behaviour of circuits. In switch level simulation, transistors are treated as voltage–controlled switches and circuits reduced to the analysis of RC networks. These approximations can be used for the analysis of some types of circuit implemented in GaAs. For example SWITCAP is a switched capacitor analysis program which models linear finite gain operational amplifiers, switches and ideal capacitors to produce time and frequency domain analyses [21]. Recently the idea of behavioural modelling of analogue circuits has been promoted as a means of reducing simulation times. Here, part of the model is represented by an analytical model which can be evaluated more rapidly than the full circuit description. As always in analogue simulation, the trade-off is between speed and accuracy.

As we have already seen, when circuits are physically designed, additional parasitic components are inevitably introduced into the circuit because of coupling between parts of the circuit which are physically adjacent. For a fully accurate simulation of the circuit, these parasitic components should be included in a re-simulation of the circuit to ensure that the desired performance has not been compromised.

4.4.3 Digital Simulation

By designing electrical sub-circuits to have well-defined digital logic functions such as AND, OR, etc and then using these in turn to design larger circuits, a great simplification in design and analysis is possible. These sub-circuits can be represented by very simple models which hide their detailed electrical operation. Using these digital gate models, complete systems can be simulated at gate level with a large time saving over analogue simulation at the transistor level. Provided sensible design procedures are adopted, no significant cumulative loss of accuracy in the simulation occurs.

Gate-level digital simulation is now an essential tool of the digital designer, particularly as he aims to achieve maximum system performance. A well characterised library of gate models is required which specifies for each gate its logical behaviour, the propagation delay of signals through the gate and the delay associated with driving a specific load. Using this library, and the designer's circuit, a digital simulator will apply a set of given signals to the circuit and predict the circuit outputs as a function of time. It will also normally detect violations of setup and hold times and other timing hazards which are specified in the gate models. As with analogue simulation, it is important for accurate results that effects of the physical circuit layout, such as additional delays caused by lumped track capacitance or transmission line effects at higher frequencies, are taken into account in the simulation.

In some cases, the calculations of logical and timing behaviour are separated for synchronous systems. The logical values of signals are simulated as described above. The timing behaviour is calculated by adding the delays associated with each gate for each signal path through the circuit and checking that no timing violations occur and that signals are stable within the specified clock period. Maximum and minimum gate delays as well as typical values are often used to check for worst case situations.

As circuits get larger, simulation at logic gate level becomes very slow. A solution to this problem is to recognise frequently used collections of gates which have a well-defined function and to create a functional model for this set of gates which directly calculates the outputs in terms of the inputs without having to simulate the detailed gate-level operation. A typical digital simulator is System HILO from GenRad. This supports both gate and functional level modelling and logic and timing simulation with maximum and minimum delays.

4.4.4 Polygon Layout

Having designed and simulated the circuit to the point where it is believed that it would work, the designer's next step is to proceed with its physical implementation. For circuits to be built in monolithic form in GaAs, this involves designing the layout of all the masks which will be used to fabricate the circuit. Each mask represents a patterning step in the fabrication process in which all the geometrical shapes or polygons to be created in that step, for example on the metal layer, are defined. These shapes must obey a number of so-called design rules which are intended to ensure that no fabrication problems are created by shapes being, for example, too narrow or too close together.

The traditional method of doing this is to draw the polygons on each mask by hand. Normally an interactive graphical editing program is used for this. Such a program, which would be run on a medium to high resolution colour graphics workstation, provides the designer with a range of commands to create and manipulate polygons and to modify his view of the design by means of operations such as zoom and pan. Colour is an essential aid to distinguishing features on the different mask layers. For GaAs design it is important to be able to draw polygon edges at 45 degrees as well as parallel to the axes in order to be able to represent FET gate structures.

As described earlier, hierarchy is necessary to handle the complexity of large designs. In a layout editor this is supported by allowing the design of individual cells which can then be re-used as many times as required. Cells can be created which contain references to other cells and so a layout hierarchy is created to match the design hierarchy. Where cells are likely to be useful in other designs these can be put into a library to protect their integrity and make future access easy.

A typical IC layout editor is PRINCESS from Silvar-Lisco. Plate II shows a PRINCESS screen display with the layout of a GaAs amplifier cell. On the right side of the screen are a number of menu boxes which the designer uses to control operation of the program. PRINCESS allows users to create cells and store them away for future use. It also contains a powerful macro command feature which allows sets of editing commands to be parameterised and then saved for subsequent use. By supplying values for the requested parameters, the designer can invoke a complex sequence of commands to build, for example, a multi–layer structure such as a spiral inductor.

Being the lowest level representation of a circuit, the mask data tends to be the largest in volume. This presents interesting problems of data management in order to cope with the complexity while preserving efficiency of operation. Programs designed to handle the layout of silicon ICs which can contain millions of primitive polygon shapes, tend not to have a problem dealing with the complexity of the average GaAs design.

4.4.5 Symbolic Layout

The normal process of IC layout forces the designer to address many issues at once, both electrical and geometrical. Among these are the rules which determine the detailed construction of active and passive devices from their constituent mask features. Symbolic layout aims to remove this latter concern by allowing the designer to use simple symbols to represent the possibly quite complex sets of mask polygons making up a device. These symbols are placed to indicate the approximate relative position of the devices and they are interconnected to achieve the correct circuit topology.

This process is usually then followed by an automatic process conventionally called compaction. Here a program, which has been told about the layout design rules, moves the devices until they are at their minimum legal separation while preserving the connectivity of the circuit. A third step then generates the full mask shapes for all the devices and interconnection structures in the circuit, thereby creating correct mask manufacturing data of, theoretically, minimum size.

In reality, however, this tends to be a somewhat simplistic process and does not in general produce results as good as those of a good designer. It also does not take account of the many issues which affect circuit performance besides layout density. In particular, for high frequency circuits, the proximity of certain parts of the circuit can have a crucial effect on performance because of potential parasitic coupling. These issues still remain the province of the experienced designer.

As an example of a symbolic layout system, Plate III shows the screen display of the MAGIC program for the same amplifier cell shown in Plate II. MAGIC was developed at the University of Washington and is in the public domain. It does not include a compaction process [22].

One advantage of symbolic layout over mask polygon layout is that in symbolic layout the designer is working at a level closer to the circuit description and therefore extraction of electrically meaningful information such as device parameters or circuit connectivity for use in subsequent verification is more straightforward.

4.4.6 Layout Verification

Manufacturing integrated circuits is a lengthy and costly process. Designers are anxious therefore that their circuits should be free of design errors which would cause malfunction and require a redesign. Simulation provides confidence that the circuit design is likely to be functional. Layout verification provides confidence that the physical implementation of the design is also likely to be functional and that it does in fact implement the required circuit.

Fabrication houses supply both layout and electrical design rules which designers must follow to ensure working circuits. There may be 60 or more rules for a given process and these can be quite complex, sometimes relating shapes on up to three mask layers simultaneously. Designers naturally find it difficult to remember to follow all of the rules all of the time. To check for errors by hand is slow, laborious and unreliable. For this reason, layout verification software is used to check that a design obeys all the rules before it is committed for manufacture.

Layout design rules are usually expressed in a special language which then controls the checking program to apply each of the rules in sequence to the layout. Simple examples of these rules are to check that shape widths are not too small or that shapes are not too close together. The results of these layout checks are usually displayed graphically on the original layout to highlight the fault and allow the designer to correct it.

It is very convenient if layout design rules can be checked while a design is in progress as it is usually very much easier to correct a fault before it becomes deeply embedded in the layout. For this reason interactive layout verification is becoming available within layout editing programs as a menu option. Because of the extensive computation involved in carrying out these checks, however, interactive checks are usually limited to modest amounts of layout, for example a small group of cells. Final layout checks on a large design are performed off-line as a batch procedure, usually on a powerful mainframe computer.

As mentioned above, layout rules are concerned with the correct construction and arrangement of the polygons making up each mask from the point of view of ease of manufacture. Electrical rules are concerned with whether the polygons will result in correctly formed electrical devices and whether these devices will be correctly connected together to form a circuit. To apply these checks, electrical rule checking software attempts to recognise electrical devices and extract the electrical circuit implied by the layout polygons. Errors which it finds are reported in the same way as for layout errors. To do this, the software clearly has to have been given a good definition of the technology being used. Again this is conveyed to the program in a language format. GaAs circuits contain structures not normally encountered in designs on silicon such as spiral inductors. Special procedures must therefore be developed to recognise these devices.

Although the layout may represent a valid electrical circuit, it may not be the same circuit which the designer originally captured schematically and verified by simulation. For example, a gate to invert a signal may be inadvertently omitted. A further layout verification program checks the extracted circuit obtained from the layout with the original schematic circuit and indicates any discrepancies. Also, having extracted all the devices from the layout, it is possible to deduce their electrical parameter values, for example the value of a capacitor from its area knowing the capacitance per unit area of the process. The next step is then to produce this information in the format required by a circuit simulator such as SPICE. The expected performance of the circuit which would result from fabrication using these mask layouts can then be found. Occasionally this detects a potential failure to meet a specification which would otherwise only have been found when the circuit was fabricated.

Together, all these layout verification checks give the designer confidence to proceed with fabrication knowing that all possible steps have been taken to ensure a correctly working circuit. Examples of layout verification software are the Design Verification Products suite from Silvar–Lisco and the Dracula suite from Cadence.

4.4.7 Data Standards

When it is necessary to transfer electronic design data from one physical location to another, for example to fabricate an integrated circuit, the problem arises as to which format should be used to represent the data. The transfer is either by some machine-readable magnetic medium such as magnetic tape, or increasingly, directly over a communications network.

A pragmatic solution has arisen in the case of mask layout data where the Calma GDS2 Stream format has been widely adopted. In other areas, such as netlist transfer, no clear industry standard existed. This led to the development of the Electronic Design Interchange Format (EDIF) [23]. This defines standards for the exchange of schematic, netlist and mask layout data, with future plans to confer printed circuit board and test data. Many commercial CAD systems are now providing EDIF interfaces both in and out.

Most CAD systems have proprietary data bases in which all types of design data for a given design is stored. This makes the integration of additional CAD tools into such a system difficult. Recently discussions have been taking place among a number of the leading CAD vendors with a view to making such interfaces easier to implement.

4.5 Acknowledgements

Many thanks are due to numerous people involved in the industry, for the provision of material or for their permission to reproduce data. Thanks are particularly due to GaAs Code Ltd., Jansen Microwave, Coss–Mic Ltd., Barnard Microsystems Ltd., Plessey III/V and Plessey Research (Caswell) and the Universities of Kent and Leeds.

4.6 References

1 Gupta, Garg and Chadha: Computer Aided Design of Microwave Circuits, Publishers: Artech

2 J. K. Fidler and C. Nightingale: Computer Aided Circuit Design, Publishers: Nelson

3 F. W. Grover: Impedance Calculations, Van Nostrand Princeton N.J. 1946 reprinted by Dover Publications New York 1954

4 Walter R. Curtice: A MESFET Model for use in the Design of GaAs Integrated Circuits, IEEE MTT–28 No. 5 May 1980 pp448–456

5 Walter R. Curtice: A Non–Linear GaAs FET Model for use in the Design of Output Circuits for Power Amplifiers, IEEE MTT–33 No. 12 December 1985 pp1383–1393

6 Adam Jastrebski: Parameter Extraction for Non–Linear MESFET models, IEEE colloquium on Large Signal Device Models and Parameter Extraction for Circuit Simulation 19th April 1988

7 Andrzej Materka and Tomasz Kacprzak: Computer Calculations of Large Signal GaAs FET Amplifier Characteristics, IEEE MTT–33 No. 2 February 1985, pp129–134

8 G. J. Gardner et al: Design Techniques for GaAs MESFET switches, IEEE MTT–S Digest June 1989 Long Beach California

9 E.Oran Brigham: The Fast Fourier Transform Publishers: Prentice Hall Inc

10 C. S. Burrus and T. W. Parks: DFT/FFT and Convolution Algorithms, Theory and Implementation, Publishers John Wiley and Sons

11 Stephen A. Maas: Microwave Mixers, Publishers, Artech House

12 A. R, Kerr: A Technique for Determining the Local Oscillator Waveforms in a Microwave Mixer, IEEE MTT–23 p828 1975

13 A. K. Jastrebski and M. I. Sobhy: Analysis of Non–Linear Microwave Circuits using the State–Space Approach, proceedings of the International Symposium on Circuits and Systems, Montreal pp1119–1122 May 1984

14 Rolf H. Jansen: A Novel CAD Tool and Concept Compatible with the Requirements of Multilayer GaAs MMIC Technology, IEEE MTT–S St. Louis MO 1985 pp711–714

15 Rolf H. Jansen, Ronald G. Arnold and Ian G. Eddison: A Comprehensive CAD Approach to the Design of MMICs up to mm–wave frequencies, IEEE MTT–36 No. 2 February 1988 pp208–219

16 Recent Advances in the Full-Wave Analysis of Transmission Lines for Application in MIC and MMIC design, SMBO Int. Microwave Symp. (Rio de Janeiro) July 1987 paper M-111

17 Peter Ladbrooke: MMIC Design: GaAs FETs and HEMTs, Publishers: Artech House

18 Christopher M. Snowden: Introduction to Semi-Conductor Modeling, Publishers: World Scientific

19 Peter A. Sandborn, Jack R. East and George I. Haddad: Quasi-Two-Dimensional Modeling of GaAs MESFETs, IEEE ED-34 No. 5 May 1987 pp985-991

20 C. M. Snowden and Renato R. Pantoja: Quasi-2-D MESFET Simulations for CAD, IEEE ED-? September 1989

21 S. F. Fang, Y. P. Tsividis and O. Wing: SWITCAP: A Switched Capacitor Network Analysis Program, IEEE Circuits and Devices Magaine, Vol 5, No. 3, September 1983

22 J. K. Ousterhout et al: MAGIC: A VLSI Layout System, ACM-IEEE Proc. 21st Design Automation Conference, 1984, pp152-159

23 EDIF Steering Committee, EDIF Version 2.0.0.,Electronic Industries Association 1987z

Chapter 5

Technology Comparison: Gallium Arsenide vs. Silicon

Peter H. Saul

5.1 Introduction

This chapter compares Gallium Arsenide and Silicon technologies for both analogue and digital applications. The bases of comparison are physics, engineering and economics. Gallium Arsenide is shown to be limited in digital applications, but advantageous for analogue. Silicon bipolar is fastest and capable of higher complexity in digital applications. Where speed is less important, CMOS will continue to dominate. Finally, silicon motherboard, multi–technology "flip–chip" techniques have an important role.

There are many misconceptions of the relative roles of silicon and gallium arsenide in integrated circuits. Many authors are too intent on pressing their own case, while "proving" that the alternative technology could not achieve the same result. This chapter will attempt to bring together the relevant data and to make some definitive statements on the respective merits, seen from the viewpoint of the integrated circuit designer, of the various technologies available. The section headings have been chosen to give the reader a background to the basic concepts in physics, in engineering and in outline at least, in economics.

5.2 Physics

GaAs is claimed to be "five times faster than silicon". This is a distortion of the facts, which has arisen from comparison of the mobilities in the two materials. The physical properties of the materials are comprehensively documented in Ref. 1; Table 5.1 is a selection of the properties retabulated for comparison purposes. Figure 5.1 is also from Ref. 1, and illustrates both the origin of the five times claim and the reason why it is unlikely to be valid in real circuits. The horizontal scale, for electric field, has been relabelled in volts per micron from the original in volts per centimetre. In the region below 0.1 $V.\mu m^{-1}$, the carrier drift velocity lines for electrons in GaAs and Silicon are parallel and differ by a factor of about five up to the peak of the GaAs curve at 0.3 $V.\mu m^{-1}$. Silicon does not peak, but tends monotonically to a limit of slightly over 10^5 m s^{-1}.

Table 5.1 Properties of Si and GaAs at 300K

Properties	Si	GaAs
Linear coefficient of thermal expansion, $\Delta L/L\Delta T$ (°C^{-1})	2.6 x 10^{-6}	6.86 x 10^{-6}
Melting point (°C)	1415	1238
Minority carrier lifetime (s)	2.5 x 10^{-3}	~10^{-8}
Mobility (drift) (cm^2/V–s)	1500	8500
	450	400
Thermal conductivity at 300 K (W/cm–°C)	1.5	0.46
Thermal diffusivity (cm^2/s)	0.9	0.44
Dielectric constant	11.9	13.1
Effective Mass, m^*/m_o	m^*_l = 0.98	0.067
Electrons	m^*_t = 0.19	
	m^*_{lh} = 0.16	m^* = 0.082
Holes	m^*_{hh} = 0.49	m^* = 0.45
Electron affinity, χ(V)	4.05	4.07
Energy gap (eV) at 300 K	1.12	1.424

Current practice in GaAs MESFETs is to use a gate width of 0.5 μm to 1μm. If the voltage across the device is 1 Volt (rather lower than usual), then the field is very roughly 1V.μm^{-1}, i.e. past the peak velocity. The exact state of the GaAs device is not described by this simplistic view, since pinch effects increase the local fields close to the drain possibly by an order of magnitude, thus apparently making the device even slower. The drift velocity is therefore lower than the peak for GaAs, although how much lower is debatable, since ballistic transport effects (ref. 2) probably occur locally within the device. A limiting saturation velocity for GaAs has been reported (in Ref. 2) as 1.4 x 10^5 m s^{-1}. Thus the GaAs : Si advantage, at least on an electron velocity argument, is approximately 1.4 : 1. The hole velocity in GaAs is actually lower than silicon and is one of the reasons why no CMOS equivalent in GaAs has been produced; if p–channel GaAs devices were used as active switches they would be slower than silicon.

Silicon FETs now have gate widths comparable to those of GaAs, so that under the same field conditions GaAs has an advantage in the region of

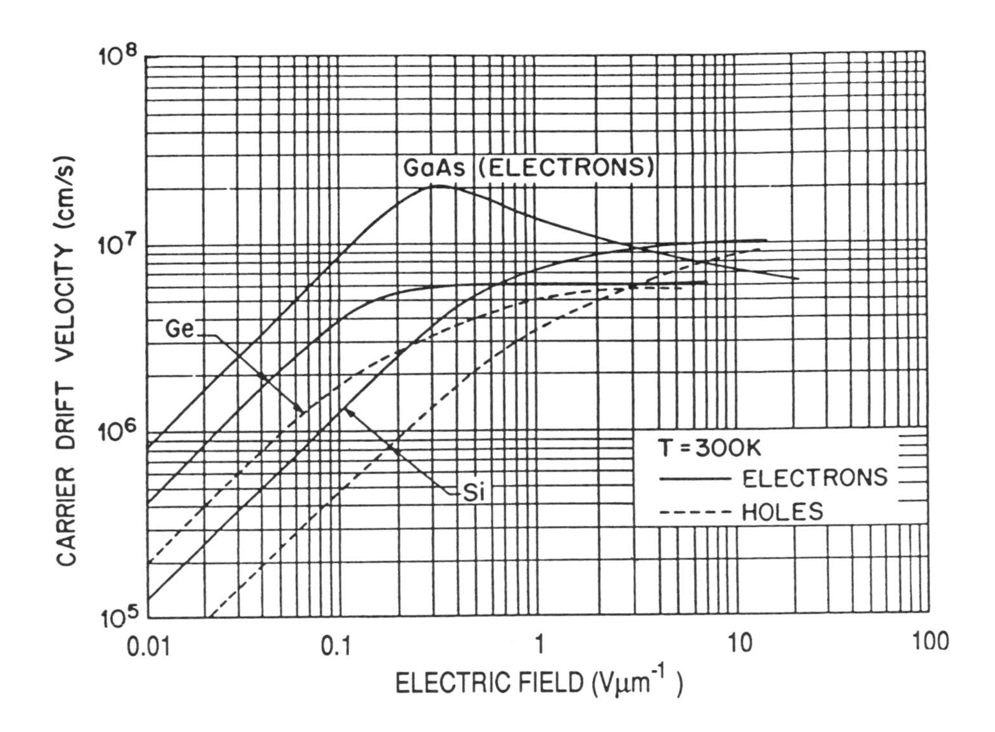

Figure 5.1 Drift velocity in semiconductors (after Sze[1])

2 : 1. Silicon bipolar devices have base widths of 0.2 μm or less, so the fields are even higher; the relative advantage of GaAs is therefore around 1.4 : 1.

Silicon is an indirect bandgap semiconductor, with only a single effective bandgap (1.12 V at 300 K), hence the monotonic increase of the carrier drift velocity in Figure 5.1. GaAs has a direct bandgap (1.424 V at 300 K) with a secondary sub–band only a further 0.31 V higher. Under high field conditions, electrons are excited into the sub–band, where they have higher effective mass and hence lower velocity. This is the reason for the reduction in carrier drift velocity at high fields; it is a measure of the relative distribution of carriers in the main band and the sub–band. True velocity saturation reduces the carrier drift velocity to below that of silicon at fields greater than 3 V μm^{-1}, although the effects mentioned above give a practical figure of 1.4×10^5 m s^{-1}.

Other relevant physical properties are the melting point, the minority carrier lifetime, the thermal conductivity and the coefficient of thermal expansion.

High temperature operation is a desirable characteristic both for military, and increasingly for civilian devices. GaAs has a higher melting point than silicon (1415°C against 1238°C) but this is offset by the dissociation of the base material which occurs in GaAs above approximately 600°C. The wide bandgap of GaAs is an advantage for high temperature operation, but the limitation in present generation devices in both Si and GaAs is metal diffusion into the semiconductor. Refractory metal technologies are being developed to overcome this problem.

Thermal conductivity is important for many circuits, especially where high power dissipation is encountered, such as in complex high speed circuits. Silicon, at 1.5W/cm-°C, is three times better than GaAs, so that more care is needed in the thermal design of GaAs chips. Again, offsetting this, some GaAs devices can be designed to dissipate less than their Si counterparts. GaAs wafers are customarily prepared very much thinner than silicon, which helps thermally, but is a disadvantage in mechanical handling and would lead to difficulties on large chips with significant thermal mismatch to the package. An interesting compromise recently proposed is the use of GaAs chips solder–bump bonded to silicon "motherboards"; in the appropriate design, this can offer many of the advantages of both technologies.

In practical circuits, the main advantage of GaAs over silicon is the semi–insulating substrate which can now be reliably produced in quantities and without chromium doping. This has a major impact on circuit design, since the device to device and device to substrate parasitic capacitance terms are very small. When compared with silicon technologies, this is an enormous advantage, both for digital circuitry, where the reduction in parasitics improves speed and power–delay product, and in analogue/microwave circuitry where inductors are a sensible component on GaAs, and not so on conventional silicon substrates due to the very low effective "Q" which would result. Recently, there has been moves in silicon technology to either a high resistivity substrate or to silicon–on–insulator. The latter technique has been in use for many years in silicon–on–sapphire form, which is in general very expensive, but more recently implanted SOI substrates have become available commercially. These latter, while suitable for surface devices such as CMOS, are not yet available for bipolar work. It may be that a very high resistivity substrate produced by fairly conventional means is better for bipolar. In this context, the diagram of Figure 5.2 is interesting. This has been taken from Ref. 3. This figure shows that, for conventional substrates in the region of 1 Ω–cm, operation of all devices up to several GHz is clearly in the

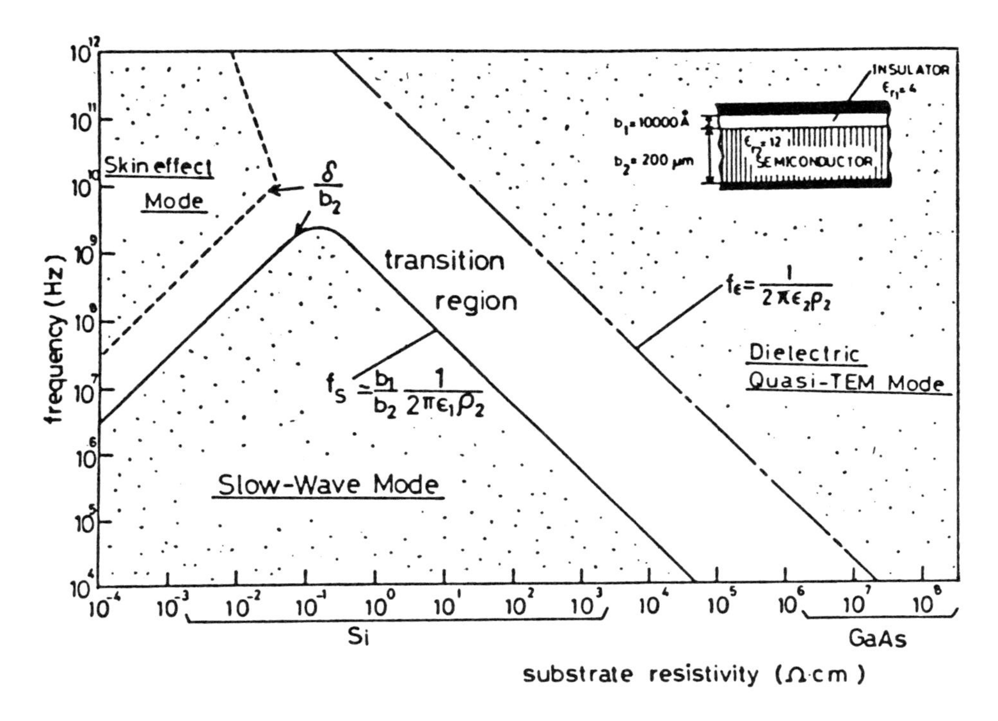

Figure 5.2 **Resistivity–frequency mode chart of the MIS microchip line. δ is the skin depth in semiconductor and ρ is the semiconductor resistivity**

"slow–wave" mode, i.e. propagation of a signal is through the bulk of the semiconductor, which appears as a lossy medium. Transmission in the skin effect mode is very unlikely in practical semiconductors. However, transmission in the "dielectric quasi–TEM" mode is possible, if the substrate resistivity and/or the frequency is high enough.

The threshold from ref. 3 is 3 GHz for 50 Ω–cm, 1.5 GHz for 100 Ω–cm and 900 MHz for 170 Ω–cm. Although substrates are commercially available up to 3000 Ω–cm, a practical upper bound is 1000 Ω–cm after processing, i.e. dielectric quasi–TEM mode would be available from 150 MHz upwards. Of course, the lines on Figure 5.2 are rather artificial; the transition would be usable with some losses, while desirably operation would be well inside the dielectric quasi–TEM region for lowest loss. The advantage of this mode is that, while propagation is still through the bulk of the material, losses are low. GaAs is always operated in this region.

Finally in this section, GaAs has the enormous advantage of light emission. Silicon, with an indirect bandgap, and only a single possible electron transition, cannot emit light. GaAs has a direct bandgap and an

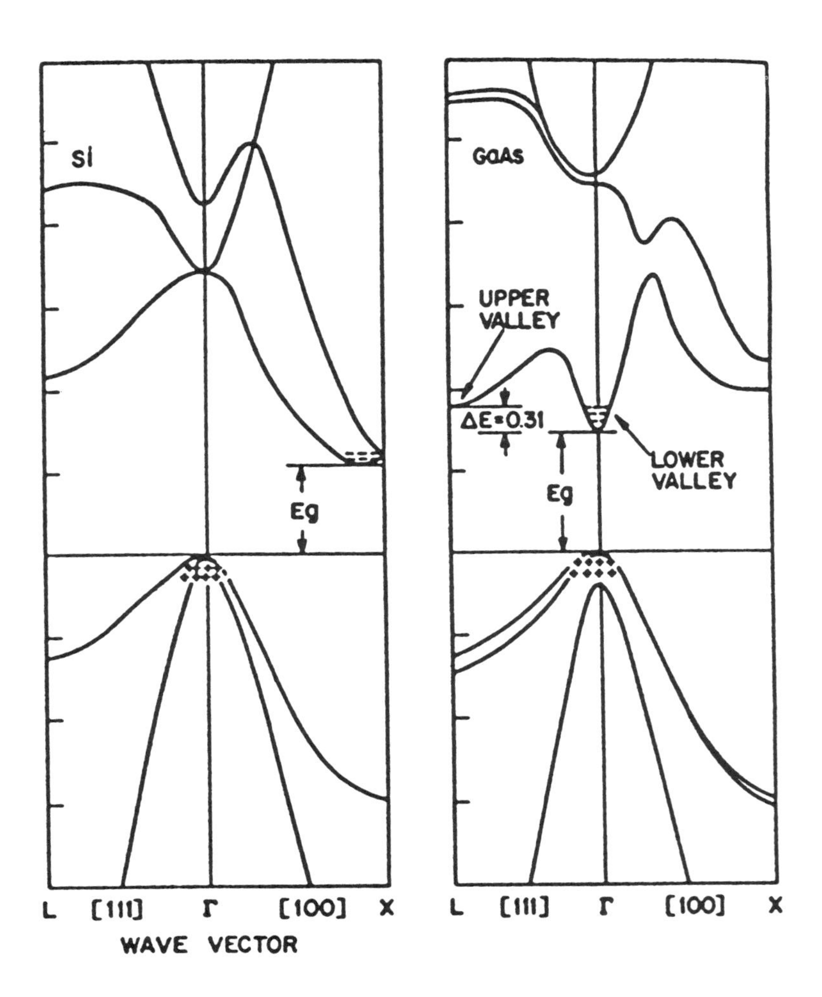

Figure 5.3 **Energy–band structures of Si and GaAs, where Eg is the energy bandgap. Plus (+) signs indicate holes in the valence bands and minus (–) signs indicate electrons in the conduction bands. (After Sze, Ref. 1)**

upper valley sub–band (Figure 5.3, also from Ref. 1). Electrons existed into the sub–band decay with the emission of light; actually in the infra–red for pure GaAs, while ternary compounds with, for example phosphorous give visible light. This property can, in principle, give rise to the fully integrated electro–optic device.

5.3 Engineering

5.3.1 Digital Circuits

For relatively low speed circuits, CMOS technology in silicon is unlikely to have any serious rivals in the near future; certainly none is visible today. The main reason for this is that very large chips become interconnection limited and speed becomes independent of basic transistor technology. Therefore, CMOS, with its inherently low power consumption, is the best choice in most instances. It is possible to achieve lower power delay products on some bipolar processes; but not at low clock rates. Variants on the CMOS theme include BiCMOS (Refs. 4 and 5) and CMOS–SOS or CMOS–SOI. The bipolar devices on CMOS processes offer better drive capability, better memory sense amplifiers and better analogue functions, although always at the cost of increased process complexity. Probably the most optimised BiCMOS processes will emerge as basically CMOS like, with bipolar assisted peripheral and analogue circuitry. The use of silicon–on–sapphire or silicon–on–insulator technology gives major speed and sometimes packing density improvements. The very fastest devices of this class can challenge all but the best bipolar processes on speed.

The main area of overlap between silicon and GaAs is in very high speed circuits, typically over 1 GHz. Figures 5.4 and 5.5 show logic gates in silicon bipolar and GaAs MESFET technology respectively. The former offers a three or four level logic stack capable of realising a complex logic function at high speed. Emitter follower level shifters are used prior to the next stage. GaAs circuits can perform equivalent functions, but the requirement in buffered FET logic (BFL) for a negative supply can be a limitation. Newer GaAs configurations use Direct Coupled FET Logic (DCFL), (Figure 5.6) similar in principle to NMOS. This requires an enhancement type FET device, which has proved difficult to reproduce, but sufficient devices are now available that these problems must have been largely overcome by the major suppliers of GaAs logic. An alternative logic configuration, source coupled FED logic (SCFL) has been proposed to overcome the limitations of very accurate process control needed for DCFL. The configuration is very similar to ECL (Figure 5.7), but with the source follower and level shifted output similar to BFL. Operating speed is similar to BFL on similar power levels. There is some possibility of improvement by reduction of FET threshold voltage; a good choice for Vt is –0.5 Volt, which enables a small voltage swing of about 1 volt as opposed to the 3 to 4 voits used on BFL. This threshold also is more easily reproduced than the very low forward threshold needed for DCFL. The basic difference between GaAs MESFETs and silicon MOSFETs is that the former will take a very significant gate current if the gate–source junction is forward biased. This occurs at about 0.6 Volts on

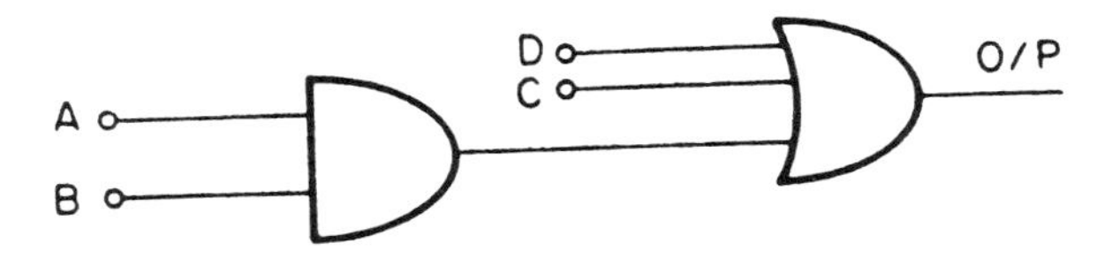

(a) logic diagram

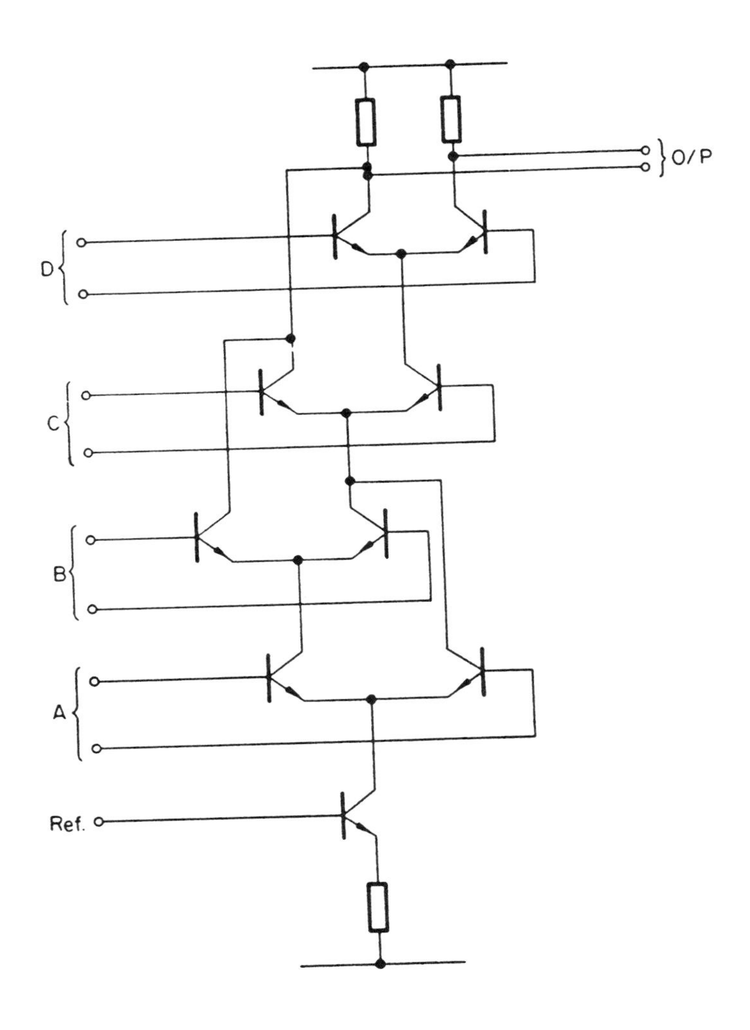

(b) circuit diagram

Figure 5.4 Four–level logic stack implemented in Silicon:

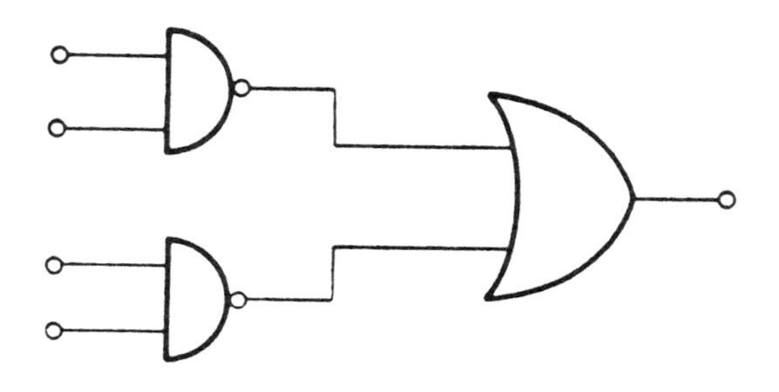

(a) logic diagram

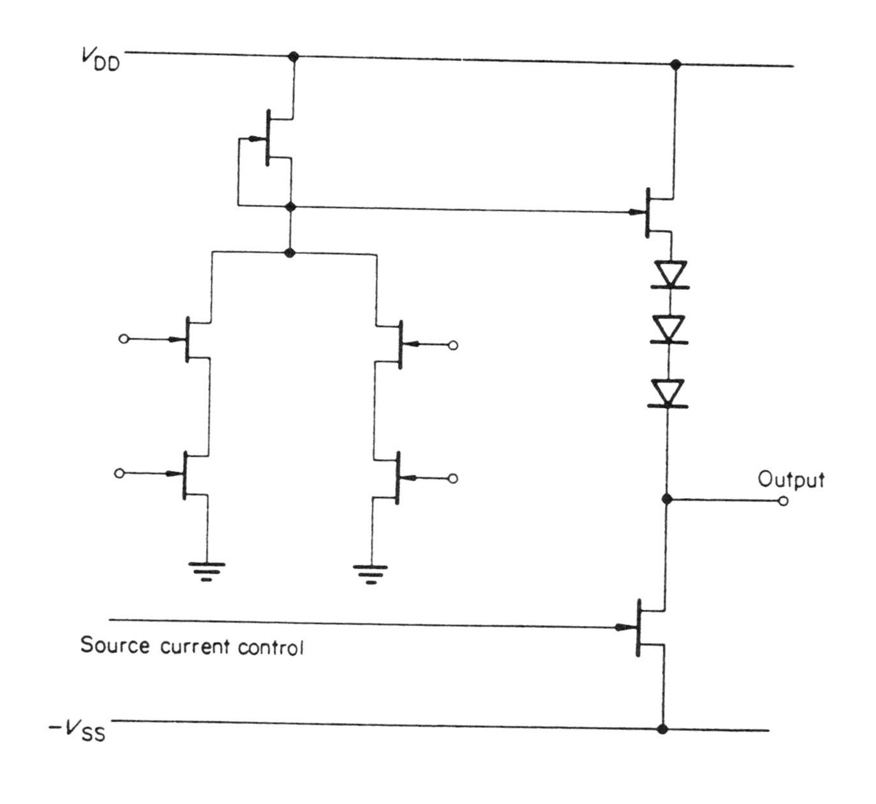

(b) circuit diagram

Figure 5.5 GaAs gate (BFL) performing similar logic functions to the silicon circuit in Figure 5.4

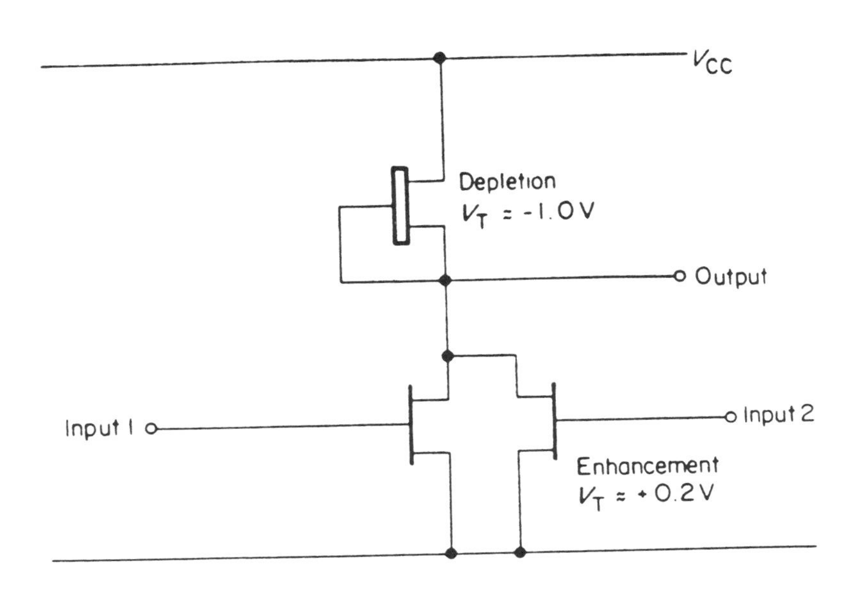

Figure 5.6 Direct coupled FET logic (DCFL)

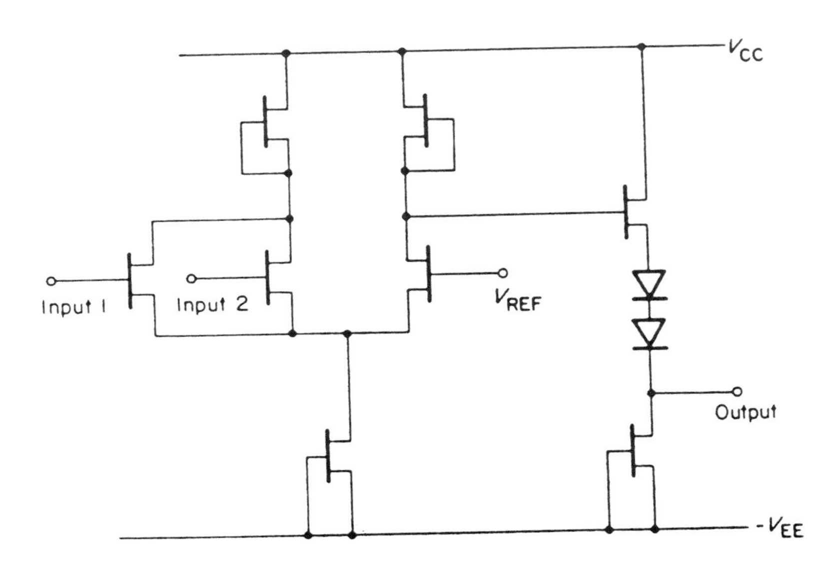

Figure 5.7 Source coupled FET logic (SCFL)

the Schottky gate. Hence, for DCFL, the MESFET Vt must be above zero, to give a noise margin, but below the turn–on voltage of the gate. Statistical predictions indicate that, for complex circuits, very tight control of Vt is needed.

A further possibility in GaAs is the use of heterojunction bipolar transistors (HBTs, or HJBTs). This involves the use of complex processing using Molecular Beam Epitaxy (MBE) or Metal Organic Vapour Deposition (MOCVD) to achieve the GaAs/AlGaAs heterojunctions and offers the possibility of extremely high Ft s, over 100 GHz having already been reported. First generation devices using this technology are already showing good performance, since they combine the advantages of the very fast bipolar transistor with the semi–insulating GaAs substrate. Indeed, this starts to point the way to the "ideal" fast logic form. As stated above, there is little doubt that CMOS, or its BiCMOS variant, will be the dominant logic family for low speed systems. The definition of low here is very variable; CMOS gate delays of 200 ps have been reported, well into the area considered the preserve of bipolar ECL only a few years ago. At higher speeds, say below 100 ps, there is much debate between the protagonists of GaAs and silicon. A more constructive approach would consider bipolar versus FET active devices. On this basis, some interesting generalities can be made independently of the process technology.

The first of these is that junction FETs, which effectively includes GaAs MESFETs, are a very non–ideal logic switching element. The reasons for this are that the threshold, if negative, requires the use of a dual power supply and a difficult level shift, as shown in Figure 5.5. It is emphasised that this is true for any junction FET, GaAs or silicon, or any other. Low positive threshold FETs suffer the tolerancing problems for the threshold described above. The Schottky gate used on most DCFL is particularly sensitive since it has a low turn–on voltage; this is the reason why some manufacturers have turned to true junction FETs.

In contrast with this is the position of bipolar devices. The threshold is very well defined at the forward base–emitter junction voltage. This is usually very reproducible on a given process. Forward base current is not often a problem, and the DC level of the threshold can be arbitrarily set by the use of a long–tailed pair input (ECL families) or by diodes (TTL families). The device has a high transconductance set primarily by the current through it, independent of the device dimensions, so that a very high ratio of transconductance to capacitance ratio can be achieved; this is particularly important in switching applications.

One measure of the current state–of–the–art is to compare reported results from the major players in the field. This is possible using either catalogue products, which tend to lag the leading edge of research, but are

Table 5.2 GaAs Gate Arrays

Manufacturer	Process/ logic family	Gate length (μm)	Number of gates	Gate delay (μm)	Power/gate (mW)	Number of gates dissipating 1 W	Funtional throughput rate (gate Hz W^{-1})
Fujitsu	GaAs DCFL	1.2	1520	217	2.6**	385	4.4×10^{11}
Harris ***	GaAs DCFL	1.5	170	130	35	29	5.6×10^{10}
Honeywell	GaAs SDFL	1	432	110	1.5	667	1.5×10^{12}
NEC	GaAs BFL	1.4	3000*	177	2.3	435	6.1×10^{11}
OKI	GaAs SBFL	1	1000	185	0.27	3700	5.0×10^{12}
Tektronik	GaAs DCFL	1	1224	225	0.25	4000	4.4×10^{12}
Texas Instruments	GaAs HI^2L	3	4000	400	1.0	1000	6.3×10^{11}
Toshiba	GaAs DCFL	1	1050	350	0.20	5000	3.6×10^{12}

* Only 2000 can be used due to the prohibitive power consumption
** At 77 K with delays of 158 ps
*** Product curently available

(© *1984 IEEE)*

real, or by taking publications in the leading technical conference digests. These latter present the leading edge of research; not all the devices reported make it into production, and a particular disadvantage is the lack of intercomparability of data. However, tables can be drawn up which give some comparison between the processes. The author presented two such tables, for GaAs and silicon gate arrays in Ref. 6. These are reproduced in Tables 5.2 and 5.3. An update on a slightly different basis is shown in Table 5.4. Unfortunately, CMOS gate array manufacturers seem not to publish even typical or average power consumptions, presumably on the assumption that the typical chip will have only a fraction of the gates operating at full speed and hence full power. For bipolar and GaAs, figures are published, so the final column in the figures can be constructed, that of functional throughput rate. The units are "gate–Hz per Watt", i.e. the number of gates times the operating frequency (taken as 1/delay) per unit power. This indicates the power efficiency of the gate. In general, slower gates will show a better figure, while gates which stretch the limits of a given technology will be poorer. It is interesting therefore to compare the technologies on this basis, and against time. The respective time windows for data gathering are approximately two years apart (spring 1987 vs. spring 1989).

Table 5.3 Silicon Gate Arrays

Manufacturer	Process/ logic family	Minimum feature size (μm)	Number of gates	Gate delay (ps)	Power/gate (mW)	Number of gates dissipating 1 W	Funtional throughput rate (gate Hz W^{-1})
Fairchild	Si bipolar	1.5	2840	250	2.0	500	5×10^{11}
NEC	Si bipolar	1.5	2000	700	1.9	520	1.9×10^{11}
Honeywell	Si bipolar	1.5	5000	600	1.0	1000	4.2×10^{11}
Siemens	Si bipolar	2.0	9000	600*	2.2	450	5.6×10^{11}
NTT	Si SST-1 bipolar	1.0**	2500	80	2.6	380	1.2×10^{12}
Plessey	Si bipolar	3.0	600	400	10	100	6.3×10^{10}
Plessey	Si bipolar	3.0	2500**	250/ 350	1.2	830	5.9×10^{11}

* Figure adjusted for comparability
** Emitter 0.5 μm
*** Semi-custom chip so not strictly comparable

On silicon bipolar, gate delays have halved, from a best figure of 80 ps to 40 ps, while available chip complexities have gone from a few thousand gates to 54000, although this latter figure is certainly not typical. In the same period, functional throughput rate has increased by an order of magnitude, from 1.2×10^{12} to 1.6×10^{13}. These figures are from a single manufacturer, and therefore ought to be reasonably comparable. The very best results are 4.2×10^{13}, although this is for slower gates and is based on ring oscillator results; broadly, the state of the art should be taken as a few times 10^{13}.

In GaAs, the very best 1987 results are 3.5×10^{12}. In 1989, the figure is only improved by a small factor to 8.3×10^{12}, although this is for a faster gate, and the average to average gain over the two year period is closer to the order in silicon. Most interesting is the appearance of HBT gate arrays with throughput rates about half that of DCFL; but the GaAs devices remain a factor between two and five times poorer than silicon. In fairness, the very fastest custom designed ring oscillators are in HBT technology at the present time.

All comparisons of this kind suffer some deficiencies, since the data base cannot be strictly comparable; but there is definite indication from the data that silicon digital devices are faster than GaAs in practice, and that the gap has widened during the last two years. Although there have been

Table 5.4 1988/1989 Comparison of Gate Array 'State of the Art'

Manufacturer	Process	Minimum feature	Number of gates	Gate delay	Power /gate	Funtional Throughput gate-Hz/W
NTT	BiCMOS	0.8	2000	450ps	–	–
IBM	CMOS	0.5	2000	200ps	–	–
Hitachi	CMOS	0.8	130k+38k	350ps	–	–
Mitsubishi	CMOS	0.8	67000	405ps	–	–
Hitachi	CMOS	0.8	65000	800ps		
NTT	Bipolar	1.0	2100	43ps	1.5e–3W	1.6e13
Fujitsu	Bipolar	1.0	54000	40ps	7e–4W	3.4e13
Mitsubishi	Bipolar	1.0	12k+36k	110ps	1.8e–3W	5e12
IBM Rockwell	GaAsHBT	1.4	1100	71ps	4.5e–3W	3.1e12
NTT	GaAsFET	0.4	250	50ps	2.4e–3W	8.3e12
AMCC/ Plessey	Bipolar	1.0	13000	100ps	2.8e–4W	4.2e13

Data:– CICC 1988, ISSCC 1988, ISSCC 1989

great strides made in GaAs MESFETs, and particularly in HBTs, silicon devices have been improved even more, as the market for really high speed logic becomes more apparent. It is also worth noting that CMOS and BiCMOS devices have been produced which are comparable in delay to many of the silicon bipolar and GaAs arrays of only two years ago. An approximation to the throughput rate on the same basis as for the published results would probably show CMOS as best at about $1x10^{14}$.

5.3.2 Analogue Circuits

Analogue circuits can be broadly divided into two areas, i.e. classic analogue circuits such as operational amplifiers, voltage references, multipliers etc., and R.F. functions such as R.F. amplifiers, mixers, oscillators and filters.

Since the very earliest days of integrated circuits, silicon bipolar devices have dominated analogue integrated circuit design. This is because bipolar parts have a high and predictable transconductance and a low "offset". Offset is the random variation of input threshold between two nominally identical devices; it is a critical parameter of most op–amps and comparators. In addition, in many analogue processes, complementary transistors are available, giving the facility for very high stage gains and push–pull outputs. It can also be exploited to achieve amplifiers which can swing almost rail–to–rail between the power supplies. In contrast, GaAs MESFETs have a relatively poor transconductance for small devices; the transconductance to capacitance ratio is fixed as mentioned above. Threshold mismatch (i.e. offset) can be very poor, at least an order of magnitude worse than bipolar. Gain per stage, at least in GaAs with its lack of a real complementary device, can be very low. In addition, the l/f noise is high in GaAs at frequencies below 100 MHz. Clearly in the range up to about 100 MHz silicon is the dominant analogue technology, usually in bipolar form, but increasingly CMOS circuit tricks are used to obtain good performance with MOSFETs. GaAs is used above 100 MHz in op–amp like structures, but usually where the offset and total gain parameters are not critical.

GaAs does come into its own through the second category of circuits, that of R.F. stages. The advantages of GaAs are two–fold; in noise performance, above the l/f knee, where an FET will generally beat a bipolar device, and in the semi–insulating substrate, where it becomes possible to integrate inductors on the chip as well as capacitors, resistors and active components. This leads to the typical "MMIC" (Monolithic Microwave Integrated Circuit) of today, where the bulk of the active area is used up in passive components such as filters and power splitter/combiners, with a few active components, the whole performing a function which could not be realised in any other way. Plate I shows such an MMIC; this device is extensively described in Ref. 7. Although some advances in silicon R.F. and microwave circuits are expected, due to the increasing use of high resistivity substrates which will enable some filters to be achieved on chip, GaAs will probably remain the dominant technology in performance terms; it remains to be seen whether sufficient high volume applications can be found where the performance of silicon is adequate. It should be remembered that a definition of good engineering might be the achievement of an objective at the minimum cost; ultimate performance may not be the goal.

5.4 Radiation Hardness

Radiation hardness is of major importance to the military systems engineer and in certain specialist application such as satellite operations. Table 5.5, which is taken from Ref. 6 illustrates the advantage of GaAs over silicon; the best silicon technology reported extensively is CMOS/SOS, although it is to be expected that CMOS/SOI will turn in broadly similar results. Of course, in many applications, silicon bipolar technologies are sufficiently hard for many applications. This has been historically true for the "Collector Diffusion Isolation" (CDI) processes from Ferranti

Table 5.5 Radiation Hardness

Technology	Total dose	Neutrons
CMOS/SOS	10^6	10^{15}
Bipolar	10^6	10^{14}
GaAs	10^7	10^{15}

Electronics (now part of Plessey Semiconductors). More recently, the moves to oxide isolated, and especially to very small dimension processes has improved the radiation tolerance of many advanced bipolar processes.

The mechanisms for radiation degradation vary between devices and imposed conditions. In bipolar devices the main effects are a decrease in current gain (h_{FE}) due to generation/recombination activity in the base and emitter–base depletion region and an increase in the transistor saturation voltage due to a decrease in the carrier mobility. In MOS devices there is a change in threshold voltage; a reduction of the n–channel threshold and an increase of the p–channel threshold, due to increased oxide charge.

In the past this has led to the relative "softness" of MOS processes, but newer CMOS processes are made harder by reduction of the oxide thickness. None of the above factors significantly affect GaAs. Failure due to transient upset is also better for GaAs; under extreme conditions, silicon devices can pass large photo–currents, leading to latch–up effects and burnout. In silicon, avoidance of latch–up is a major issue; it was the reason for the apparent radiation hardness of integrated injection logic (I2L), a silicon technique which operates from a supply of less than 1 volt, i.e. parasitic thyristors cannot be tuned on. Also, CDI structures contain such a very low gain PNP that they again are very unlikely to latch up. In CMOS, the problem is more complex, and many anti–latch–up structures have been proposed. Very thin epitaxial processes have been the most commercially successful, but the complete freedom from latch structures afforded by CMOS/SOS is a decided advantage. Silicon–on–insulator (SOI), when more readily available, will probably be as good as SOS, at a (predicted) much lower price.

5.5 Economics

Silicon wafers in a range of sizes from 100 mm diameter to 200 mm diameter are routinely available from several suppliers. The quality of the basic material is very good and supplies are reasonably assured. In GaAs,

Table 5.6 Cost Comparisons

Technology	Cost
CMOS Bulk	Lowest
CMOS SOI	↑
Silicon Bipolar	
Silison BiCMOS	
GaAs MESFET	↓
GaAs HBT	Highest

this has not been historically the case, although the situation is much improved in recent years, with fairly ready availability of 75 mm and 100 mm slices.

The base material of GaAs is usually estimated as ten times the cost of silicon, but this ratio is being improved. General processing costs are slightly lower on GaAs, because the process has fewer stages; in silicon, where stepper technology is used to achieve fine lines, mask cost can be very high. A typical GaAs process uses conventional photolithographic techniques for most layers, with a stepper stage, or even E–beam, for the critical gate stage. This is generally a much cheaper route than for silicon, but does not compare like with like; real reductions in GaAs chip area will also require the use of steppers and the transfer of much of the silicon technology to GaAs.

GaAs has the advantage of light emission as mentioned above, although true electro–optic integrated circuits are still at the very small scale integration level since the processes are fairly incompatible.

A very realistic prospect is for growth of GaAs layers on a silicon substrate to combine the advantages of GaAs and silicon. This applies equally to analogue and some digital circuits. This combined process would of course be very complex, although it has been demonstrated. In the short term, and with excellent prospects for growth, is the "solder–bump" method of mounting together chips of differing technologies. It is possible to connect optimised chips in several processes together with very short path lengths and very many paths. Assemblies with 100,000 solder bumps on a single chip have been demonstrated [8] .

It is impossible to give a precise view of the costs of all the relevant processes; even a relative view is very subjective and dependent on the supplier and the type of product. Instead, the summary in Table 5.6 shows the cost range from relatively low to relatively high. Most of this table is unlikely to change in order, although the actual costs will vary considerably. The conclusions are that silicon CMOS, probably in bulk but possibly on an insulating substrate, will remain the cheapest technology. Most expensive of the processes immediately in view will be HBT, with bipolar silicon and GaAs HBT in the mid-ground.

5.6 Conclusions

Overall, some very simple conclusions can be drawn in comparing silicon and GaAs technologies. The first is that, at low speeds, say less than 200 MHz clock rate, CMOS, possibly in BiCMOS or SOI form, will predominate. Silicon bipolar will dominate all higher speed digital markets. Only the very fastest prescalers and similar small scale circuits will go to GaAs in HBT form. Even then, the competition from silicon will be strong. From this perspective there seems to be little or no place for GaAs MESFETs in high speed digital applications. Such devices as are currently commercially available could be realised in silicon and indeed would be if the markets were sufficiently large. Better still would be higher levels of integration in very high speed silicon, beyond the capability of GaAs.

In analogue applications, at frequencies below 1 GHz, silicon bipolar will continue to dominate over GaAs. CMOS or BiCMOS may well take a significant part of the market, especially at lower frequencies. Above 1 GHz, we can expect to see increasing use of silicon in R.F. amplifiers, mixers and oscillators. However, GaAs will retain the fundamental advantage of lower noise. New substrates in silicon may make possible on-chip inductors and filters; but only at the cost of more complexity. GaAs FETs are likely to remain the best choice in the higher frequency regions; the cross-over will depend on the total criteria of performance, cost and market size.

It seems that, for the integration of complex systems, the most likely route will be to optimise several technologies, then bring them together in some form of multi-chip arrangement such as the solder-bump motherboard technique mentioned above. This would offer the advantages of all technologies in a compact and ultimately low-cost package.

5.7 References

1 S.M.SZE, "Physics of Semiconductor Devices" (2nd ed.) Wiley, New York, 1981

2 RE.C.EDEN "Comparison of GaAs device approaches for ultrahigh-speed VLSI" Proc. IEEE Vol. 70 No.1 January 1982 pp 5-12

3 H.HASEGAWA, S.SEKI "Analysis of Interconnection Delay on Very High-Speed LSI/VLSI Chips using an MIS Microstrip line mode" IEEE Trans. Microwave Theory and Techniques Vol. MTT-32 No. 12 Dec. 1984

4 C. SUNG et. al. "A 76 MHz Programmable Logic Sequencer" ISSCC 1989 Digest of Technical Papers pp 118-119

5 T.HOTTA et. al. "A 70 MHz 32b Microprocesor with 1.0 μm BiCMOS Macrocell Library" ISSCC 1989 Digest of Technical Papers pp 124-125

6 P.H.SAUL "technology choice for High-speed applications" Microprocessors and Microsystems Vo. 11 No. 8 October 1987 pp 428-442

7 A.A.LANE "High Packing Density Gallium Arsenide circuits for Radar" Plessey Research and Technology Annual Review 1989 pp 83-92

8 D.PEDDER "Flip-chip solder bonding for precision structures" Plessey Research and Technology Annual Review 1989 pp 69-81

Chapter 6

GaAs Digital Circuits and Systems

B. W. Flynn, H. M. Reekie and J. Mavor

6.1 Introduction

High speed logic circuits, fabricated in gallium arsenide have shown great promise for applications such as computing and signal processing since the first results were published in the late 1970's and early 1980's. Unfortunately, the early predictions were often hopelessly over optimistic and made no allowance for the extreme immaturity of GaAs fabrication technology as compared to the existing silicon technology. This overstatement of the case for using GaAs[1] slowed its acceptance by loosing credibility among system designers. It is only now, nearly a decade later, that there are signs that the technology is maturing to an extent whereby systems can be fabricated which fully capitalise on the potential speed advantages that this technology has to offer. In order to gain the full speed advantages of using GaAs for device fabrication it is necessary to engineer a design at each stage, from architecture right through to packaging and testing. It is not sufficient simply to take a design in silicon and convert it into GaAs.

The object of this chapter is to review the development of GaAs digital IC technology and to consider the current state of the art, in order to discuss likely future developments. A further aim is to give the reader some insight into applications and systems where the appropriate use of GaAs LSI and VLSI circuits will give a real advantage.

From the earliest studies of the semiconductor properties of GaAs its potential as a material for the fabrication of high speed/high frequency devices was realised. Table 6.1 compares the principal material properties

Table 6.1 Room Temperature Properties of GaAs and Si

PROPERTY	GaAs	Si
Electron mobility	500cm/V.s	800cm/V.s
Hole mobility	$250cm^2/V.s$	$350cm^2/V.s$
Maximum electron drivt velocity	$2 \times 10^7 cm/s$	$1 \times 10^7 cm/s$
Onset of high-field transport	$2 \times 10^3 cm/s$	$1 \times 14^3 cm/s$
Energy gap	1.43eV (direct)	1.12eV (indirect)
Maximum resistivity	10^9 ohm.cm	10^5 ohm.cm
Schottky barrier height	0.7–0.9V	0.4–0.6V

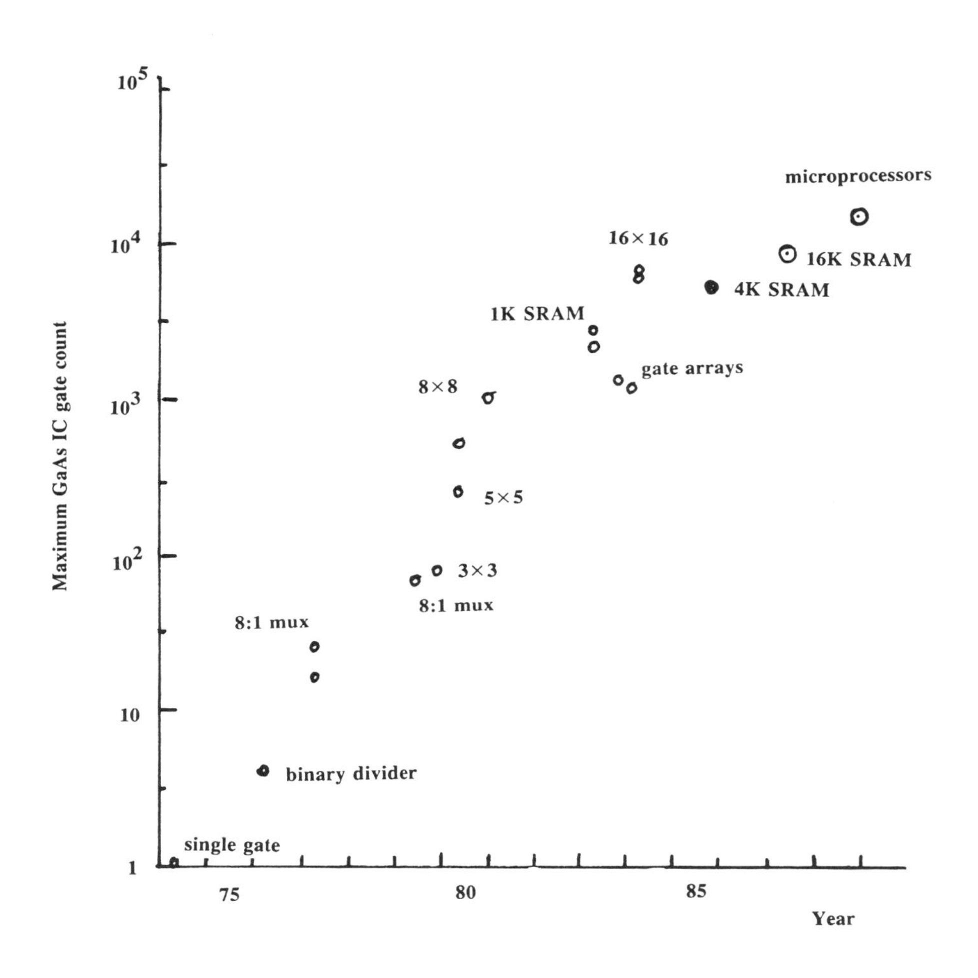

Figure 6.1 Maximum gate count versus year GaAs ICs

affecting device performance for GaAs and silicon at room temperature. it can be seen that GaAs has a considerably higher electron mobility than silicon, but the maximum drift velocity is only twice as large. To set against this, high field transport occurs at lower fields in GaAs. Generally GaAs is accepted to be capable of a two to three times speed advantage over silicon.

Early attempts to make transistors using GaAs were frustrated by the difficulty of working with a compound material, which tended to dissociate at high processing temperatures. There is no native oxide (a key feature of silicon!) and it was found difficult to control the surface states on the material. Additionally, it was not generally possible to use diffusion to place dopant materials into substrates, as is done in silicon planar processing. The solution to this problem was to use ion–implantation, but this technology was not always available. The result was that early discrete devices could operate at high frequencies, but had characteristics that drifted with time. It was not until the mid 1970s that these problems were solved [2] and discrete transistors became commercially available for small–signal microwave applications.

Around this time work started on the development of integrated circuit technologies. The earliest devices used FETs which had active areas produced by epitaxial growth, with isolation between devices achieved by a mesa–etching process. Since the mid 1970s GaAs integrated circuits have developed at an astounding rate, particularly assisted by the introduction of planar processing. Figure 6.1 shows this development illustrating a doubling in complexity every 7–8 months. This is a far higher rate of development than for silicon but, in part, can be explained by the availability of well established semiconductor manufacturing processing techniques and equipment development for use in the silicon industry, and also by the fact that the staring base was at a very low level of integration.

6.2 GaAs Devices for Digital Logic

6.2.1 The MESFET

The transistor structure in widest use in GaAs IC logic technologies is the N–channel metal semiconductor FET (MESFET) [2]. This comes about for a variety of reasons. The structure is relatively easy to fabricate on account of its simplicity and secondly, the electron mobility in GaAs is high and thirdly, the high metal–semiconductor Schottky barrier makes a metal–semiconductor diode a convenient gate structure in the absence of a native oxide on GaAs. Figure 6.2 shows a typical GaAs MESFET structure. The source and drain are commonly formed by implantation and gold is used for the interconnect metal. The gate metallisation is formed from a refractory metal such as titanium or tungsten to form a Schottky contact on the GaAs. Controlling the doping concentration in the channel determines the threshold voltage of the device, allowing the production of "normally on" depletion devices, ($V_T < 0$ V) and "normally off" enhancement devices, ($V_T > 0$ V). The electrical characteristics of each type are shown in Figure 6.3. Both types can be operated up to a gate–source potential equal to the Schottky barrier voltage of the gate diode. Beyond this point gate current will flow and upset the operation of the driving stage. The earlier forms of GaAs logic circuitry only used

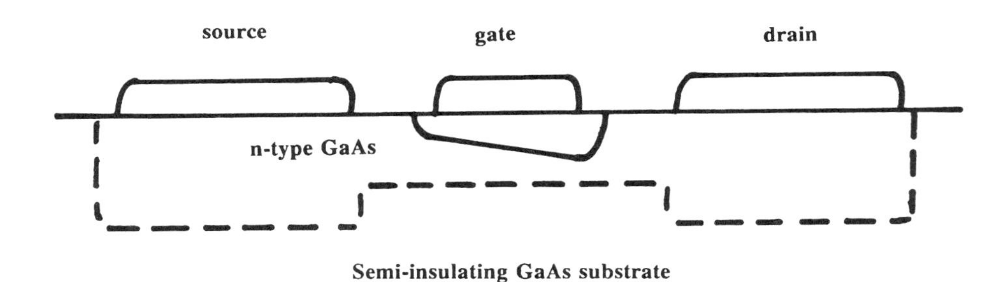

Figure 6.2 GaAs MESFET

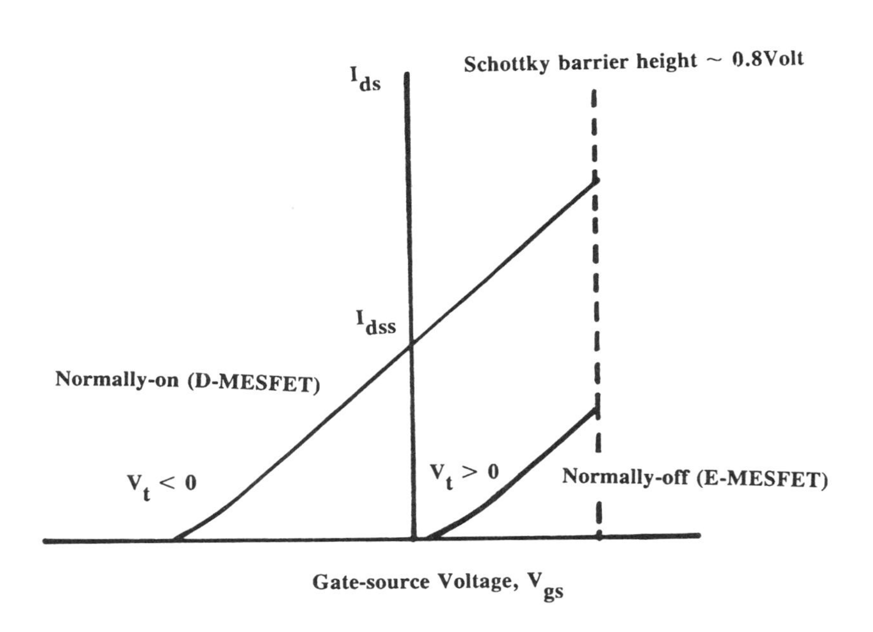

Figure 6.3 Transfer characteristic for E and D–MESFETs

depletion devices as it was difficult to obtain the degree of control on threshold voltage, V_T, required to make enhancement devices with reproducible characteristics.

6.2.2 The High Electron Mobility Transistor (HEMT)

The structure of the high electron mobility transistor [3,4], (HEMT) is very similar to that of the MESFET and its development has extended useful transistor performance to around 100 GHz. This device is known by a variety of names and is sometimes referred to as a "ballistic" transistor or a modulation–doped field effect transistor (MODFET). The speed of the simpler MESFET is limited by the mobility of electrons in the n–type channel region. The mobility is reduced by increased channel doping which in turn slows the device down. When the channel material is doped n–type by the addition of donor impurity atoms to give an excess of electrons, donor sites must also be introduced into the semiconductor lattice. Each donor site acts as an impurity centre which scatters charge carriers as they move through the crystal lattice, reducing the carrier mobility. In contrast in the HEMT carriers move in an undoped layer of GaAs beneath the gate region. As explained the mobility of electrons in the undoped material is many times that in the doped material, resulting in a greatly increased speed of operation.

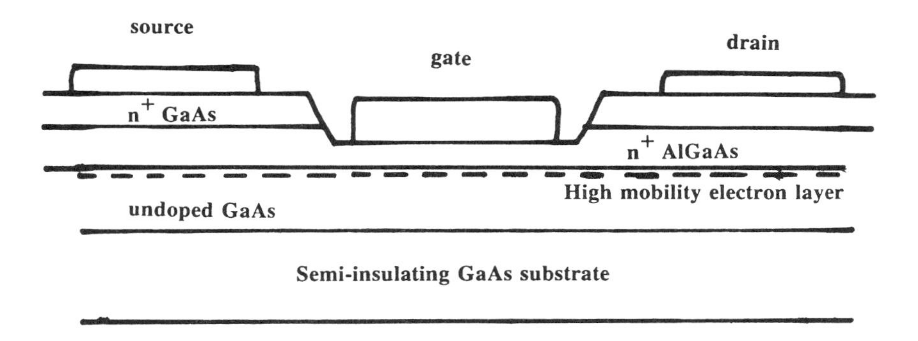

Figure 6.4 HEMT structure

Enhancement and depletion devices can be fabricated in HEMT technology. Figure 6.4 shows the HEMT structure. It is similar to the conventional MESFET except for the layers introduced beneath the gate. These are produced by molecular beam epitaxy (MBE) or metal–organic chemical vapour deposition (MOCVD). The critical area is the interface between the underlying undoped GaAs and the heavily doped AlGaAs. Free electrons in the AlGaAs diffuse into the undoped GaAs where they can move with high mobility because of the lack of impurity scattering centres. Conduction takes place in this interface layer and is controlled by the gate potential. HEMTs have a higher forward turn on voltage than MESFETs giving them a higher noise margin in circuit applications. This factor may be crucial in their choice for designs.

6.2.3 The Heterojunction Bipolar Transistor (HBT)

Another active device that is used in GaAs IC design is the HBT [5,6]. This is an npn bipolar transistor formed from dissimilar semiconductor materials for the emitter and the base regions. The base is formed from p–type AlGaAs while the emitter and collector are n–type GaAs. When the emitter–base junction is forward biased electrons are injected into the base region and travel by means of diffusion across the base to the collector. As noted earlier the mobility of carriers in the p–type material is low therefore the drift velocity is relatively small. The technique used to overcome this problem is to alter the doping across the base region to vary the bandgap over the base region. This produces a pseudo electric field across the base which greatly assists the passage of electrons, speeding up the operation of the device. The graded doping profile across the base is produced by means of so–called "band–gap engineering" using molecular beam epitaxy.

HBTs are attractive for digital circuitry and are used in emitter coupled logic circuit configurations. They offer the low impedance advantages of bipolar structures and appear to be ultimate for high speed logic.

6.2.4 The Junction Field Effect Transistor (JFET)

The final device worthy of mention in this brief review of GaAs devices for logic circuitry is the junction FET (JFET) [7]. These are similar to the MESFET structure but incorporate a p–n junction as the gate region, rather than a Schottky diode. This gives a higher gate turn–on voltage and can again lead to better noise margin performance than for MESFETs. Developments in MESFET technology including better control of threshold voltages have lead to wider noise margins. This has limited the adoption of the JFET as a device in GaAs IC designs.

6.3 GaAs MESFET Logic Technologies

6.3.1 Buffered FET Logic (BFL)

Buffered FET logic [8] is an all–depletion device logic family which was the earliest type to be used for circuit design. Figure 6.5 shows a typical

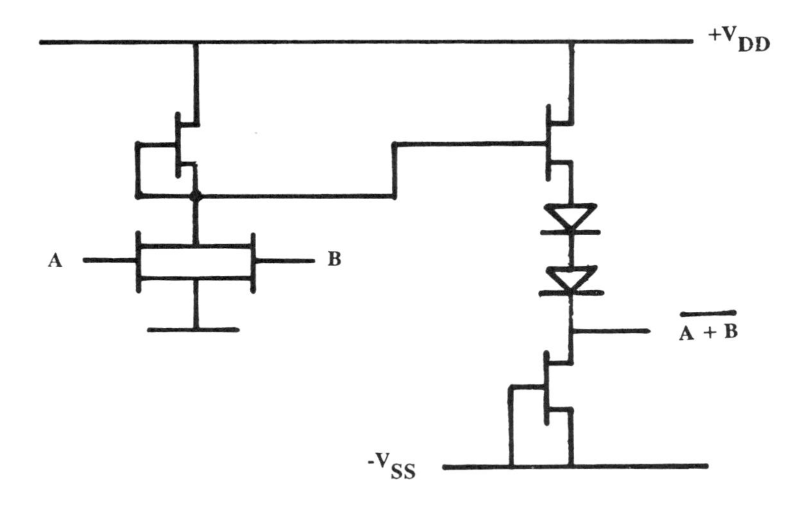

Figure 6.5 Typical BFL NOR gate circuit

BFL two-input NOR circuit. The main feature is the use of a level-shifting buffer stage after the logic circuitry. This operates with a second negative power rail ($-V_{SS}$) and shifts the voltage levels sufficiently negative to drive the succeeding depletion FET stage to threshold. The logic part of the circuit is similar to that of silicon nMOS with the exception that it uses all depletion transistors with the result that the gate inputs must be driven negative to below the transistor threshold voltage in order to be able to turn them off, hence the output level shifting buffer with the negative power supply rail.

BFL is the fastest form of MESFET logic with gate propagation delays of around 50 ps being reported for 0.5 micron FETs. Unfortunately this speed is obtained at the expense of power consumption, with a typical BFL circuit taking around 10 mW per gate. These gates use a fairly large number of FETs and a substantial layout area per gate. This, combined with its large power consumption, makes it unsuitable as a contender for LSI and VLSI circuit designs.

BFL is used for the MSI logic parts that are commercially available from Gigabit Logic and Harris Semiconductor. Its power consumption renders it unsuitable for more complex circuits but for raw speed it is unrivalled.

6.3.2 Schottky Diode FET Logic (SDFL)

This is an all-depletion device logic family [9] which uses forward-biased Schottky diodes to perform level shifting. Figure 6.6 shows a typical SDFL gate circuit. Here a NOR function is effected with only one switching FET (T1), plus one active pull-up (T2) and one pull-down (T3). The diodes D1 and D2 perform the actual logical OR operation using D3 and T3 to perform the level-shift required to switch the depletion transistor T1, which in turn provides an inversion of the signal. SDFL gates dissipate less power than BFL at the expense of speed and fan-out capability. Typical gate delays are around 150 ps at a power dissipation of 2.5 mW per gate.

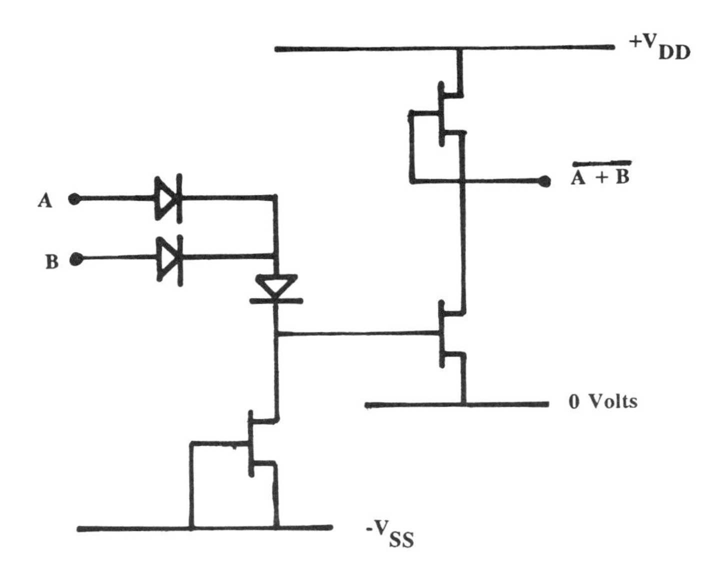

Figure 6.6 Typical SDFL NOR gate circuit

6.3.3 Capacitive Coupled FET Logic (CCFL) and Capacitor Diode FET Logic (CDFL)

An obvious approach to the level shifting problem is to use capacitive coupling between stages. Two approaches to this have been reported. The first [10,11] employs a reverse-biased Schottky diode to couple successive stages as shown in Figire 6.7. This scheme has the advantage of simplicity

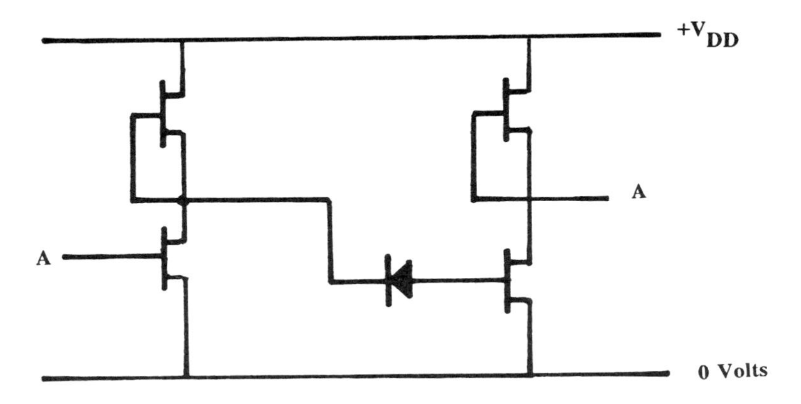

Figure 6.7 Typical CFFL inverter circuits

and low power consumption, but will not clock at low speeds. This approach has been pioneered by British Telecom.

An alternative [12] is to use a combination of reverse and forward-biased diodes in the capacitor diode FET logic (CDFL) configuration, as shown in Figure 6.8. The forward-biased diodes (D1–3) act as level shifters while the reverse-biased diode (D4) is used as a "speed up" capacitor to couple the switching waveform edges. Because the speed of operation does not depend on the level shifting diodes the current through them can be made very small, allowing a low power consumption per stage. This type of circuitry is used by Gigabit Logic in their Picologic range of products. These have logic levels that are compatible with silicon ECL parts and are available in complexities of up to around a thousand gates. Capacitively coupled logic exhibits a gate delay of around 100 ps for a 0.36 mW per gate power consumption.

6.3.4 Direct Coupled FET Logic (CDFL)

As GaAs IC fabrication technology has developed and matured better control of device threshold voltages has been obtained, making it possible to fabricate enhancement devices into circuits in a reproducible way. This has allowed designers to dispense with the level shifting buffer stages and use directly coupled stages. Figure 6.9 shows a typical DCFL stage. The

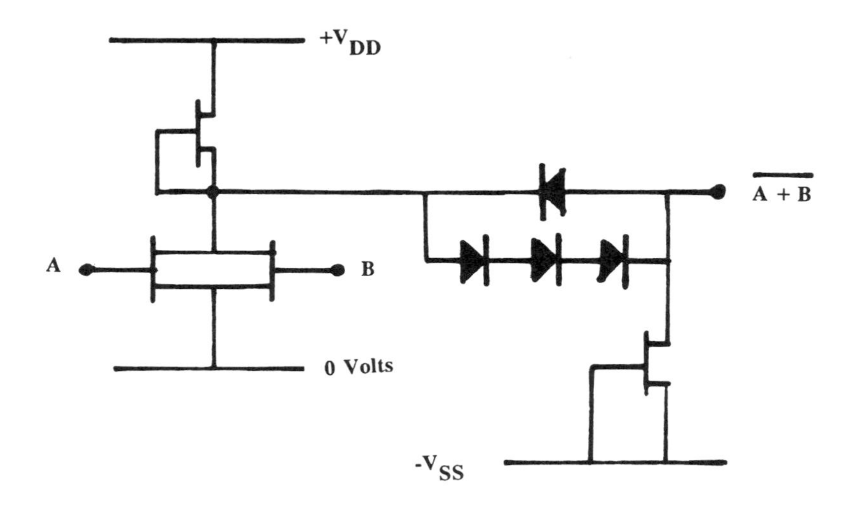

Figure 6.8 Typical CDFL NOR gate circuit

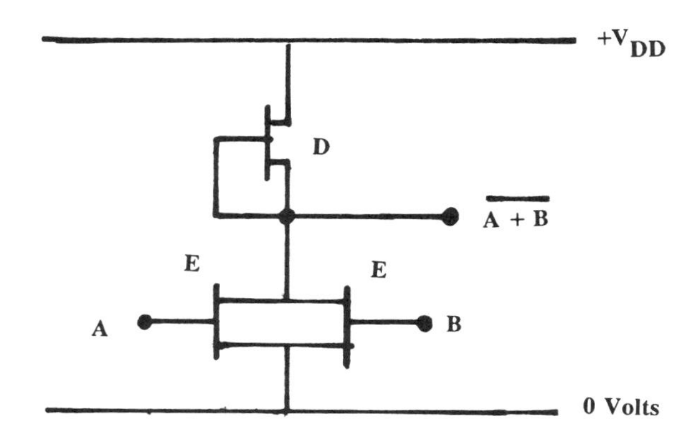

Figure 6.9 Typical DCFL NOR gate circuit

circuit configuration is similar to silicon nMOS. DCFL has the lowest power consumption of any GaAs logic family at around 100 μW per gate and a propagation delay of around 100–150 ps. Additionally the lack of level shifting circuitry allows a greatly increased packing density. DCFL is generally acknowledged to be the preferred logic design type for GaAs VLSICs at the present time.

6.3.5 *Performance of Si and GaAs*

The most vaunted advantage of GaAs over silicon is that of speed, however, by trading-off speed considerable power consumption advantages can be obtained. Additionally GaAs circuits show considerably improved radiation hardness over silicon, making them attractive in military and aerospace applications.

The main disadvantages of GaAs stem from the material problems which can lead to a low yield of working circuits, which in turn limits the feasible circuit complexity. Current gate count for large GaAs ICs is around 100,000 [13] while that for silicon is around 4 million.

Figure 6.10 shows a plot comparing the speed and power performance of a range of silicon and GaAs IC technologies. Generally the GaAs technologies are faster than silicon but recent advances in small geometry MOS devices mean that the gap is closing between silicon and GaAs enhancement-depletion (E-D) MESFET devices. Ultimately the greatest advantages may be gained by using HEMTs or HBTs as a circuit technology. Currently several Japanese companies [14] are pursuing this approach very vigorously.

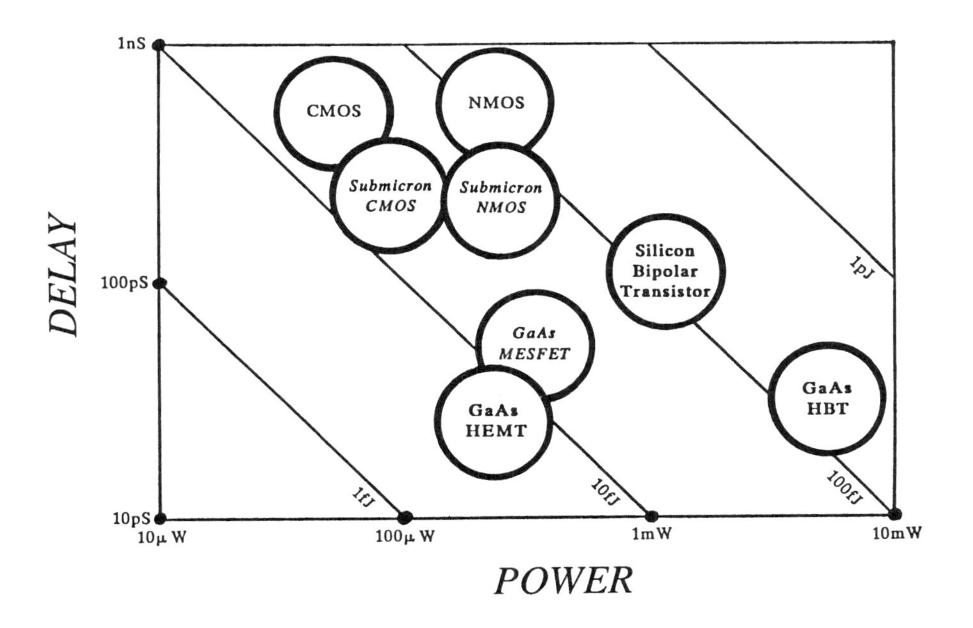

Figure 6.10 Power-delay comparison of important transistor technologies

6.4 Small-scale Static and Dynamic Circuits

The progression of GaAs circuits from a laboratory curiosity to practical circuits has been greatly influenced, perhaps even driven, by the yield of fabrication processes available. The first GaAs logic ICs were all depletion mode which led to the need for level shifters. As discussed earlier, this need was met by BFL, SDFL, CCFL and others. In the usual way, each had its advantages and disadvantages and each was used for some small-scale circuit design. In this section it is proposed to discuss briefly representative cells of each of the main all-depletion technologies. However, more time will be devoted to an examination of DCFL circuits as it is probable that enhancement transistors will be required if power requirements are to be kept low enough, and circuit layout densities are to be kept high enough, to allow even MSI circuits to be commercially realisable.

6.4.1 General Considerations Relevant to all GaAs Digital Technologies

No matter which GaAs fabrication technique is employed, some factors are common to all GaAs MESFET designs.

(i) The Gate is a Forward-biased Schottky Diode

The metal to semiconductor gate terminal contact forms a Schottky barrier junction. This is of particular note in DCFL, in which case the gates are forward biased (the enhancement transistor threshold voltage is positive so forward bias is essential if enhancement transistors are to be turned ON). These gates can draw significant amounts of current when they are forward biased by more than a few hundred milli-volts.

A possible Schottky diode characteristic is shown in Figure 6.11. Noticeable gate current is drawn for forward voltages significantly less than the Schottky barrier voltage. Note, also, that the rise in current after the Schottky barrier voltage has been reached is nothing like so great as would be expected from a silicon diode. This is due to the greater intrinsic resistance of the device and to the physics of the device.

(ii) Gate Charge Storage Time

Gate terminal storage time, i.e. the length of time a charge dumped on a gate capacitance remains there before leaking away, is short owing to the gate current drain. Therefore dynamic circuits which rely on charge storage on nodes (such as those commonly used in NMOS and CMOS) cannot be used. This is a major limitation which is very restrictive in practice.

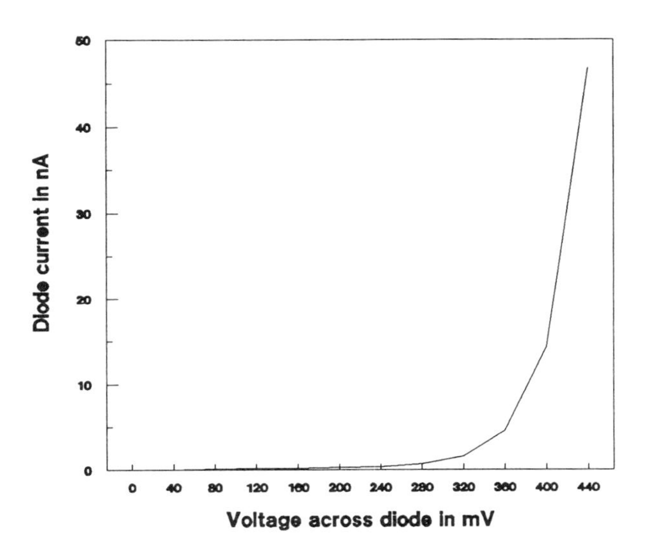

Figure 6.11 Schottky diode characteristic

(iii) Thermal Conductivity

The thermal conductivity of GaAs is several times worse than silicon. This, can be a major limiting factor in the design of circuits. In particular, circuit techniques such as Buffered FET Logic, which are rather power hungry, are limited in scale by this problem. Sometimes manufacturers attempt to circumvent this difficulty by lapping the reverse of a processed wafer in order to thin it. Such thinning can assist in the transfer of heat from the site of its production (the transistors on the top of the wafer) to where it can be dissipated (the package).

The poor thermal conductivity can also cause thermal gradients to be set up across the chip which can cause tracking errors in supposedly matched sections of circuitry.

(iv) Speed

Apart from the obvious speed advantage GaAs has over silicon due to higher electron mobility, GaAs has a speed advantage due to low loading capacitance. Because of the semi–insulating GaAs substrate, capacitances to ground are virtually negligible and, in fact, interconnection capacitances are dominated by line to line capacitances.

We will now consider in turn each of the major GaAs circuitry technologies.

6.4.2 Buffered FET Logic (BFL)

BFL was the first on the scene; a NOR gate is shown in Figure 6.5. It has the great advantages of (relative) simplicity and speed and uses MESFETs with relatively high threshold voltages (about –2.5 V). Such circuits may operate at around 10 GHz and are relatively insensitive to fanout loading and load capacitance. However, BFL is a low–density technology and is very power–hungry (tens of mW per gate).

It is also possible to use MESFETs with a lower threshold voltage (about –1 V) and only two level shifting diodes in which case we have Low Powered Buffered Fet Logic [15]. Such modifications can result in circuits with a noticeably lower power dissipation per gate (around 5 mW) and only slightly reduced speed. However, this approach has not been particularly successful as it requires tighter control of the transistor threshold voltages, something which is difficult to achieve in GaAs.

The elimination of the source follower of the buffer circuit can result in significantly reduced power dissipation [16] but the resulting circuit is more sensitive to the effects of increasing fan–out. Even with this reduced power consumption, the circuit still requires some 12 mW per gate and gives a propagation delay of about 60 ps [17]. This lower power dissipation still limits BFL to MSI densities.

6.4.3 Schottky Diode FET Logic (SDFL)

A simple Schottky Diode FET Logic NOR gate has already been shown in Figure 6.6. The extremely fast Schottky diodes available in a GaAs process are used to good effect, both as level shifters and as logic elements (to carry out the OR function). Speed is comparable to BFL but circuits consume [18] much less power (between 0.2 mW and 2 mW per gate). The SDFL gate structure allows virtually unlimited fan–in at the first (positive OR) logic level (up to 8–input NOR gates have been described in the literature [19]) but no more than two transistors may be wired in series in the NAND part of the gate if dynamic performance is to be maintained. Fan–out is limited to about three unless the gate output is buffered (though SDFL is not as sensitive to fan–out as, for example, Direct Coupled FET Logic). Circuit size is very much less than with BFL as large FETs have been replaced by very small Schottky diodes (typically 1 μm x 2 μm) and, also, the fact that diodes are two terminal devices makes for a more efficient layout.

6.4.4 Capacitor Coupled FET Logic (CCFL)

In a further attempt to reduce the power levels associated with level–shifting circuitry, use has been made of reverse–biased Schottky diodes as capacitors [11, 12]. Such a gate is shown in Figure 6.7. This type of coupling has the advantage that no power is consumed in the capacitors as once they are charged the charge is merely transferred between the FET gate and the diode. As the capacitor is in series with the FET gate, the loading capacitance this circuit presents to the driving circuit is reduced. However, there is an additional stray capacitance presented by the relatively large coupling diode. These circuits only require one supply rail but have the disadvantage that they will not operate below some specified minimum frequency, about 20 kHz. These circuits are also fairly tolerant of variations in processing parameters because of the ac coupling between stages. Power consumptions and propagation delays are about 3.5 mW and 55 ps per gate [20, 21]. CCFL circuits are sometimes also referred to as Schottky–coupled Schottky–barrier FET Logic (SSFL).

6.4.5 Direct–Coupled FET Logic (DCFL)

All the above mentioned GaAs technologies used only depletion FETs so giving rise to the need for level shifters. Each of the early technologies realised the level shift in different ways, producing different trade–offs and requiring different manufacturing and design techniques. Clearly, a large number of difficulties would be avoided if enhancement FETs were available. Such devices started to come on stream in the late seventies but, as has been mentioned previously, required extremely tight control on certain processing parameters and so had a very low yield. As time went by, new techniques were developed which made designing with enhancement MESFETs a viable proposition.

The first circuits designed using enhancement transistors were very similar to those designed using silicon nMOS technologies. A typical NOR gate is shown in Figure 6.8 and can be seen to be identical in form to a standard enhancement/depletion nMOS gate.

Figure 6.8 is based on a 1 μm gate non self–aligned enhancement–depletion process. The ratio of the pull–down to the pull–up transistors is usually about 8 : 1 and nominal threshold voltages are about 0.2 V and –1 V for the EFET and the DFET, respectively. Typical inverter delay is approximately 150 ps for a fan–out of one.

Non self–aligned processes that make use of conventional fabrication techniques, such as the one shown, have a severe yield/performance trade–off limitation. In DCFL where the switching device is the EFET, performance is limited primarily by the parasitic resistance between the gate and the source and drain ohmic contacts.

This parasitic resistance is made up of two major parts. The first is the ohmic contact resistance itself (the resistance between the metal source or drain connection and the active area beneath the contact). The second is the resistance between the channel under the gate and the active area under the ohmic contact. In the EFET, the latter dominates and has a high value because the thin active channel layer is almost totally depleted. This can be reduced by decreasing the gate to ohmic contact separation. Conventional alignment techniques cannot cope effectively with the very tight tolerance required without seriously affecting the yield. A number of self-aligned GaAs techniques have been successfully developed which realise the very high speed potentials of GaAs with good yields [22].

Consider the simple DCFL inverter drawn in Figure 6.12(a). Assuming the input voltage is very low, the output should be logic "1". In this case, because the output is not connected to anything, the actual voltage will be 1.7 V, equal to the positive power supply rail. However, in real life it is not usual to design inverters with open-circuited outputs and a more normal situation is as shown in Figure 6.12(b), where there are two inverters in cascade. Now the output of the inverter is, in effect, connected through a Schottky diode to ground. Therefore the voltage on the gate will be about 0.7 V, the Schottky barrier voltage. The current into the gate is provided by the pull-up device of the inverter.

Suppose two gates are now connected to the output of the inverter, as shown in Figure 6.12(c). The sum of the current into the two driven gates is approximately the same as before, because it is provided by the same pull-up transistor in the inverter, which is acting as a current-source. Assuming that they are matched, each receives half the current of the original driven gate in Figure 6.12(b). Because each gate receives half the current, it has less voltage across it. To determine the actual value of the voltage on a driven gate, the Schottky diode forward characteristic must be known, but it can be as little as 0.67 V. If the inverter is used to drive three gates, each of the driven gates will receive one third of the current and the logic "1" voltage would reduce to around 0.64 V.

Clearly, the addition of extra gates will reduce the logic "1" voltage seen by each gate. As the logic "1" voltage decreases the noise margin reduces and the rise and fall times increase. The practical limit for the number of gates to a single gate output, the **fan-out**, is about four.

Note that in the above argument the gates are assumed to be of equal sizes. If, however, the pull-down transistors of a driven gate are, for example twice the normal size, such gates will add two to the fan-out.

Another unfortunate characteristic of MESFETs is that they have a rather low OFF resistance. In other words, they conduct significant leakage current when they are supposed to be OFF. Consider the inverter shown in Figure 6.13. Assume that the input voltage is below the pull-down transistor threshold by a sensible amount (the noise margin,

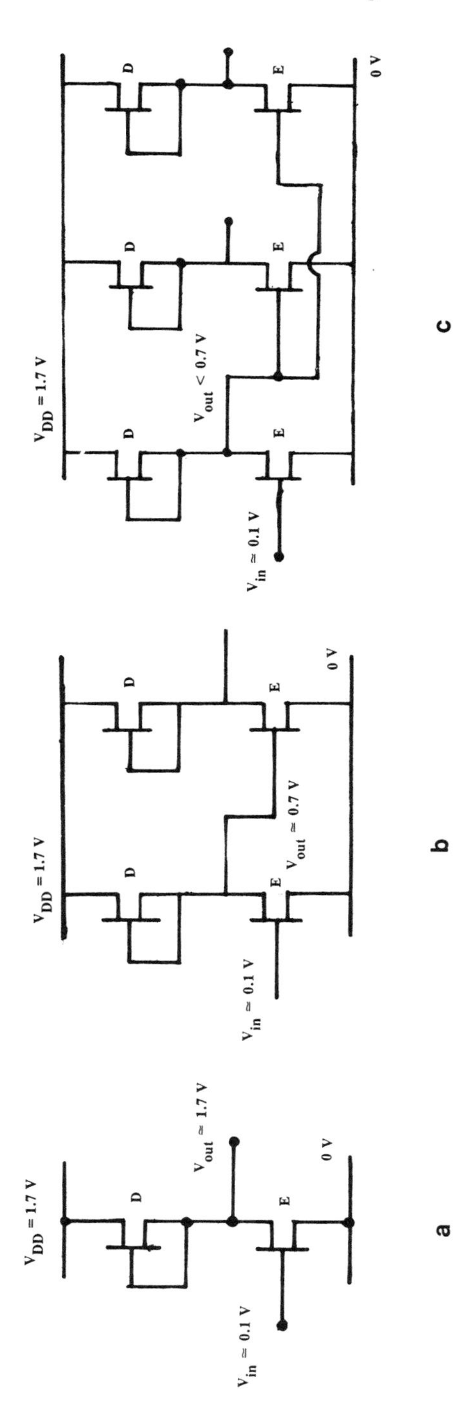

Figure 6.12 Inverter circuits
(a) Simple DCFL inverter (b) Cascaded DCFL inverter (c) Two gates being driven by one DCFL inverter

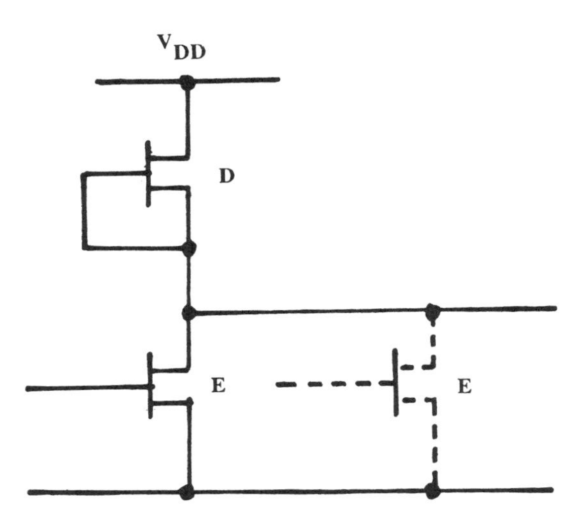

Figure 6.13 DCFL fan–in limitation

usually about 100 mV). A small current will pass through this supposedly OFF device and will pull the logic "1" value down.

Suppose another pull–down transistor is added to the inverter, converting it into a two–input NOR gate. Again, even if both pull–down devices are supposed to be OFF, they will conduct a small current which will reduce the logic "1" voltage output by the gate. As further pull–down devices are added the logic "1 value output by the gate will decrease further. Each reduction will cause a degradation of the noise margin and a decrease in drive to the following gate. Also, as further pull–down devices are added, the parasitic capacitance on the output of the gate increases so decreasing operating speed. For these two reasons the maximum number of pull–down transistors on a single gate, the **fan–in** , is limited, usually to about six.

These limits on fan–in and fan–out are the major problems in designing with GaAs MESFETs. However, these are not the only constraints on a designer and the following points should also be taken into account.

(i) NAND Gates

The use of NAND gates is to be avoided. Apart from the fact that their pull–down transistors are large (N times larger than normal for an N–input NAND gate) and add greatly to the driving gate's fan–out, the switching voltages of the upper and lower pull–down transistors are significantly different (by some tens of mV) which reduces noise margins. Two input

NAND gates can be used if absolutely necessary but NAND gates with more than two inputs are usually forbidden.

(ii) OR Gates

OR gates, which are not available in nMOS, can be made in GaAs DCFL. The circuit for a two input OR gate is shown in Figure 6.14 and can be seen to be based on a source follower. As with a source follower, the gain is less than one, so two such gates should not be cascaded as this can result in a significant reduction in logic levels and hence reduced noise margins. Also, for this gate to work, its input logic "1" must be of the order of two Schottky barrier voltages. For this reason, gates driving OR gates cannot drive other NOT, NOR or NAND gates at the same time as this would clamp the driving gate's logic "1" output voltages to one Schottky barrier voltage.

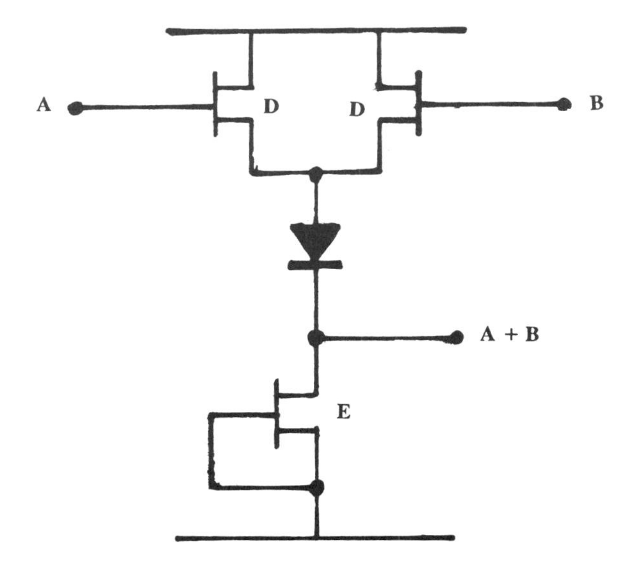

Figure 6.14 DCFL OR gate circuit

(iii) Complex Gates

Complex gates are best avoided altogether though simple versions can be used at the expense of operating speed.

6.4.6 Comparison Between Logic Families

Comparisons between the logic families are not often made directly as it is hard to compare like with like. If comparisons are to be made they are typically between ring oscillators which are simple circuits consisting of chains of an odd number, N, of inverters or gates. The propagation delay, τ_D, of the inverters is related to the frequency of oscillation, f, by $\tau_D = [^1/_2 fN]$. The values obtained by measuring ring oscillator performance are about as good as can be obtained because each inverter has a fan–out of one and loading capacitances are minimised. Because of this the figures obtained are somewhat misleading as they do not reflect how a real circuit would work in practice using that technology. However, ring oscillator based measurements are often used to evaluate a particular process.

About seven years ago (1982) comparisons were being made between the various GaAs technologies then available. To aid in making this comparison ring oscillators were fabricated using the available GaAs technologies. The results obtained are presented in Table 6.2. Though this information is now rather dated it is still relevant as a comparison. Also, it is rather difficult to obtain such information for up–to–date processes as most work today is being carried out using either pure DCFL or combined DCFL and BFL and the structures reported in the literature are rather more complex than ring oscillators!

Table 6.2 Ring oscillators speed–power performance for several GaAs IC technologies (extracted from [23])

Source	Approach	Gate Length and Width (μm X μm)	Prop. Delay (pS)	Speed–Power Product (pJ)	Fan–in/ Fan–out
Hughes [24]	BFL inverter	0.5 X 50	34	1.4	1/1
HP [8]	VFL NOR	1 X 20	86	3.9	2/2
Rockwell	SDFL NOR	1 X 10	120 52 †	0.04 0.063 †	2/1 2/1
Thomson CSF [25]	BFL	0.75 X 20	68	2	1/1
FUJITSU [26]	E–D MESFET	1.2 X 20	170	0.12	1/1
FUJITSU [27]	DCFL (*)	1.5 X 30	50	0.287	
NTT [28]	DCFL	0.6 X 20	30 17.5 †	0.057 0.616 †	1/1 1/1
MCD [29]	EFET/Pseudo Complementary	1 X 1	150	0.06	1/1
Thomson CSF [30]	EMESFET/quasi normally–off	1 X 35	105	0.23	1/1

† Measured at 77°K (*) Self–Aligned Gate Process

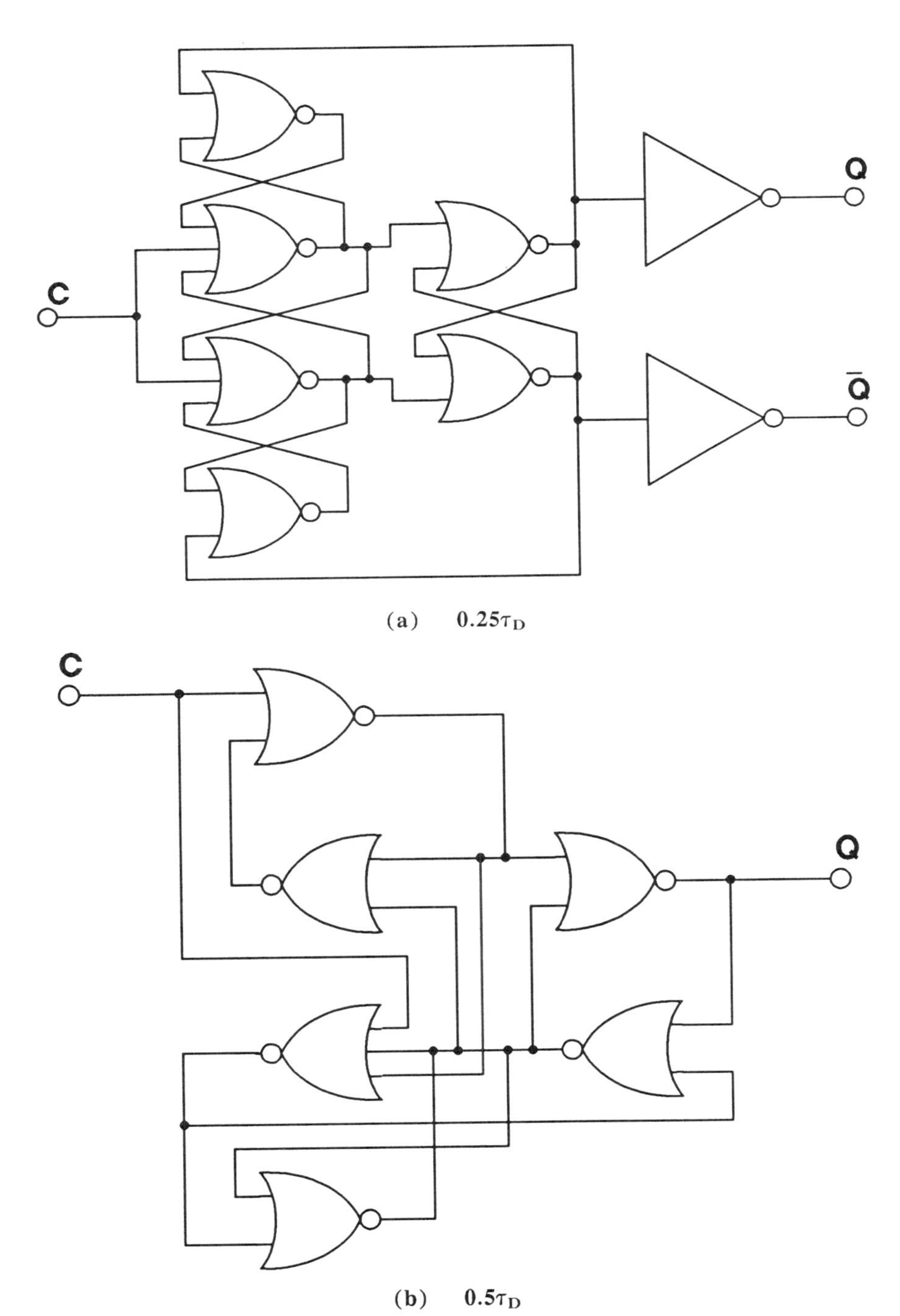

Figure 6.15 Single-clocked flip flops

Interpretation of Table 6.2 is involved because performance will vary with gate length, gate width, parasitic capacitances and resistances, temperatures, etc. Shorter gate lengths should increase transistor transconductance and so decrease propagation delays. The same effect can be produced by increasing gate widths, though in this case the effect is lessened by the consequent addition of parasitic capacitance. It is probably more meaningful to compare performances of a circuit such as a binary frequency divider because then the values of fan–in, fan–out and loading capacitance are more likely to be typical.

Let us compare the performances of four types of frequency divider that have been implemented in GaAs and are shown in Figures. 6.15(a–d). Again, this is old data but the points made remain valid.

Figure 6.15(a) shows a $0.25\tau_D$ D–type single clock flip–flop, Figure 6.15(b) shows a $0.2\tau_D$ D–type single clock flip–flop, Figure 6.16(a) shows a $0.25\tau_D$ master–slave, complementary–clocked flip–flop and Figure 6.16(b) shows a $0.5\tau_D$ master–slave, complementary–clocked flip–flop.

The results are given in Table 6.3, which presents a comparison of speed and power for a variety of approaches to frequency division. By examining the maximum dividing frequencies of these circuits the propagation delays can be calculated and they are fairly close to those obtained from ring oscillator evaluation. This indicates that the speed of

Table 6.3 GaAs IC Frequency divider performance (extracted from [23])

GaAs IC Technology	Circuit Approach	Theoretical Maximum Tog. Freq	Measured Maximum Tog. Freq	Equiv. τ_d(ps)	Power Diss. (mW/gate)	$P_D\tau_d$ (pJ)
1 μm SDFL Rockwell [19]	DFF/2 (NOR Gate)	$0.2\tau_D$	1.9 GHz	105	2.5	0.26
0.7 μm SDFL TCSF [25]	DFF/2	$0.2\tau_D$	3 GHz	67	40	2.68
1 μm BFL Hughes [31]	DFF/2	$0.2\tau_D$	2.2 GHz	91	78	7.1
1 μm BFL HP [8]	NAND/NOR Com.Clock	$0.5\tau_D$	4.5 GHz	111	40	4.4
0.6 μm BFL LEP [32]	NAND/NOR Com.Clock	$0.5\tau_D$	5.5 GHz	91	40	3.6
0.6 μm DCFL NTT [28]	DFF/8 (NOR)	$0.25\tau_D$	3.8 GHz	66	1.2	0.079
1.2 μm DCFL NEC [33]	Comp Clock / 2 NOR	$0.25\tau_D$	2.4 GHz	100	3.9	0.39

logic gates in depletion mode circuits is not greatly reduced by fanouts of three. The higher toggling rates of the BFL circuits produced by HP [8] and LEP [32] are due to their use of complementary–clocked master–slave flip–flops rather than any fundamental speed advantage over the circuits produced by TCSF [25] or Hughes [31].

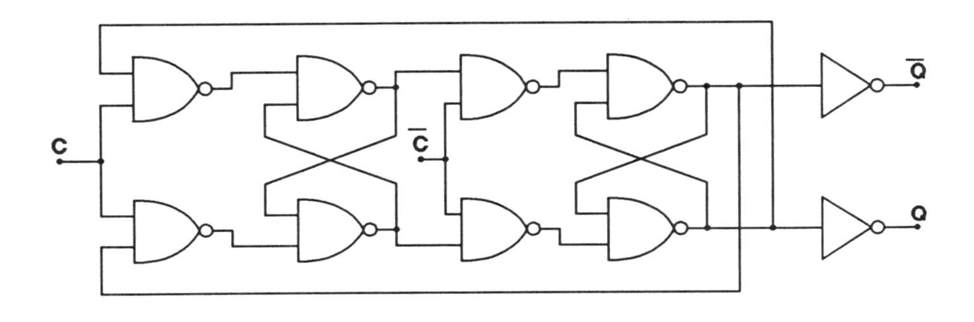

(a) $0.25\tau_D$

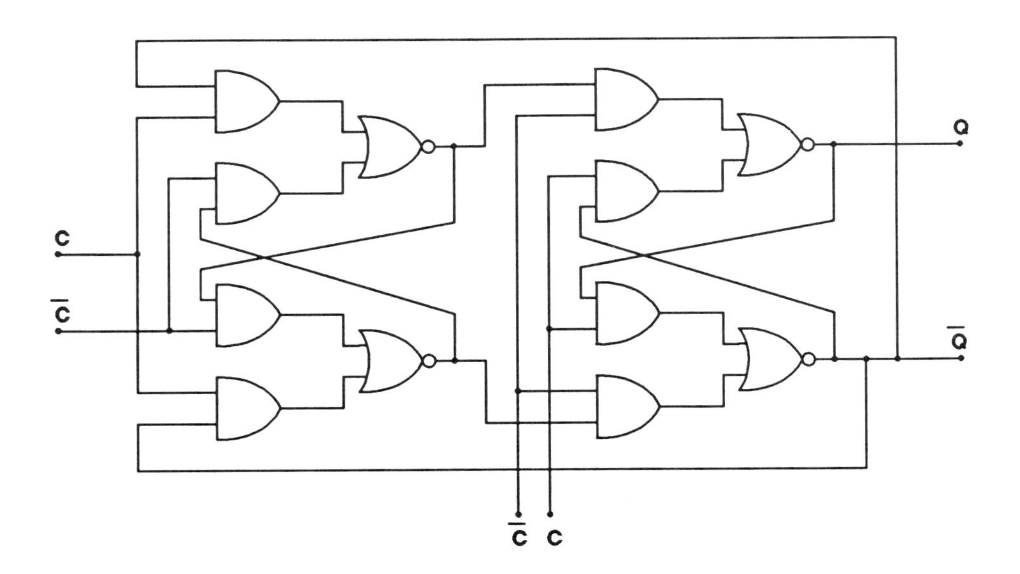

(b) $0.5\tau_D$

Figure 6.16 Master–slave, complimentary–clocked clock flip flops

The gates in frequency divider circuits which use enhancement mode FETs have a rather higher propagation delay than those demonstrated in the ring oscillators. This is most probably due to the greater affect of increasing fan-out of DCFL logic. However, use of 0.6 μm gate lengths has allowed the fabrication of a divide-by-eight counter clocking at 3.8 GHz [28] with a power dissipation of 1.2 mW per gate. This is a good result even by the standard of the late 80's.

If such comparisons were to be made today the results would not be greatly different to those obtained in 1982. The last few years have seen manufacturers concentrating on increasing yield and repeatability rather than on further reducing gates lengths or transistor transconductance.

6.4.7 Achievements in the Production of GaAs ICs

BFL Circuits

Early demonstration BFL circuits of about 20 gates were reported [8] in the literature in 1977. Since then BFL has been used in a number of applications where speed is of the essence and power consumption is less critical [34, 35]. Typical examples would be pre-scalers for counters. The first frequency divider result offered 4.5 GHz frequency division [8] and other workers later achieved 5.5 GHz [36] and 5.7 GHz [37]. D-type flip-flops have been reported working at 3 GHz and over [38]. Other offerings such as an 8:1 multiplexer circuit designed for use in a 5 GHz pattern generator [17] and a 4-bit ALU [39] have also been reported but there is little doubt that the high power consumption of BFL will limit it to MSI densities. Perhaps one of the best, latest, offerings is reference [34] which reports on a 1.5 GHz programmable divide-by-N GaAs counter. This device is significant because it is fully engineered. It will operate over a temperature range of −60°C to +110°C, has ECL compatible inputs and outputs, and has been designed taking into account long term reliability considerations, resulting in a mean time to failure of 1.8 million hours.

SDFL Circuits

SDFL circuits have increased in complexity with early (1980–81) offerings including variable modulus dividers [40], and 8:1 multiplexer and a 3-bit multiplier [41], a 5-bit multiplier [42] and an 8-bit multiplier with over 1000 gates [45].

By early 1988 SDFL was being used in the production of a gate array chip containing 2000 cells [44]. Each cell had a power dissipation of 108 μW and a propagation delay of 1.6ns. The cells were configured to give an eight-bit adder with an addition time of 11ns at a power dissipation of 236 mW.

DCFL Circuits

In the early 80s it was thought that problems with the fabrication of enhancement transistors would prove insuperable if high yields were to be achieved. This view led to companies such as Gigabit Logic putting a great deal of effort into all depletion technologies such as SDFL, and a variant which combines some of the features of SDFL and CCFL. This variant, sometimes known as CDFL (Capacitor Diode FET Logic), places reverse biased diodes (capacitors) across the level shifting forward biased diodes used in SDFL. These "speed–up" capacitors couple edges between gates faster than if they had simply passed through the forward biased level shifting diodes. However, this has probably not been the correct way to proceed and most effort seems to have been put into DCFL. The main problems involved in the fabrication of enhancement GaAs transistors appear to have been solved and fairly large chips are now available and are produced with high yields.

One example of a recently reported result is a 12x12 multiplier [45] which carries out its multiplication in 82.5ns with a power dissipation of 1.7W. This has been constructed from a gate array consisting of 19000 devices in 6000 gates. This is not a laboratory–only chip, having been designed to work up to 160°C and fabricated at a 10% yield.

Another example is a 1K–bit static RAM with 1.3ns access time [46]. This is constructed from a mix of DCFL and BFL. The use of BFL carries a power dissipation penalty, but if it is employed where speed is critical the overall speed of the chip can be greatly increased without the power dissipation becoming totally unreasonable. In this case the chip draws 1.4W with only 40 mW required by the actual memory array while the remainder is taken by the support circuitry, in particular the output buffers. This RAM is fully ECL compatible so it delivers its output into transmission lines terminated with 50Ω resistors.

DCFL has also been used to design a modified Booth's Algorithm 4x4 two's complement multiplier with a 2.5ns multiplication time [47]. This was fabricated on a 1 μm self aligned gate process and had gate delays of about 120 ps.

Another DCFL single chip, produced on a self–aligned process, contained both a 4x4 multiplier and an 8 x 8 multiplier/accumulator [48]. The 4 x 4 multiplier operated at 1.2 GHz (940 ps multiplication time) at a 131 mW power dissipation. Perhaps more interestingly, this part of the chip had a yield of 93%. The 8x8 multiplier/accumulator multiplied in 3.17ns, the fastest yet reported, using 4278 FETs in 1317 gates. Its yield was 43%, a very respectable value.

These examples demonstrate that DCFL is far from being a laboratory curiosity. It has matured to the point where real, useful, circuits are being produced at a respectable yield.

6.5 System Applications for GaAs Digital ICs

Ever since GaAs ICs were first developed in the late 1970s and the early 1980s enthusiastic claims have been made for how they would find widespread use in high speed system applications. In fact these early claims were never fulfilled and currently very few are in use in products or systems that are commercially available.

Several reasons can now be discerned for this with the advantage of hindsight. Firstly, many of the results reported in the technical literature dealt with technology demonstrator circuits which were not designed with a specific application in mind. Secondly, the yield of working circuits was generally low, because of material problems, making production uneconomical. Thirdly, system designers were understandably wary of committing to a new and relatively immature device technology. The result was that there existed a credibility gap with the silicon effort which could only be met by making bold claims which were probably counter-productive in the long term.

This situation is now changing with the identification of specific application areas where GaAs offers unique and significant advantages which justify the costs of its use. These are:

- military systems
- digital signal processors
- supercomputers
- telecommunication systems

A number of microprocessors fabricated in GaAs have been reported in the literature. All represent the current state-of-the-art in terms of complexity. The first such circuit is the bit-slice processor chip set developed by Vitesse Semiconductors [49]. This consists of three circuits; a 4-bit-slice multifunction ALU with some on-board memory, a 12-bit programmable sequencer and a carry look-ahead generator for a 16-bit wide data path, all implemented in enhancement-depletion MESFET technology. The architecture of these devices is based on the Advanced Micro Devices (AMD) 2901, 2910 and 2902 circuits. The performance of these parts used to implement a 16-bit microprocessor is three times faster than a conventional silicon implementation. The ALU uses 4856 transistors, the programmable sequencer uses 6875 transistors and the high–speed carry look–ahead generator uses 625 transistors. This chip set is claimed to be the first commercially available GaAs microprocessor.

The other two microprocessor developments [50] are at McDonnell Douglas and Texas Instruments. Both are funded by the US Defense Advanced Projects Research Agency (DARPA) as part of a project to produce a space–based signal processor system. Signal processing microprocessor systems such as this are required in ;military applications where ;they must operate at very high data rates, over a wide range of temperatures, consume little power and are radiation hard. The objective of the project is to design a 32–bit reduced instruction set computer (RISC). The completion of this advanced project is due in 1989–1990. The goals for the project are [52]:

- a circuit complexity of around 10,000 gates
- clock rate of 200 MHz
- a custom, main processor and a floating point co–processor
- a data processing rate of 100MIPS

The McDonnell Douglas chip uses JFET logic and contains approximately 23,000 transistors in an area of 8.5 mm by 11.1 mm. The power consumption is 6.5 watts at 60 MHz. The Texas Instrument chip [53] uses a form of integrated injection logic (I^2L) and it contains 12,895 gates in a chip area of 11.3 mm by 10.5 mm. They expect it to consume around 10 Watts at 200 MHz clock rate.

Another example of a VLSI circuit implemented in GaAs for digital signal processing applications is the 16 by 16 complex multiplier chip shown in Plate IV, which has been implemented in 1 μm MODFET technology by Honeywell [54]. Using NOR logic implemented in a type of DCFL which uses a special temperature compensated load structure. This chip is clocked at 500 MHz and can perform a full 16 bit complex multiplication every 8ns. It consists of one multiplier and an accumulator. The multiplier is used four times in each multiplication cycle and the products are stored in an output register. To facilitate testing it incorporates a self test mode, whereby the product can be fed back recursively to produce a series of rotating vectors at the output, as the chip is clocked. The design employs a modified Booth's algorithm which forms the real part and the imaginary part of the product, in sequence, by multiplying the multiplicands and adding the results in an accumulator.

Another area where GaAs can give significant advantages is in supercomputers. Currently Cray Research [55] is using GaAs MSI parts for the CPU and control logic in the Cray–3 machine, giving it a cycle time significantly less than the Cray–2. The Cray–3 has a computing power of 10 gigaflops while that of the silicon ECL–based Cray–2 is around 1.2 gigaflops. This large increase in performance is obtained both from the inherent speed advantage of using GaAs and by careful design of the packaging and interconnect wiring between chips to minimise off-chip propagation delays. Both machines consume around the same amount of power. The technology used in this machine is GaAs MESFET depletion mode logic with a complexity of around 200–300 equivalent gates per chip. The chip fabrication is being carried out on a foundry basis by Gigabit Logic. The use of GaAs technology has allowed ultra high speed performance at moderate power levels. Japanese companies are known to be actively developing GaAs circuitry for large mainframe computer applications [56].

Telecommunications is another area where GaAs can offer distinct system advantages. These stem from the opto-electronic properties, which make it useful for light emitting and detecting devices for fibre optic systems. Additionally the serial nature of most communication media necessitate a single high bandwidth data path as the front end to most systems. GaAs ICs fit this application well combining low complexity with high speed.

Hitachi [57] has introduced a set of GaAs ICs that provides a complete transmit–receive interface for fibre optic communication systems at data rates up to 2.4 Gbits/second. The set consists of a multiplexer and laser diode driver for the data transmission side and five other circuits that provide signal conditioning and demultiplexing at the receiver end of the fibre.

6.6 Conclusions

GaAs ICs cannot be expected to supplant silicon in most applications. The two materials should be seen as complementary to each other with silicon being ideally suited for lower speed high complexity circuitry and GaAs most useful for high speed low complexity applications. GaAs is useful for applications that are difficult for silicon because of extremely exacting requirements on power consumption, speed, temperature range and radiation hardness. Initially the application areas will only occur where significant system advantages can be obtained in order to justify and bear the cost of developing the technology further towards maturity.

As clock rates on systems using GaAs ICs increase, considerable engineering effort will be required to be directed toward the packaging of the devices to minimise the interconnect delays between devices. Efforts must be made to ensure that signal connections behave as transmission lines and are reasonably well terminated to avoid reflection and the resulting distortion of the pulses. Consideration must be given to system architectures in the context of packaging to ensure that propagation delay effects are minimised, for example systems should be partioned so that the high speed signal interconnects are minimised and techniques such as pipelining of signals must be adopted to accommodate delays.

6.7 References

1 Eden, R. C.: "The Development of the first LSI GaAs Integrated Circuits and the Path to the Commercial Market", Proc. IEEE, 1988, 76, (7), pp765–777.

2 Liechti, C. A.: "Microwave Field-Effect Transistors-1976", IEEE Trans. MTT, 1976 MTT-24, (6), pp279–300.

3 Morkoc, H and Solomon, P M : "The HEMT: a Superfast Transistor", IEEE Spectrum, Feb. 1984, pp28–35.

4 Beresford, R.: "Heterostructure FETs", VLSI Systems Design, Nov 1988, pp92–104.

5 Kroemer, K.: "Heterostructure Bipolar Transistors and Integrated Circuits", Proc. IEEE, 1982, 70, (1), pp13–25.

6 Asbeck, P. et al : "Emitter Coupled Logic Circuits Implemented in Heterojunction Bipolar Transistors", IEEE GaAs IC Symp. Tech. Dig., 1983, pp174–177.

7 Zuleeg, R.7 Notthoff, J. K. and Lehovec, K. : "Femtojoule High-Speed Planar GaAs E-JFET Logic", IEEE Trans. El. Devices, 1978, 25, (7), pp628–639.

8 VanTuyl, Rl L., Liechti C., Lee R. E. and Gowen E.: "GaAs MESFET Logic With 4 GHz Clock Rate", IEEE J. Solid State Circuits, 1977, SC–12, (10), pp485–496.

9 Eden, R. C. et al: "Planar GaAs IC Technology: Applications for digital LSI", IEEE J. Solid State Circuits, 1978, SC–13, (4).

10 Mellor, P. T. J. and Livingstone, A. W.: "Capacitor Coupled Logic Using GaAs Depletion-mode FETs", Electronic Letters, 1980, 16, (19), pp749–750.

11 Livingstone, A. W. and Mellor P. T. J.: "Capacitor Coupling of GaAs Depletion Mode FETs", IEE Proc. I, Solid-State and Electron Devices, 1980, 127, (5), pp297–300. pp419–425.

12 Eden, R. C.: "Capacitor Diode FET Ligic (CDF) Circuit approach for GaAs D-MESFET ICs", IEEE GaAs IC Symp. Tech. Dig., (Boston), 1984, pp11–14.

13 Abe, M., Mimura, T., Yokoyama, N. and Ishikawa, H., "New Technology towards GaAs LSI/VLSI for computer Applications", IEEE Trans. MTT, MTT–30, 1982, pp992–998.

14 Ishii, Y et al, "Processing Technologies for gaAs Memory LSIs", IEEE GaAs IC Symp. Tech. Dig., 1984, p121.

15 Damay-Kavala, F., Gloanec, M., Peltier, M., Nuzillat G. and Arnodo G.: "Speed Power performance in Sequential GaAs Logic Circuits", Int. Symp. GaAs and Related Compouns Tech. Digest, Oiso, Japan, Sept 1981, Paper 15.

16 Barna, A. and Liechti, C. A.: "Optimisation of GaAs MESFET Logic Gates with Subnanosecond Propagation Delays", IEEE Journal Solid-State Circuits, 1979, Vol SC–14, pp708–715.

17 Baldwin, G. et al., "A GaAs MSI Word Generator Operating at 5GBits/sec Data Rate", IEEE GaAs IC Symposium Tech. Dig., 1980.

18 Eden, R. C., Welch B. M. and Zucca, R.: "Lower Power GaAs Digital Ics Using Schottky Diode-FET Logic", Int. Solid State Circuit Conf. Dig. Tech. Papers, 1978, pp68–69.

19 Long, S. I., Lee, F. S., Zucca, R., Wlech, B. M. and Eden, R. C.: "MSI High-Speed Low-Power GaAs ICs Using Schottky Diode FET Logic", IEEE Trans. Microwave Theory Tech., 1980, MTT–28, pp466–471.

20 Hashizume, N., Yanada, H., Kojima, T., Matsumoto K. and Tomizawa, K.: "Low-power Gigabit Logic by GaAs SSFL", Electronic Letters, 1981, 17, (16), pp553–554.

21 Hashizume, N., Yanada, H. and Tomizawa, K.: "Schottky-barrier Coupled Schottky-barrier Gate GaAs FET Logic", Electronic Letters, 1981, 17, (1), pp51–52.

22 Joho, A., Toyoda, N., Mochizuki, M., Mizoguchi T. and Nii, R.: "Polar E/D-type GaAs ICs by Pt Buried Gate Technology", IEEE GaAs Ic Symposium Research Abstracts, 1981, Paper 11.

23 Long, S. I. et al., "High Speed GaAs Integrated Circuits", Proc IEEE 1982, 70, (1), pp35–45.

24 Lundgren, R. E., Kdrumm C. F. and Pierson, R. L.: "Fast enhancement-mode GaAs MESFET Logic", 37th Annual Devices Research Conference, Boulder, CO., 1979.

25 Nuzillat, G., Damay-Kavala, F., Bert G. and Arnodo, C.: "Low Pinch-Off Voltage Fet Logic (LPFL): LSI Oriented Logic Approach Using Quasi-normally Off GaAs MESFETs", Proc. IEE, 1980, 127, Pt 1, (5), pp287–296.

26 Suyama, K., Kusakawa H. and Fukata, M.: "Design and Performance of GaAs Normally-off MESFET Integrated Circuits", IEEE Trans. Electron Devices, 1980, ED–27, pp1092–1097.

27 Yokoyama, N., Mimura, T., Fukuta M. and Ishikawa, H.: "A Self-Aligned Source/Drain Planar Device for Ultrahigh Speed GaAs MESFET VLSIs", 1981 Int. Solid State Circuits Conf., Dig. Tech. Papers, Feb 1981

28 Mizutani, T., Kato, N., Osafune K. and Ohmori, M.: "Gigabit Logic Operation with Enhancement Mode GaAs MESFET ICs", IEEE Trans. Electron Devices, Vol. ED–29 No. 2 pp199–204 Feb. 1982

29 Lehovec, K. and Zuleeg, R.: "Analysis of GaAs FETs Integrated Logic", IEEE Trans. Electron Devices, 1980, ED–27, pp1074–1091.

30 Nuzillat, G., Bert, G., Ngu T. P. and Gloanec, M.: "Quasi-normally-Off MESFET Logic for High Performance GaAs ICs", IEEE Trans. Electron Devices, 1980, ED–27, pp1102–1108.

31 Greiling, P. T., Lundgren, R. E., Krumm, C. F. and Lohr R. F.: "Why Design Logic With GaAs and How?", MSN, Jan 1980, pp48–60.

32 Cathelin, M., Gavant M. and Rocchi, M.: "A 3.5 GHz Single-Clocked Binary Frequency Divider on GaAs", Proc. IEE, 1980, 127, Pt I, No 5, pp270–277.

33 Katano, F., Furutsuka T. and Higashisaka, A.: "High Speed Normally-Off GaAs MESFET Integrated Circuits", Electronic Letters, 1981, 17, No. 6, pp236–239.

34 Kane, M. G., Chan, P. Y., Cherensky S. S. and Fowlis, D. C.: "A 1.5 GHz programmable Divide-by-N GaAs Counter", IEEE Journal Solid-State Circuits, 1988, 23, (2), pp480–484.

35 Ekroot, C. G. and Long, S. I." "A GaAs 4-bit Address-Accumulator Circuit for Direct Digital Synthesis", IEEE Journal Solid-State Circuits, 1988, 23, (2), pp573–580..

36 Cathelin, M., Gavant M. and Rocchi, M.: "A 3.5 GHz Self-Aligned Single-Clocked Binary Frequency Divider on GaAs", IEE Proc. I, Solid-State and Electron Devices, 1980, 127, (5), pp270–277.

37 Gloanec, M., Jarry, J. and Nuzillat, G.: "GaAs Digital Integrated Circuits for Very High-Speed Frequency division", Electronics Letters, 1981, 17, (20), pp763–765.

38 Gloanec, M., Nuzillat, G., Arnodo G. and Peltier, M.: "An E-beam Fabricated GaAs D-type Flip-flop IC", IEEE Trans. Microwave Theory and Tech., 1980, MTT-28, pp472–478.

39 Suyama, K., Kusakawa H., Okamura, S., Yamamura, S. and Fukata, M.: "GaAs LSI for High Speed Data Processing", IEEE GaAs IC Symposium Research Abstract, 1980, Paper 4.

40 Walton, E. R., Lee, F. S., Shen, Y. K. and Dikshit, r.: "High Speed SDFL Divider Circuit", IEEE GaAs IC Symposium Research Abstract, 1981, Paper 3.

41 Welch B. M. et al., "LSI Processing Technology for Planar GaAs Integrated Circuits", IEEE Trans. Electron Devices, 1980, ED–27, pp1109–1115.

42 Lee, F. S. et al.: "High Speed LSI GaAs Digital Integrated Circuits", IEEE GaAs IC Symposium Research Abstract, 1980, Paper 3.

43 Long, S. I., Welch, B. M., Zucca, R. and Eden, R. C.: "Gallium Arsenide Large Scale Integrated Circuits", J. Vac. Sci. and Technology, 1981, 19, pp531–536.

44 Vu, T. T. et al. "Low Power 2K-Cell SDFL Gate Array and DCFL Circuits Using GaAs Self Aligned E/D MESFETs", IEEE Journal Solid-State Circuits, 1988, 23, (1), pp224–238.

Chapter 7

High-Speed Analog-to-Digital and Digital-to-Analog Conversion with GaAs Technology: Prospects, Trends and Obstacles

Lawrence E. Larson

7.1 Introduction

GaAs integrated circuit technology has historically been portrayed as the "technology of the future". Although it has not attained the wide applicability that its early promoters predicted, it is gradually becoming an accepted and commercially available source of high-speed integrated circuits. However, a key question that needs to be addressed is its suitability for one of the most demanding of integrated circuit applications – the A/D converter. Many researchers hope to be able to exceed the 8-bit 1 GHz sample rate barrier in the near future with this technology. In addition, high-speed high-resolution D/A converters are also planned for implementation in the technology. This paper will address the promises and limitations of GaAs technology for A/D and D/A converter applications.

The principle motivation for the use of GaAs technology for A/D and D/A conversion is the high intrinsic speed of the transistors. For a given gate length or base width, the resulting f_T – or unity current gain frequency – of the device is typically 30 to 70 % higher than an equivalent silicon transistor. Comparable gains can also be achieved in the transistor's f_{max} – or unity power gain frequency. In addition, the availability of lattice matched wide band-gap AlGaAs allows the device designer an extra degree of freedom for performance optimisation. The is especially attractive in the case of GaAs Heterojunction Bipolar Transistors (HBTs), where a wide band gap emitter allows the base to be heavily doped, decreasing base resistance, without degrading current gain. Finally, since undoped GaAs is semi-insulating, the parasitic capacitance of wire interconnections is lower than in silicon technology.

The three dominant transistor types in GaAs technology are the Metal-Schottky FET (MESFET), the Modulation-Doped FET (MODFET) and the HBT. MESFET and HBT technologies are the most promising for A/D converter applications, since GaAs MODFETs do not exhibit a significant performance advantage compared to MESFET technology, and their fabrication technology is relative immature. HBT fabrication technology is even less mature.

Table 7.1 Comparison of Dynamic Performances of Silicon and GaAs transistors

	f_T (GHz)	gm(mS/mm)
SILICON BIPOLAR	25	3000
SILICON MOS	20	200
GaAs MESFET	80	400
GaAs MODFET	120	600
GaAs HBT	110	3000

Table 7.1 lists the comparative high-speed performance of state-of-the-art transistors in silicon and GaAs technology; the advantage of GaAs for high-speed applications is evident.

However, the *intrinsic* advantages of GaAs must be weighed against its *practical* disadvantages, which include a low level of achievable integration, as well as backgating, hysteresis, high 1/*f* noise and poor uniformity in the case of MESFET and MODFET approaches. These drawbacks have limited the use of GaAs technology to low-resolution A/D applications - typically 4–bits or less. Substantial improvements will be required in the manufacturability of GaAs integrated circuits before substantially higher resolutions can be routinely obtained.

In order to evaluate the impact of GaAs technology on A/D and D/A performance, a number of issues must be addressed, including the effect of transistor speed on A/D and D/A building-block performance, fundamental limitations on converter bandwidth and resolution, and optimum architectures for A/D and D/A realization in GaAs technology.

7.2 Limitations on converter performance

The performance of A/D and D/A converters is limited by a number of fundamental factors, including architecture and transistor speed, noise and matching. Each of these factors affect the achievable resolution and signal-to-noise ratio of the resulting output. This section outlines the specific technological factors that limit overall converter performance.

7.2.1 Fundamental limitations on A/D performance

Before an accurate assessment can be made of the potential of GaAs technology for A/D applications, it is important to understand the fundamental physical limits on A/D performance, which are somewhat independent of the technology used to implement the circuit. Traditional limitations like comparator offset voltage and hysteresis are not covered here since, in principle, their effects can be minimized by a variety of calibration schemes. Only those limitations that are inherent in the A/D conversion process are discussed in this section.

There are many ways of evaluating the performance of an A/D converter. For the purpose of signal acquisition, one of the most popular criteria is the *signal-to-noise ratio*, or SNR. For an ideal A/D, the maximum SNR of the output is given by the expression:

$$SNR = (6N + 1.8)dB \tag{1}$$

where N is the number of bits of resolution.

However, there are a number of factors that can degrade the SNR from its ideal value, including thermal noise, aperture jitter, and comparator uncertainty, in addition to algorithmic and dynamic errors of the A/D circuitry itself.

The maximum resolution of an A/D converter is *fundamentally* limited by the thermal noise of its environment, since there is little use for a converter whose quantization noise is less than the noise of the incoming signal. The maximum resolution of the A/D can be approximated from Nyquist's theorem, the maximum input signal, and the desired input bandwidth by the following expression:

$$N_{max} \approx log_2 \left(\frac{V^2_{max}}{8KTR\Delta f} \right)^{1/2} \tag{2}$$

where V_{max} is the maximum input amplitude, K is Boltzmann's constant, T is the temperature, Δf is the input bandwidth, and R is the impedance level of the input signal.

This equation demonstrates the well-known tradeoff between dynamic range and bandwidth. For example, assuming a maximum input amplitude of 1.0 V and an input bandwidth of 1 GHz, then the maximum achievable resolution of *any* A/D converter is approximately 14 bits in a 50Ω environment at room temperature. At 100 MHz, the maximum resolution rises to 16 bits. This limit has not been approached yet by commercially available devices, but the improved performance achievable with GaAs technology may allow designers to approach it in the future.

Another fundamental limit is related to random variations in the sampling interval; this phenomenon is known as aperture jitter. An A/D converter samples the input waveform at a periodic rate and converts it into a digital signal. Random variations in this rate degrades the resulting

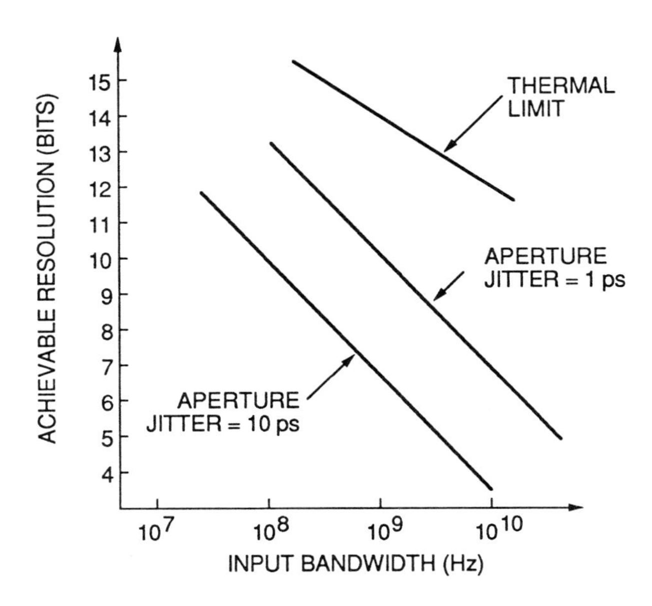

Figure 7.1 Achievable A/D resolution for different aperture jitters

performance of the converter. The allowable uncertainty in the sampling time decreases with sampling frequency and resolution. This uncertainty can be caused by low-frequency noise modulation of the incoming clock signal, temperature drift, or other instabilities. The degradation of the converter's signal-to-noise ratio in response to aperture jitter depends on the statistical properties of the jitter process and the incoming signal. However, it it useful to assume that the one-sigma aperture noise should be less than the one-sigma quantization noise for random inputs [1]. If this limit is adhered to, then:

$$2^N f_{sig}\tau_a \leq 1 \tag{3}$$

where N is the resolution of the converter, f_{sig} is the input signal frequency, and τ_a is the rms aperture jitter.

The effect of this constraint on A/D resolution is shown graphically in Figure 7.1 for different values of τ_a , along with a comparison with fundamental thermal limits. Clearly, the combination of high resolution and high bandwidth will require significant improvements with respect to timing accuracy repetition. For example, in order to digitize a 1 GHz input signal with 8-bit accuracy, a timing jitter of less than 5ps is required.

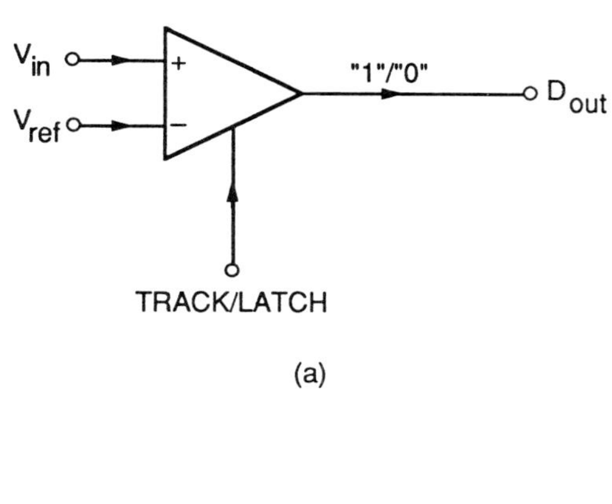

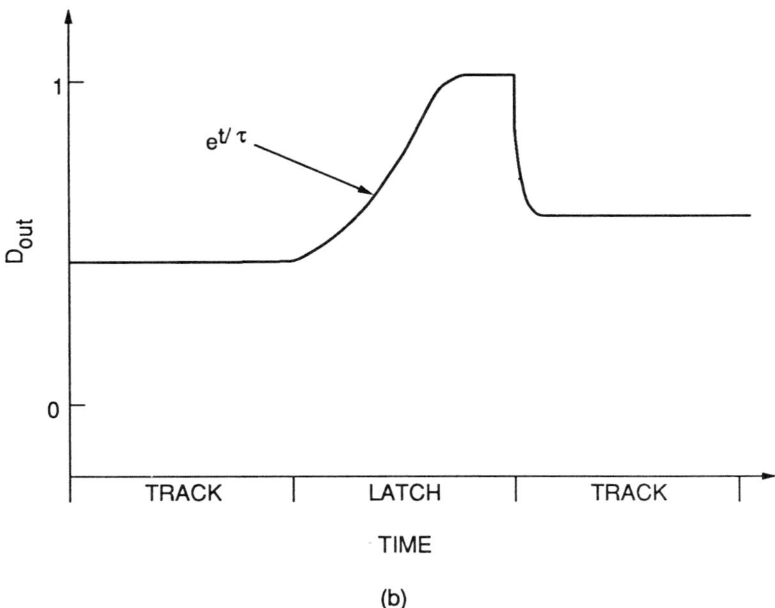

Figure 7.2 (a) **Simplified comparator model showing**
(b) **Time domain response**

A/D converters typically function by comparing the input signal against some known reference, and producing a digital output depending on the difference. Another limitation on A/D dynamic range arises from the uncertainty of the decision that the comparator makes during this process. A simplified model of the comparator function appears in Figure 7.2. The comparator is essentially an amplifier and a latch. In the "track" mode, the comparator amplifies the input voltage difference. In the "latch" mode,the comparator multiplies the amplified input voltage by a growing exponential, $e^{t/\tau}$, until the amplifier saturates, or until the latch mode ends. At the end of the latch period, a digital output is available for decoding.

The digital circuit following the comparator must be able to unambiguously determine whether the comparator output is a "one" or a "zero" at the end of the latch period. For input signals uniformly distributed in voltage, the probability that there will be an indecisive comparator output at the end of the latch period is given by [2]

$$p_i = \frac{2V_L}{A_o q} e^{-t/r} \tag{4}$$

where V_L is the minimum signal that the digital circuit can unambiguously accept, A_o is the gain of the comparator in the track mode, q is the A/D quantization level, t is the latch period, and τ is the comparator regeneration time constant.

Clearly, τ should approach zero for high accuracy operation, although there is a tradeoff between achievable speed and τ . By assuming a simplified two-transistor model for the comparator, and assuming that interconnect capacitance within the comparator is insignificant, then the comparator regeneration time can be approximated by

$$\tau \approx \frac{1}{2\pi f_T} \tag{5}$$

where f_T is the unity current gain cut-off frequency of the transistor used to realize the comparator. However, implementations of a comparator usually exhibit regeneration times higher than that predicted by this model, because of interconnection capacitance and the greater complexity of a typical comparator circuit.

Comparator indecision can effect A/D performance in a number of ways. It can be demonstrated that the variance (σ^2) of the quantization noise of an ideal A/D converter is $2^{-2N}/12$ [3], and that the total variance that results from combination of comparator indecision and quantization noise can be approximated by [4]

$$\sigma_{ex}^2 \approx \frac{2^{-2N}}{12} [1 + p_i 2^{N+1}] \tag{6}$$

For example, the quantization noise of an 8-bit A/D converter would double if p_i were 0.2%.

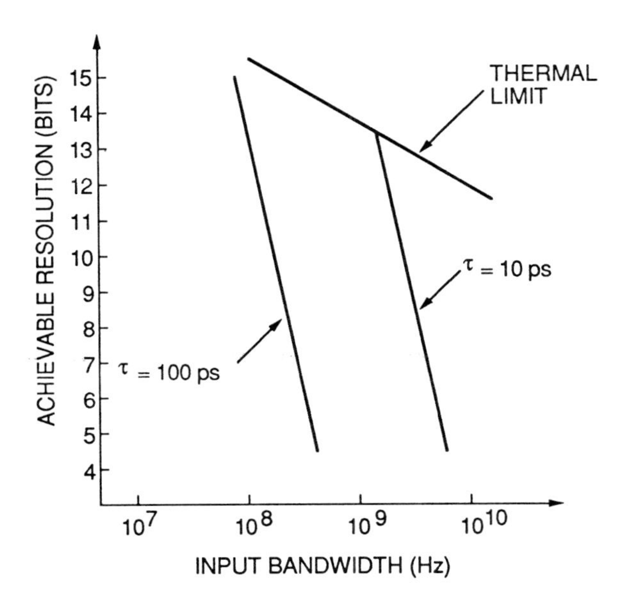

Figure 7.3 Achievable A/D resolution for different comparator regeneration times

Figure 7.3 plots achievable resolution as a function of input bandwidth, for different comparator regeneration times, assuming that the excess quantization noise is limited to 10% of the A/D quantization noise. This plot demonstrates the necessity for extremely short regeneration times in high-speed high-resolution A/D convertor applications. A regeneration time of less than 30 ps is required for applications in the 1 GHz range. This implies that the f_T of the transistor should be at least 33 GHz.

Certain A/D architectures are less sensitive to the effects of aperture jitter and comparator uncertainty than others. In particular, feedback approaches such as the $\Sigma - \Delta$ appear to "average out" these effects, although the magnitude of the improvement is unclear, and more research is required. Futhermore, certain A/D decoding schemes can be employed to minimize the effects of comparator indecision [5].

Figures 7.1 – 7.3 demonstrate the necessity for careful system-level considerations for the realization of high-speed, high-resolution A/D converters. For example, a 10-bit 1 GHz bandwidth A/D converter will require a sampling process whose aperture jitter is less than 1 ps and a comparator regeneration time of less than 20 ps. This level of performance is not achievable with currently available silicon or GaAs technology.

7.2.2 Fundamental limitation on D/A performance

High-speed D/A converters are typically implemented using the binarily weighted current switch approach, although some "low-glitch" designs may employ unarily weighted current sources instead [19]. The maximum signal-to-noise ratio of the output signal is the same as given in equation (1). This value is limited by static and dynamic errors. Static errors include offset, gain, non-linearity, and non-monotonicity. Dynamic errors include linear and non-linear settling errors.

In the current switch approach, the speed of the DAC is fundamentally limited by the speed of the transistor, which acts as a "switch". In that case, the minimum switching time can be approximately

$$\tau_{sw} \approx \frac{1}{2\pi f_T} \tag{7}$$

However, in practise, the switching time is considerably longer than this, and is limited by RC effects at the input and output of the circuit. In addition, finite slew-rate effects and switch "bucking" can cause the settling time of the D/A to be considerably longer than estimates based on the intrinsic speed of the transistor.

7.3 GaAs technology for A/D and D/A applications

GaAs technology is attractive for converter applications because of the high intrinsic speed of the transistors. Table 7.1 compares the dynamic performance of representative state-of-the-art GaAs and silicon transistors, and demonstrates the potential performance advantages of GaAs. It was demonstrated in the section on fundamental limits to A/D converters, that the comparator regeneration time constant and D/A minimum switching time is inversely proportional to the f_T of the transistor used to fabricate the circuit, and Figure 7.3 demonstrates the importance of a low regeneration time for digitizing input signals whose bandwidths exceed 100 MHz. Therefore, the high f_T obtained form GaAs transistors combined with the low parasitic capacitance, should increase the achievable speed of an A/D or D/A significantly. In the following sections, the suitability of the major GaAs technologies are discussed in relation to the requirements for a high-speed data converter.

7.3.1 GaAs MESFET technology

GaAs MESFET technology is currently capable of realizing enhancement/depletion-mode integrated circuits of LSI complexity. The

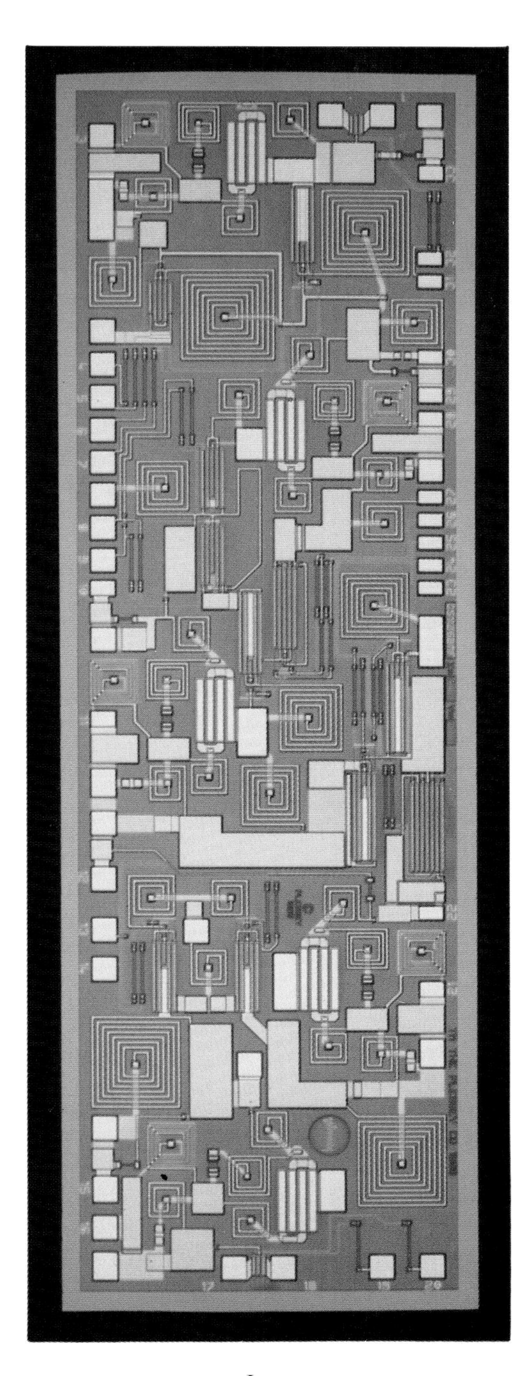

Plate I GaAs MMIC 6-bit phase shifter

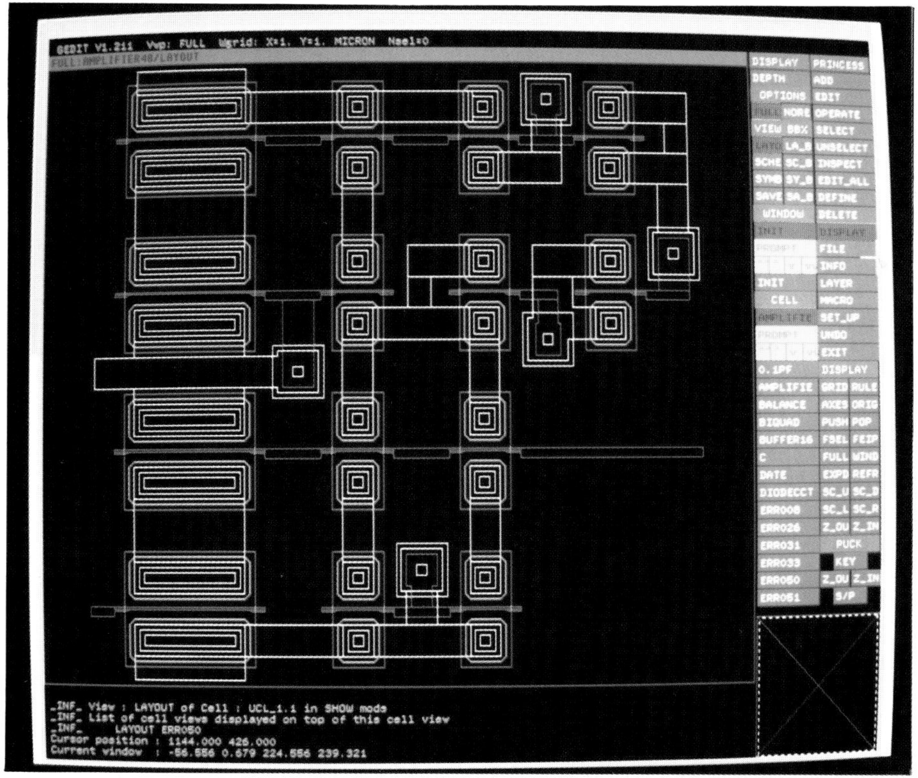

Plate II   A GaAs amplifier cell seen in the PRINCESS layout editor

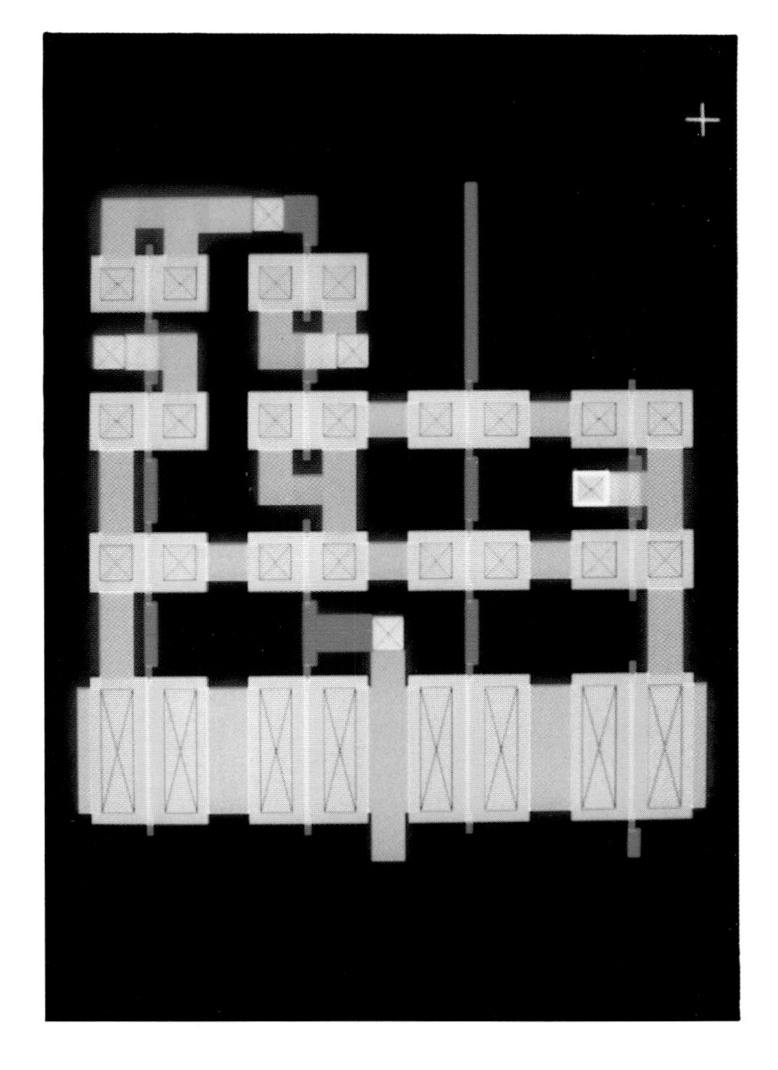

Plate III The GaAs amplifier of Plate III seen in the MAGIC symbolic layout editor

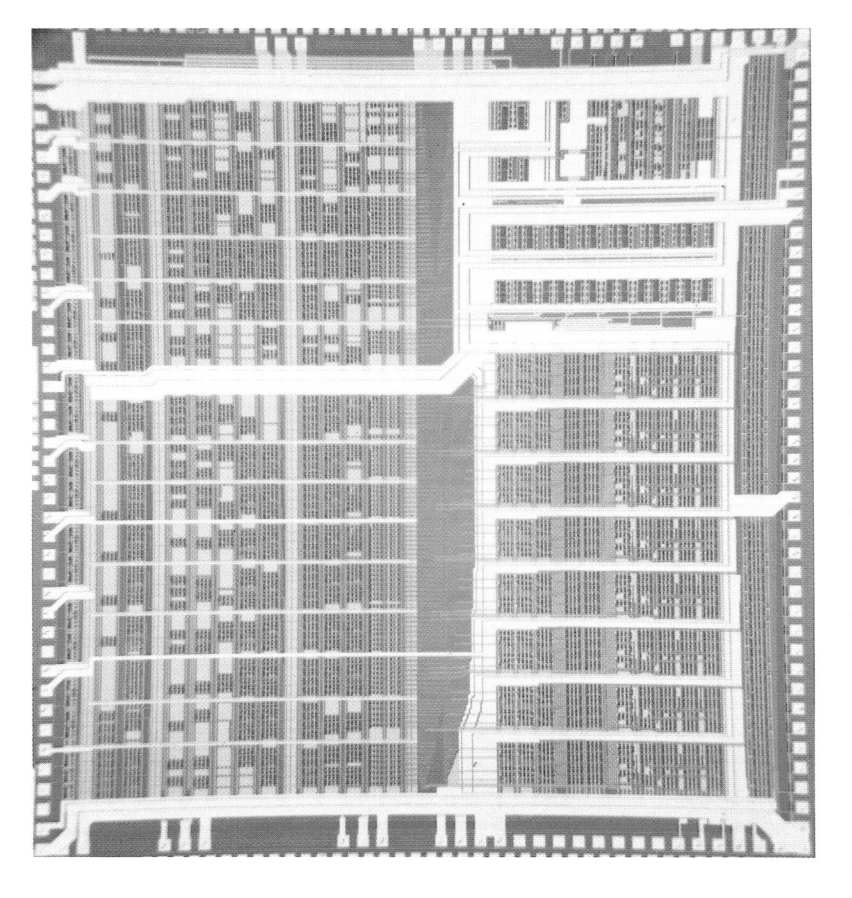

Plate IV A 16 × 16 complex multiplier implemented in 1 micron MODFET technology (courtesy Honeywell Sensors and Signal Processing Laboratory)

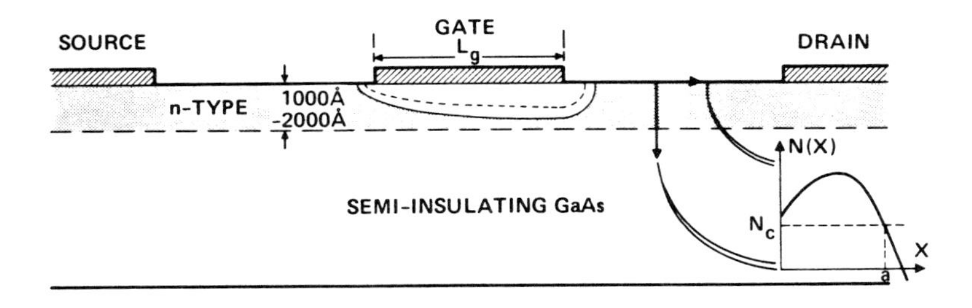

Figure 7.4 Cross-section of GaAs MESFET

earlier depletion-mode only processes were typically unable to realize acceptable A/D converters because of poor threshold uniformity and high power dissipation. A cross-section of a typical GaAs MESFET appears in Figure 7.4. The maximum f_T of commercially available MESFETs for integrated circuit applications is approximately 15 GHz, and the maximum f_T achieved by the best GaAs MESFET in a research laboratory environment is approximately 80 GHz [6]. Commercially available devices have gate lengths of approximately 1 micron, and the smallest laboratory devices have 0.1 micron gate lengths. The typical threshold voltage spread across a three-inch wafer is 30 – 40 mV, and local threshold voltage spread, which is important for comparator applications, is 15 – 20 mV [7]. Because of the Schottky barrier gate, the maximum gate-to-source voltage of a GaAs MESFET is typically limited to approximately 0.7 V.

Most commercially available GaAs MESFETs suffer from low-frequency hysteresis effects, which tend to complicate the design of high-frequency drain-source resistance of the FET when biased in the saturation region [8]. This drop in drain-source impedance is accompanied by a phase-shift, which creates a time varying offset voltage – or hysteresis – in the I–V relationship of the device. This time varying offset adds to the dc offset of the comparator, and reduces its achievable resolution. This effect can be reduced by improving circuit design procedures [9], and recent results have demonstrated that it can be eliminated altogether with improved processing techniques [10].

The low-frequency input referred noise of commercially available GaAs MESFETs is relatively high, with 1/f noise corner frequencies of approximately 100 MHz [11] and input referred noise voltages of several millivolts. In addition, low-frequency (100 Hz) drain current oscillations are often present in ion-implanted devices [12]. This presents another limitation to A/D performance, since the noise voltage of the comparator circuit, integrated over the bandwidth of interest, must be less than the quantization level. Fortunately, improved processing techniques are

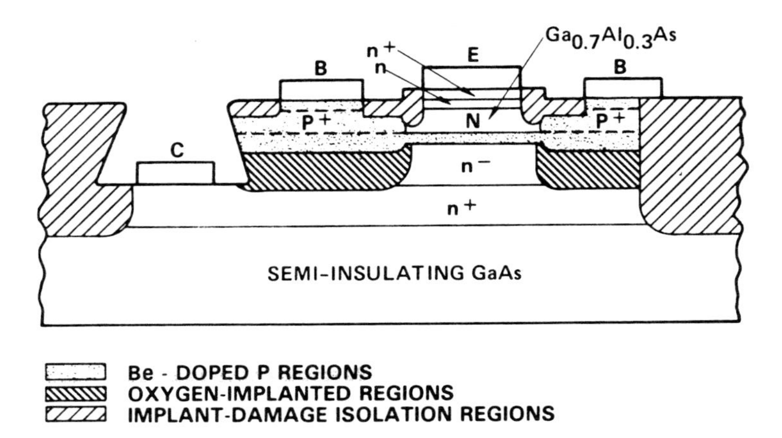

Figure 7.5 Cross–section of AlGaAs/GaAs HBT

currently being developed that minimize the low-frequency noise of GaAs MESFETs. State-of-the-art devices exhibit 1/f noise corner frequencies of less than 1 MHz [13].

GaAs MESFET technology is similar to MOS technology in its attractiveness for A/D applications. Its threshold voltage uniformity, hysteresis and 1/f noise make it suitable for high-speed low resolution applications, or for those A/Ds where some sort of digital error correction is possible for accuracy enhancement.

7.3.2 GaAs HBT technology

GaAs HBT technology is a relatively new, but extremely promising, technology for implementation of high-speed data converters. A cross-section of a typical AlGaAs/GaAs HBT appears in Figure 7.5. The device itself was proposed by Shockley, in his original transistor patent [14], and modern epitaxial growth techniques have made its realization possible. Since the emitter of the transistor possesses a wider band-gap than the the base, hole injection from the base into the emitter is suppressed by the valence band discontinuity, and the base can be arbitrarily doped without adversely affecting the transistor's current gain, or β.

This results in a device that possesses all the attractive features of a homojunction bipolar transistor, with the addition of some desirable features. Base resistance and high-level injection effects are very small because the base can be so heavily doped ($\approx 10^{19} cm^{-3}$). The f_T of a typical HBT is approximately 30 GHz, and results well over 100 GHz have been reported [15]. The base VBE uniformity of the transistors is extremely high, with offset voltages of fabricated comparators typically below 4 mV, and hysteresis is typically below 1 mV [16].

In addition, these devices suffer from few of the hysteresis effects seen in MESFET technology, and 1/f noise is substantially reduced. Finally, the key distances that govern HBT speed are vertical (base width) and controlled by epitaxial growth, rather than horizontal (gate length) and limited by lithographic considerations in the case of FET. Therefore, the f_T of HBT devices can be extended without an investment in expensive sub-micrometer lithography. The major drawback of HBT technology is its relative immaturity compared to MESFET technology. Optimised HBT processes are currently capable of fabricating circuits of at most a few hundred gates complexity. Typical HBT processes are inherently non-planer, so achievable yield is somewhat limited, although a 1K gate-array has been demonstrated [17].

Table 7.2 compares the relative performance of the two technologies. It is clear that, except for achievable complexity, HBT is the preferred technology for high-resolution high-speed converter implementations.

Table 7.2 Comparison of HBT and MESFET for A/D applications

Parameter	GaAs MESFET	GaAs HBT
MAXf_T	80 GHz	110 GHz
FOUNDRY f_T	15 GHz	15 GHz
CDS (CCS)	0.3 ff/μm	0.3 ff/μm
MAX COMPLEXITY	~10K GATES	~1K GATES
g_m	400 mS/mm	3000 mS/mm
V_{OFFSET}	15 mV	4 mV
I_{in}	1 μa	50 μa
1/f NOISE (1 GHz)	~5 mV	~1 mV
HYSTERESIS	~20 mV	~1 mV

7.4 A/D architectures for GaAs technology

The architectural considerations for A/D converters in GaAs technology are not substantially different form those of silicon technology. In general, the architecture of the A/D must be relative insensitive to expected device variations in order to maintain the desired resolution and sampling rate. The most promising architectures for GaAs A/D implementation are the flash, multi-pass, interleaved, and $\Sigma - \Delta$ approaches. Because of its superior uniformity and noise performance, HBT A/D's will require much less error correction than MESFET A/D's.

7.4.1 Flash A/D design

The flash A/D converter is perhaps the most straightforward approach for A/D realization. A block diagram of a 4-bit HBT flash A/D converter appears in Figure 7.6 [18]. This circuit exhibited a 25 GHz sample rate, and dc linearity compatible with 6-bit operation. The input voltage is compared simultaneously with 2^N reference voltages, which are generated by a resistor ladder voltage divider. The output of the comparators is a digital "thermometer" code, which is converted by digital logic into the final binary output.

The advantage of this approach is that the conversion can occur in a single clock cycle, and hence is very fast. In addition, the output vs. input transfer function can be guaranteed to remain monotonic, even under the most severe input conditions. This is a substantial advantage in high-speed implementation, where the comparators often have difficulty tracking the input signal.
Unfortunately, $2^N - 1$ comparators are required for an N–bit conversion, and the resulting power dissipation and complexity is often excessive. Furthermore, for each bit increase in resolution, the cumulative error in the comparator circuit – offset voltage, noise, hysteresis, etc. – must be cut in half. This results in very stringent comparator offset voltage and hysteresis requirements for resolutions above 4-bit. In addition, an anti-aliasing low-pass filter is required at the input of the convertor, which typically must meet very stringent performance criteria as the resolution increases. As the input bandwidth to the filter rises, the design of an acceptable anti-aliasing filter becomes almost as difficult as that of the A/D itself. Finally, the performance of a flash A/D converter often benefits from the use of a sample-and-hold (S/H) circuit at its input. This also complicates the design and fabrication of the circuit.

The implementation of the flash architecture in GaAs technology has been limited to the 4-bit level at this time [18, 20, 24], although a number

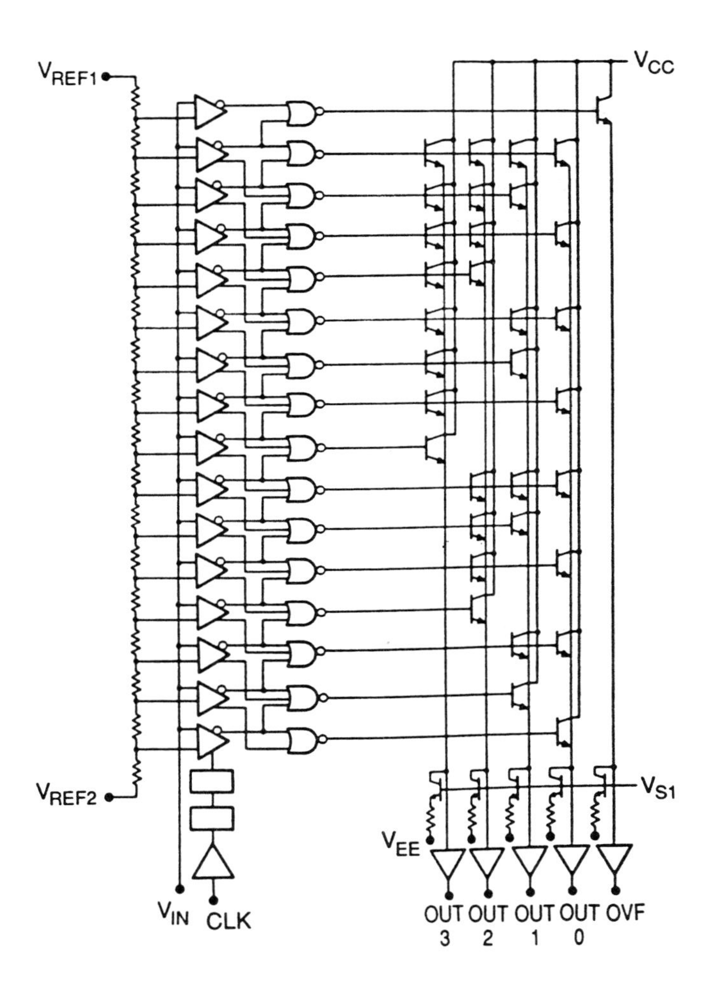

Figure 7.6 Schematic diagram of 4–bit flash A/D [18]

of groups are working on extending the resolution to 6 and 8-bits. Maximum sample rates have ranged from 1 to 2.5 GHz.

7.4.2 Multi-pass and interleaved A/D design

Multi-pass and interleaved A/D designs are general techniques capable of realizing a single high-precision, high-speed A/D with a number of lower precision, lower speed A/Ds. They are extremely attractive for GaAs

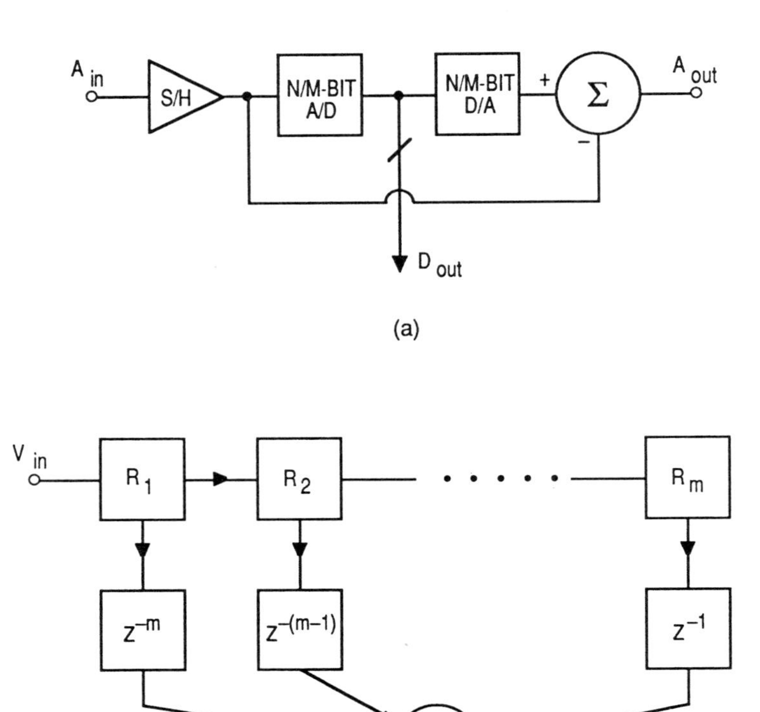

Figure 7.7 Block diagram of multi-pass A/D showing: (a) Module layout and (b) Overall schematic

technology, because the overall goal is usually to yield as high a performance converter as possible, at almost any cost.

The multi-pass A/D approach employs a number of low resolution A/Ds to create a single high-resolution A/D. Interleaved A/D converters increase the overall sample-rate by paralleling slower A/Ds, and then multiplexing the digital outputs. The former approach may be viewed simply as placing A/Ds in series, while the latter parallels the A/D's. A simplified block diagram of the multi-pass approach is shown in Figure 7.7, and the interleaved approach is shown in Figure 7.8. The

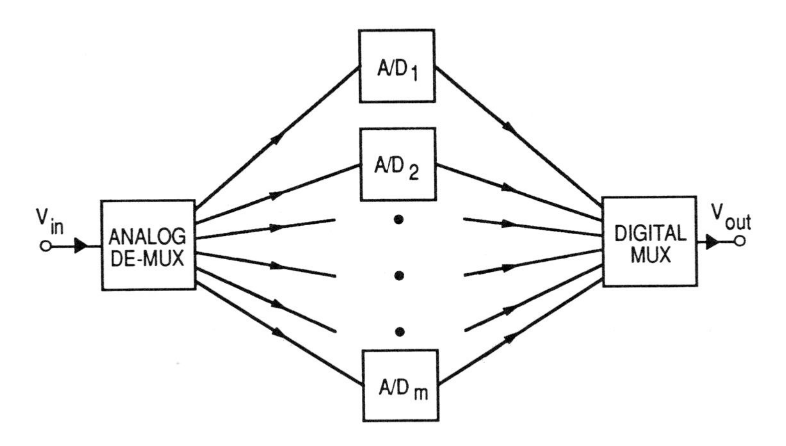

Figure 7.8 Block diagram of time interleaved A/D converter

multi-pass approach is particularly attractive for MESFET A/D converter realization, since it is unlikely that GaAs MESFET technology will be able to achieve a single high-resolution A/D in the near future. The multi-pass approach will allow a number of low-resolution A/Ds to realize a single high-resolution circuit, without paying a substantial penalty in speed.

The multi-pass A/D approach reduces the process of an N-bit A/D conversion to that of M N/M bit conversions. If each N/M bit conversion is realized with a flash A/D, then the number of comparators is reduced from approximately 2^N to $M2^{N/M}$, which can be a considerable saving. Furthermore, each N/M-bit A/D is only required to be accurate to N/M bits, so the resolution requirements of the N/M-bit A/D can be substantially eased. The overall A/D throughput is not significantly degraded from that of the low-resolution flash, since each stage of the converter can be processing a unique input. The conversion process is said to be "pipelined", and the overall latency is increased to M clock cycles, instead of a single cycle for the flash. Finally, some digital error correction can be applied at each of the M-stages, to further reduce the errors associated with the conversion process [21].

The disadvantages of the multi-pass approach include the requirement for accurate sample-and-hold, D/A and gain stages. However, these circuits are relatively simple compared to a full A/D, so the penalty is not excessive. This approach is being pursued by a number of groups in both silicon and GaAs technology, in an attempt to realize higher resolution and comparable speed compared to flash A/D.

A time interleaved approach has been employed successfully in the past in silicon technology, and was recently utilized in a hybrid GaAs/silicon A/D [22]. A very accurate analog de-multiplexer is required at the input, in order to convert a single high-speed analog input into M low-speed analog sampled-and-held outputs. These M outputs are then fed to M parallel A/D converters, whose outputs are multiplexed into a single-speed digital output. The M A/D converters can operate at a lower sample rate than the overall conversion rate. In the case of reference [22], they were realized using a silicon bipolar technology. The analog de-multiplexer at the input is required to run at the full conversion rate however, and was realized in GaAs technology. Problems can arise with this approach if mismatches occur in the M A/Ds, or in the analog de-multiplexer. These create harmonic distortion in the resulting output.

7.4.3 $\Sigma - \Delta$ *modulation A/D design*

The $\Sigma - \Delta$ architecture belongs to a class of *oversampling* A/D converters that trade-off sampling rate for resolution. The advantage of this approach is that it is quite insensitive to the typical limitations of GaAs MESFET technology, including comparator offset and hysteresis, and transfer function error. In principle, a low resolution A/D converter can be converted into a high-resolution circuit by sampling the input frequency at a much higher rate than the Nyquist rate. The basic approach is illustrated in Figure 7.9.

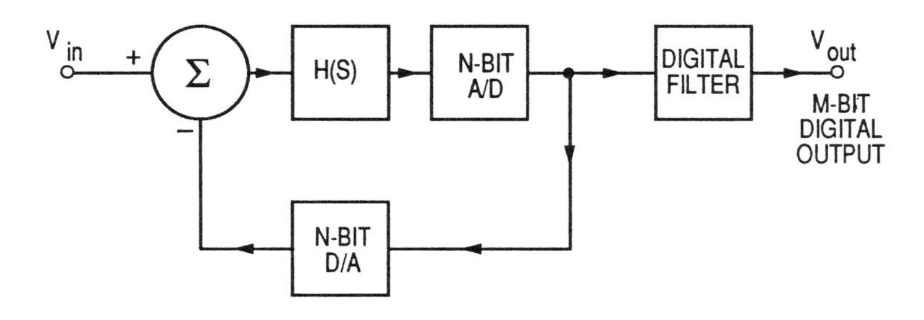

Figure 7.9 Block diagram of $\Sigma - \Delta$ A/D converter

By assuming that the A/D quantization is an additive, white, uncorrelated noise source, and that the gain of the A/D and D/A are close to one, then the quantization noise transfer function of the overall system is simply

$$H_e(s) \approx \frac{1}{1 + H(s)} \tag{8}$$

while the overall signal transfer function is

$$H_e(s) \approx \frac{H(s)}{1 + H(s)} \tag{9}$$

If H(s) is a low-pass filter with a high dc gain, then the signal transfer function is nearly unity, while the quantization noise is high-pass filtered. Thus, the high quantization noise of the low-resolution A/D has been filtered out of the frequency range of interest, and the resulting resolution is much higher. If H(s) is a first-order filter, then the improvement in resolution is 1.5 bits for each octave of over-sampling, and 2.5 bits per octave if H(s) is a second-order filter [23].

The output of the $\Sigma - \Delta$ is a high-speed digital signal that is fed into a low–pass digital filter, which filters the high–frequency quantization noise and produces a high–resolution digital output. Other advantages of this approach include the fact that an anti–aliasing filter and sample–and–hold are not needed, unlike the flash or multi–pass approaches.

This approach is particularly attractive for ultrahigh–speed GaAs technology, where a simple A/D converter, oversampled at an extremely high rate, could yield a high–resolution circuit with an impressive bandwidth.

7.5 D/A Architectures for GaAs Technology

The preferred architecture of GaAs D/A converters is quite similar to the architecture of high–speed silicon D/A converters, which is based on the weighted current switch approach. A schematic of the generic approach to the D/A design appears in Figure 7.10. In the binarily weighted current switch approach, each bit of a digital input word controls a current source. Each increasingly significant bit controls a current source that is double that of the previous bit. The various output currents are summed together at the output. This current is typically fed into a 50Ω resistor to produce a voltage output.

In the unarily weighted approach, the digital input word is first converted to a "thermometer" code, as shown in Figure 7.11. Each output of the thermometer code then controls an unique current source. All the current sources are equal valued. The advantage of this approach is the low "glitch" energy that typically results; for example, the transition from "10000" to "01111" results in switching only one current source, instead

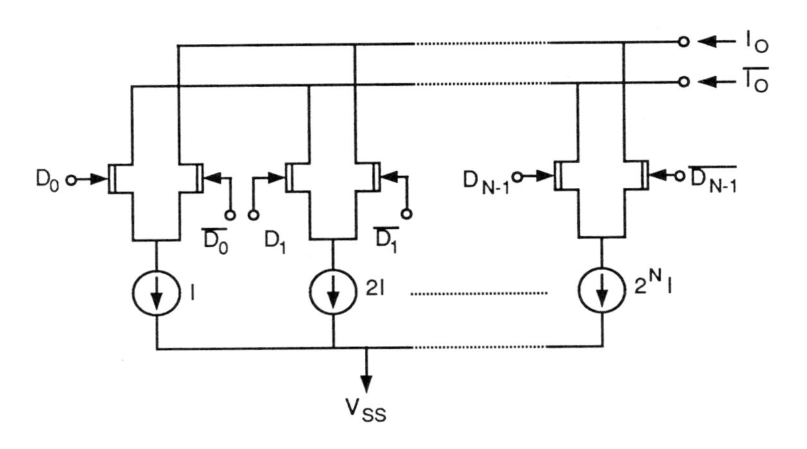

Figure 7.10 Schematic diagram of binarily weighted current-switch D/A converter

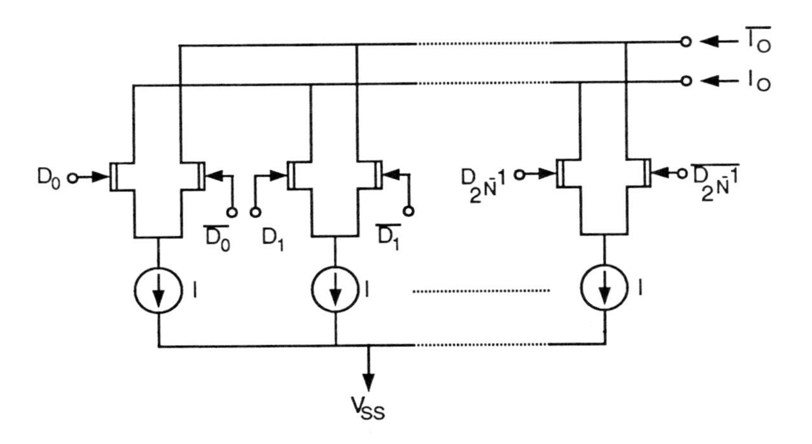

Figure 7.11 Schematic diagram of unarily weighted current-switch D/A converter

of five. Thus, the error associated with major carry transitions is substantially reduced.

Other approaches to the design of high-speed DACs are also attractive. The "classic" R-2R ladder configuration is well-established in silicon bipolar technology. In conjunction with a current summing approach, it can realise very high-speed D/A converters [25].

7.6 Conclusions

GaAs MESFET and HBT technology promise an increase in sample rate and achievable resolution for A/D and D/A conversion compared to silicon technology if a number of technological and design limitations can be overcome. GaAs MESFET technology is relatively mature, but currently suffers from poor threshold uniformity, l/f noise and hysteresis. GaAs HBT technology posses few of the limitations of MESFET technology, but the achievable level of integration is still quite low. In some ways, HBT technology represents the "ultimate" converter technology, if the obstacles to the realisation of complex circuits can be overcome. However, silicon bipolar and MOS technology continues to steadily improve, and the performance difference between GaAs and silicon technologies must be constantly re-evaluated in light of the continuing evolution of the technology.

7.7 Acknowledgements

The author would like to thank Professors Gabor Temes and Kenneth Martin of UCLA for valuable discussions. Also, the author would like to thank Dr. Paul Greiling and Dr. Ron Lundgren of the Hughes Research Laboratories for encouragement and support. The author would also like to acknowledge many valuable discussions with Mr. Joseph Jensen of the Hughes Research Laboratories.

7.8 References

1 F. A. Collins and C. J. Sickling: Properties of low-precision analog-to-digital converters, IEEE Trans. Aerosp. Electron. Syste., Vol AES-12 pp643-647, Sept. 1976

2 K. H. Konkle, C. E. Woodward and M. L. Naiman: A monolithic voltage comparator array for A/D converters, IEEE J. Solid-Stage Circuits, Vol 6, pp392-399, Dec. 1975

3 A. V. Oppenheim and R. W. Schafer: Digital Signal processing, Prentice-Hall, 1975

4 Aerospace Group, 500 MHz analog-to-digital converter, Final report, Hughes Aerocraft Co., Rep. P72-64, 1972

5 L. Curran: These chips are breaking the linear bottleneck, Electronics, pp61-64, Dec. 17, 1987

6 U. K. Mishra, R. S. Beaubien, M. J. Delaney, A. S. Brown and L. H. Hackett: MBE grown GaAs MESFETs with ultia-high g_m and f_T, Proc. Intl. Electron Devices Meeting, pp829-831, 1986

7 Y. D. Shen, M. R. Wilson, M. McGuire, D. A. Nelson and B. M. Welch: An ultra high performance manufacturable GaAs E/D process, IEEE GaAs IC Symposium - Technical Digest, pp125-128, 1987

8 L. E. Larson: An improved GaAs MESFET equivalent circuit model for analog integrated circuit applications, IEEE J. Solid-Stage Circuits, Vol SC-2, No. 4, pp567-574, Aug. 1987

9 N. Scheinberg: A hysteresis-free GaAs comparator, Abstract submitted to 1987 IEEE GaAs IC symposium

10 P. C. Canfield et al: Buried-channel MESFETs with improved small-signal characteristics, GaAs IC Symposium - Technical Digest, pp163-166, 1987

11 Triquint Semiconductor Short-Course Notes

12 N. Scheinberg: Designing high-speed operational amplifiers with GaAs MESFETs, Proc. IEEE Intl. Conf. Circuits and Systems, pp193-198, 1987

13 B. Hughes, N. G. Fernandez and J. M. Gladstone: GaAs FET's with a flicker noise floor below 1 MHz, IEEE Trans. Electron Devices, ED-34(4), pp733-741, 1987

14 H. Kroemer: Heterostructure bipolar transistors and integrated circuits, Proc. of the IEEE, Vol 70, No. 1, pp13-25, Jan. 1982

15 P. M. Asbeck et al: Heterojunction bipolar transistors for microwave and millimetre-wave integrated circuits, IEEE Trans. Microwave Theory and Techniques, MTT-35(12), pp1462-1470, 1987

16 K. Wang et al: High-speed, high-accuracy, voltage comparators implemented with GaAs/(GaAl)As heterojunction bipolar transistors, GaAs IC Symposium - Technical Digest, pp99-102, 1985

17 H. T. Yuan, H. D. Shin, W. V. McLevige and A. S. Hearn: GaAs heterojunction bipolar 1K gate array, ISSCC Digest of Technical Papers, 1984

18 K. Wang et al: A 4-bit quantizer implemented with AlGaAs/GaAs heterojunction bipolar transistors, IEEE GaAs IC Symposium - technical Digest, pp83-86, 1987

19 V. W-K Shen and D. A. Hodges: A 60 nS Glitch-Free NMOS DAC, IEEE Intl. Solid-State Circuits Symposium, pp188-18, 19839

20 T. Ducourant et al: 3 GHz, 150 mW, 4 bit GaAs analogue to digital converter, IEEE GaAs IC Symposium - Technical Digest, pp209-212, 1986

21 S. H. Lewis and P. R. Gray: A pipelined 5-Msample/s 9-bit analog-to-digital converter, IEEE J. Solid-State Circuits, SC-22(6), pp954-961, 1987

22 K. Poulton, J. J. Corcoran and T. Hornak: A 1 GHz 6-bit ADC system, IEEE J. Solid-State Circuits, SC-22(6), pp962-970, Dec. 1987

23 Y. Matsuya et al: A 16-bit oversampling a-to-d conversion technology using triple integration noise shaping, IEEE J. Solid-State Circuits, SC-22(6), pp921-929, 1987

24 J. Kleks et al: A 4-bit single chip analog-to-digital converter with a 1.0 GHz analog input bandwidth, GaAs IC Symposium - Technical Digest, pp79-82, 1987

25 R. Plassche et al: Dynamic element matching for high accuracy monolithic D/A converters, IEEE J. Solid-State Circuits, SC-14, No. 3, pp552-556, 1979

Chapter 8

Low Noise Oscillators

J. K. A. Everard

8.1 Introduction

The oscillator in communication and measurement systems, be they radio, coaxial cable, microwave, satellite, radar or optical fibre, defines the reference signal onto which modulation is coded and later demodulated. It is therefore essential to produce as pure a signal as possible to obtain low noise transmission and reception. This chapter will describe the theory required to design low noise oscillators operating from low to microwave frequencies. It will show that there is an optimum coupling coefficient between the resonator and the amplifier to obtain low noise and that this optimum is dependent on the definitions of the oscillator parameters. Low noise oscillator circuits which use inductor/capacitor, Surface Acoustic Wave (SAW) and transmission line resonators will be described in detail showing both the theoretical design and experimental results. It will be shown that the theory accurately predicts the noise performance and optimum operating conditions. Techniques to tune these oscillators while maintaining optimum noise performance will be described.

GaAs offers the designer the opportunity to design complete oscillators, including the resonator, on a microwave integrated circuit up to frequencies over 30 GHz and higher frequencies for discrete devices such as Gunn devices, HEMT's and HBT's. This is because the substrate is semi–insulating enabling the design of inductors, gyrators [13] and transmission line resonators to be fabricated on the substrate with 'Quality factors' (Q's) from 10 to 100. It is therefore possible to design a complete front end including the filtering all on a single integrated circuit (MMIC).

Initially one might imagine that because of the low noise figures obtainable from GaAs devices that they would offer better noise performance, but this is not the case due to the effect of transposed Flicker noise which is often 30 dB higher in GaAs than in silicon. This low frequency noise is modulated onto the carrier greatly degrading the noise performance of the oscillator. The accurate measurement and reduction of noise, especially Flicker noise which is high in GaAs MESFET's, will be discussed and techniques for reduction will be described as Flicker noise reduction is essential if the full potential of GaAs is to be realised.

The design of oscillators using GaAs monolithic microwave integrated circuits will be discussed.

8.2 Oscillator Noise Theories

The model chosen to analyse an oscillator is extremely important because the conclusions drawn can often vary. For this reason two models will be presented which in each case can produce different results as well as improving the understanding of the basic model. Both an equivalent circuit model and a block diagram model will be described. We will start with the equivalent circuit model originally used by the author to design oscillators with the potential for higher efficiency [18,19] and easy analysis.

8.2.1 Thermal Noise

8.2.1.1 Equivalent Circuit Model

The model is shown in Figure 8.1 and consists of an amplifier with two inputs with equal input impedance, one for noise (V_{in2}) and one as part of

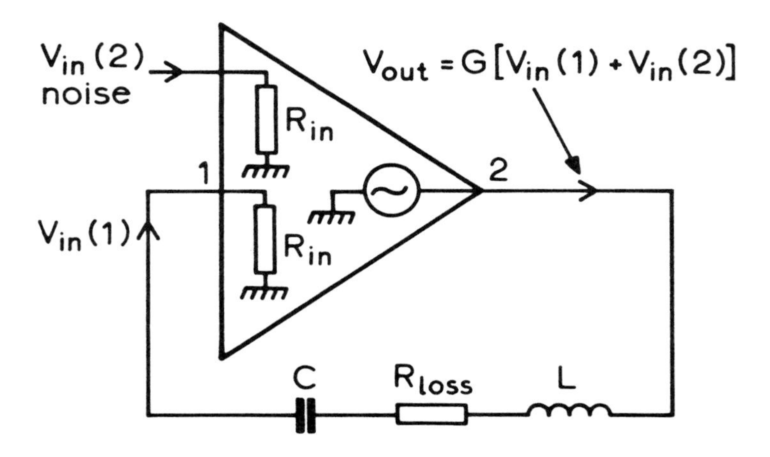

Figure 8.1 Equivalent circuit model of oscillator

the feedback resonator. The feedback resonator is modelled as a series inductor capacitor circuit with an equivalent loss resistance R_s which defines the unloaded Q (Q_o) of the resonator as $\omega l/rs$.

The operation of the oscillator can best be understood by injecting white noise at the input V_{in2} and calculating the transfer function while incorporating the usual boundary condition of $G\beta_o = 1$ where G is the limited voltage gain of the amplifier when the loop is closed and β_o is the feedback coefficient at resonance where $f_o = 1/2\pi\sqrt{LC}$.

The amplifier model had zero output impedance a known input impedance and a resonant positive feedback network. The zero output impedance of the amplifier was used because the design of highly power efficient oscillators was of interest. This also reduced the pulling effect of the load. The zero (low) output impedance of the amplifier is achieved by using a switching output stage. In fact the same theory and conclusions can be obtained for oscillators with a finite output impedance so for convenience this will be left at zero.

Input V_{in2} is used at the input of the amplifier to model the effect of noise. In a practical circuit the noise would come from the amplifier. The noise voltage V_{in2} is assumed to be added at the input of the amplifier and was dependent on the input impedance of the amplifier, the source resistance presented to the input of the amplifier and the noise figure of the amplifier. In this analysis, the noise figure under operating conditions, which takes into account all these parameters, was defined as F. This varies from amplifier and will be discussed later.

The circuit configuration is very similar to an operational amplifier feedback circuit and therefore the voltage transfer characteristic can be derived in a similar way

$$V_{out} = G(V_{in2} + V_{in1}) = G(V_{in2} + \beta V_{out}) \tag{8.1}$$

Where G is the voltage gain of the amplifier, β is the voltage feedback coefficient between nodes 1 and 2 and V_{in2} is the input noise voltage.

The voltage transfer characteristic is therefore:

$$\frac{V_{out}}{V_{in}} = \frac{G}{1 - \beta G} \tag{8.2}$$

By considering the feedback element between nodes 1 and 2, the feedback coefficient β is derived as:

$$\beta = R_{in}/[R_{loss} + R_{in} + j\ (\omega L - 1/\omega c)] \tag{8.3}$$

Where ω is the angular frequency.

As the loaded Q is $Q_L = \omega_o L/(R_{loss}+R_{in})$ and the unloaded Q is $Q_o = \omega_o L/R_{loss}$, the feedback coefficient at resonance is

$$\beta_o = R_{in}/(R_{loss} + R_{in}) = (1-Q_L/Q_o) \tag{8.4}$$

This shows that as the loaded Q is increased towards the unloaded Q the insertion loss of the resonator increases towards infinity

$$\text{for } \Delta\omega/\omega o \ll 1, \qquad (\omega L - 1/\omega c) = \pm 2\Delta\omega L \tag{8.5}$$

$$\beta = (1 - Q_L/Q_o) \;/\; (1 \pm 2jQ_L\Delta f/f_o) \tag{8.6}$$

Where f_o is the centre frequency and Δf, is the offset frequency from the carrier, in hertz.

As the noise of interest occurs within the boundaries of $Q_L\Delta f/f_o \ll 1$ equation 8.6 simplifies to:

$$\beta = (1 - Q_L/Q_o)\;(1 \pm 2jQ_L\Delta f/f_o) \tag{8.7}$$

This approximation is acceptable because the noise of an oscillator has reduced to the background noise floor when $\Delta\omega$ = the 3 dB point of the resonator.

The voltage transfer characteristic of the closed loop is therefore:

$$V_{out}/V_{in2} = \frac{G}{[1 - G(1 - Q_L/Q_o)\;(1 + 2jQ_L\Delta f/f_o)]} \tag{8.8}$$

At resonance Δf is zero and Vout/Vin2 is very large. The output voltage is defined by the maximum swing capability of the amplifier and the input voltage is noise. The denominator of equation 8.8 is approximately zero, therefore

$$G = 1/\;(1 - Q_L/Q_o) \tag{8.9}$$

This is effectively saying that at resonance the amplifier gain is equal to the insertion loss ($G\beta_o = 1$). The gain of the amplifier is now fixed by the operating conditions. Therefore:

$$V_{out}/V_{in2} = G/\;(\pm 2jQ_L\Delta f/f_o) \tag{8.10}$$

It should be noted however that this equation does not apply very close to carrier where Vout approaches and exceeds the peak voltage swing of the amplifier. As $V_{in} = \sqrt{FkTR_{in}}$ which is typically 10^{-9} in a 1 Hz bandwidth and V_{out} is typically 1 volt, G is typically 2, then for this criteria to apply $Q_L\Delta f/f_o \gg 10^{-9}$. For a Q_L of 50 $f_o = 10^9$ errors only start to occur at frequency offsets closer than 1 Hz to carrier.

As the sideband noise in oscillators is usually quoted in terms of power not voltage it is necessary to define output power.

It is also necessary to decide where the limiting occurs in the amplifier. In this instance limiting is assumed to occur at the output of the amplifier as this is the point where the maximum power is defined by the power supply. In other words, the maximum voltage swing is limited by the power supply.

To investigate the ratio of the noise power in a 1Hz sideband to the total output power, the voltage transfer characteristic can now be converted to a characteristic which is proportional to power. This is achieved by taking the square of the total output voltage. Only the power dissipated in the oscillating system and not the power dissipated in the load was of interest. The low output impedance of the amplifier ensured that the load did not affect the noise performance.

The input noise power in a 1Hz bandwidth is FkT (k is Boltzmann's constant and T is the operating temperature), where kT is the noise power that would have been available at the input had the source impedance been equal to the input impedance (R_{in}). F is the operating noise figure which includes the amplifier parameters under the oscillating operating conditions. This includes such parameters as source impedance. The dependence of F with source impedance will be discussed later. The square of the input voltage is therefore $FkTR_{in}$.

It should be noted that the noise voltage generated by the series loss resistor in the tuned circuit was taken into account by the noise figure of the amplifier. The important noise was within the bandwidth of the tuned circuit allowing the tuned circuit to be represented as a resistor over most of the close to carrier performance. In fact the sideband noise power of the oscillator reaches the background level of noise around the 3 dB point of the resonator.

The noise power is usually measured in a 1 Hz band-width. The square of the output voltage in a 1Hz bandwidth at a frequency offset Δf is:

$$(V_{out}\Delta f)^2 = \frac{G^2FkTR_{in}}{4QL^2}\left(\frac{f_o}{\Delta f}\right)^2 \qquad (8.11)$$

As this is a linear theory, the sideband noise is effectively amplified narrow band noise. To represent this as an ideal carrier plus sideband noise, the signal can be thought of as a carrier with a small perturbation rotating around it [93,99] Figure 8.2. Note that there are two vectors rotating in opposite directions, one for the upper and one for the lower sideband. The sum of these vectors can be thought of as containing both AM and PM. The component along the axis of the carrier vector being AM noise and the component orthogonal to the carrier vector being phase

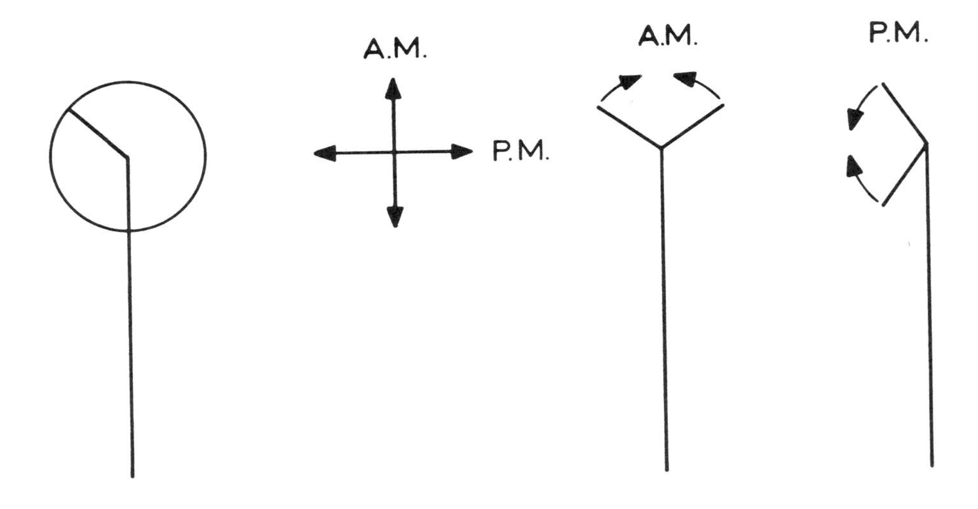

Figure 8.2 Signal representation of AM and PM noise

noise. Phase modulation can be thought of as linear modulation as long as the phase deviation is considerably less than 0.1 rad.

Eqn.8.11 accurately describes the noise performance of an oscillator which uses automatic gain control (AGC) to define the output power. However, the theory would only describe the noise performance at offsets greater than the AGC loop bandwidth. Although linear, this theory can incorporate the nonlinearities, i.e., limiting in the switching amplifier, by modifying the absolute value of the noise as described by authors [99,107–109,122,123,125]. The absolute value of the sideband noise often appears to be different from the theoretical predictions although the shaping of the noise appears to be correct.

If the output signal amplitude is limited with a hard limiter, the AM component would disappear and the phase component would be half of the total value shown in eqn. 8.11. This is because the input noise is effectively halved [122,123]. This assumes that the limiting does not cause extra components due to mixing. Limiting also introduces a form of coherence between the upper and the lower sideband which has been defined as conformability [99].

The output noise performance is usually defined as a ratio of the sideband noise power to the total output power.

If the total output voltage is $V_{outmaxrms}$ the ratio of sideband phase noise, in a 1Hz bandwidth, to total output will be $L_{(fm)}$ where:

$$L_{(fm)} = \frac{(V_{out}\Delta f)^2}{(V_{outmaxrms})^2} = \frac{G^2FkTR_{in}}{8QL^2}\left(\frac{f_o}{\Delta f}\right)^2 \quad (8.12)$$

When the total RF feedback power P_{rf} is defined as the power in the oscillating system, excluding the losses in the amplifier, and most of the power is assumed to be close to carrier, then P_{rf} is limited by the maximum voltage swing at the output of the amplifier and the value of $R_{loss} + R_{in}$.

$$P_{rf} = \frac{(V_{outmaxrms})^2}{R_{loss} + R_{in}} \quad (8.13)$$

$$L_{(fm)} = \frac{G^2FkTR_{in}}{8QL^2P_{rf}(Rloss + Rin)}\left(\frac{f_o}{\Delta f}\right)^2 \quad (8.14)$$

$$As\ G = (Rloss + Rin)/Rin \quad (8.15)$$

$$L_{(fm)} = \frac{GFkT}{8QL^2P_{rf}}\left(\frac{f_o}{\Delta f}\right)^2 \quad (8.16)$$

Eqn. 8.16 shows that $L_{(fm)}$ is inversely proportional to P_{rf} and that better noise is thus obtained for higher feedback power. This is because the absolute value for the sideband power does not vary with the total feedback power. It should be noted that P_{rf} is the total power in the system excluding the losses in the amplifier, from which:

$$P_{rf} = (DC\ input\ power\ to\ the\ system)\ x\ efficiency \quad (8.17)$$

For minimum noise the noise figure (F), and the value of $G/Q^2{}_L$ should be as small as possible. It should be noted, however, that F, G and Q_L are directly related to each other and thus cannot be varied independently.

8.2.1.2 Optimisation For Minimum Phase Noise

Eqn. 8.16 is now examined to see which parameters are interrelated so that the equation can be optimised for minimum phase noise. At resonance the gain of the amplifier is $1/\beta_o$ and Δf is 0, then:

As $$G = 1/(1-Q_L/Q_o) \tag{8.18}$$

$$L_{(fm)} = \frac{FkT}{8Q_o{}^2(Q_L/Q_o)^2(1-Q_L/Q_o)P_{rf}} \left(\frac{f_o}{\Delta f}\right)^2 \tag{8.19}$$

This noise equation is minimum when

$$\frac{dL_{fm}}{dQ_L/Q_o} = 0 \tag{8.20}$$

Minimum noise therefore occurs when $Q_L/Q_o = 2/3$. To satisfy $Q_L/Q_o = 2/3$, the voltage insertion loss of the resonator is 1/3 which sets the amplifier voltage gain to 3.

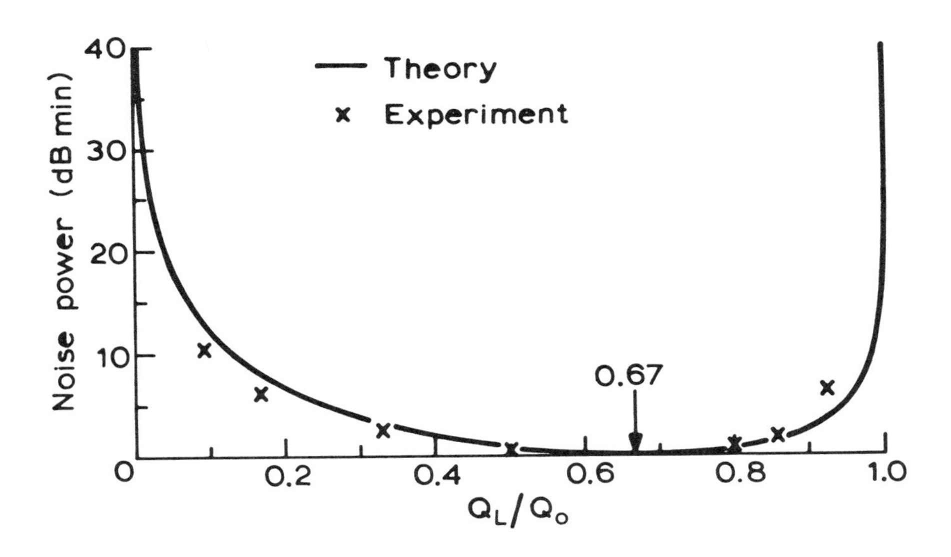

Figure 8.3 Variation of phase noise with Q_L/Q_o

A graph of the noise degradation with Q_L/Q_o is shown in Figure 8.3. An experimental oscillator was built at 1 MHz to test the theory and the results, which show close correlation, are also shown.

To summarise:

(a) the sideband phase noise to carrier ratio under optimum operating conditions, when $Q_L/Q_o = 2/3$, is therefore

$$L_{(fm)} = \frac{27FkT}{32Q_o{}^2P_{rf}} \left(\frac{f_o}{\Delta f}\right)^2 \quad (8.21)$$

(b) Q_o and P_{rf} should be as large as possible, and the noise Figure F should be as small as possible.

The constant 27/32 may be modified if the action of limiting causes noise from other frequencies to be mixed to the operating frequency.

The noise sidebands fall off at 6 dB per octave. This can be modified to incorporate the flicker characteristic as described in Section 8.2.2.

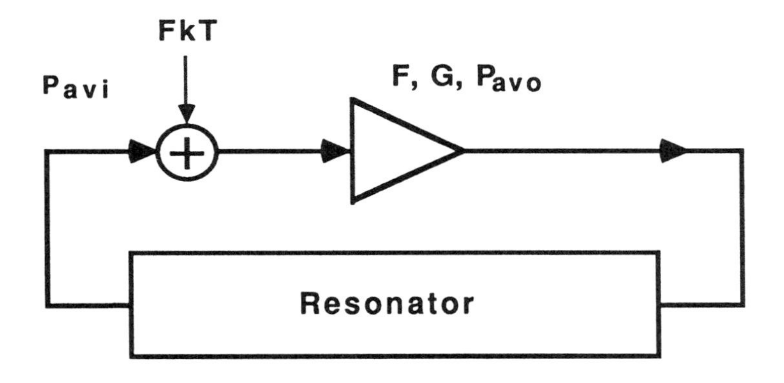

Figure 8.4 Block model of oscillator

8.2.1.3 Block Diagram Model

Another more general model is shown in Figure 8.4. Here it is assumed that the thermal noise can be modelled as a single noise source (FkT) at the input of the amplifier, where F is the noise factor. The amplifier and resonator are assumed to have both inputs and outputs matched to Z_o. If we set $G=1/S_{21}(0)^2$, which is the steady state condition for oscillation, where it can be shown that $S_{21}(0) = (1-Q_L/Q_o)$ and $S_{21}\Delta f$ = $S21(0)(1+2jQ_L\Delta f/f_o)$, then the noise power spectrum at the output of the amplifier is given by,

$$L_{(fm)} = \frac{FkTG}{4QL^2P} \left(\frac{f_o}{\Delta f}\right)^2 \quad (8.22)$$

where P is the power available at the output of the oscillator $(V_{out})^2/4R_{out}$.

If the output signal amplitude is limited with a hard limiter, the AM component will disappear and the phase component will be half of the total value shown in equation. This is because the input noise is effectively halved. It is assumed that the limiting does not cause extra components due to mixing. Limiting also introduces coherence between the upper and lower sideband when demodulated because phase modulation requires that both sidebands are rotating in opposite directions as shown in Figure 8.2. This has been described by Robbins [99] as Conformability in his book on phase noise in signal sources.

Therefore:

$$L_{(fm)} = \frac{FkTG}{8QL^2P} \left(\frac{f_o}{\Delta f}\right)^2 \tag{8.23}$$

It is extremely important to use the correct definition of power (P), as this affects the values of the parameters required to obtain optimum noise performance.

If the power is defined as the power available at the input of the amplifier P_{avi} then the gain (G) will disappear from equation 8.23. At first glance it would appear that minimum noise occurs when Q_L is made large and hence tends to Q_o. However this would require that the amplifier gain and hence output power would tend to infinity. If the power is taken as the power available at the output of the amplifier then equation 8.24 can be derived where it is assumed that $G = (1-Q_L/Q_o)^2$.

$$L_{(fm)} = \frac{FkT}{8Qo^2\ (Q_L/Q_o)^2\ (1-Q_L/Q_O)^2\ P_{avo}} \left(\frac{f_o}{\Delta f}\right)^2 \tag{8.24}$$

The minimum of equation 8.24 occurs when $Q_L/Q_o=1/2$. It should be noted that P_{avo} is constant and not related to Q_L/Q_o. The power available at the output of the amplifier is different from the power dissipated in the oscillator, but by chance is close to it. Parker [74] has shown a similar optimum for SAW oscillators.

This should be compared with the result obtained when the equivalent circuit model was used where the power is defined as the total power dissipated in the resonator and the impedances of the amplifier (P_{rf}). This would be useful if highly efficient oscillators were required [19]. Then the optimum condition occurred when $Q_L/Q_o = 2/3$. The equation for the noise performance which is derived in Section 8.2.1.1. becomes:

$$L_{(fm)} = \frac{FkT}{8Qo^2\ (Q_L/Q_o)^2\ (1-Q_L/Q_O)\ P_{avo}} \left(\frac{f_o}{\Delta f}\right)^2 \tag{8.25}$$

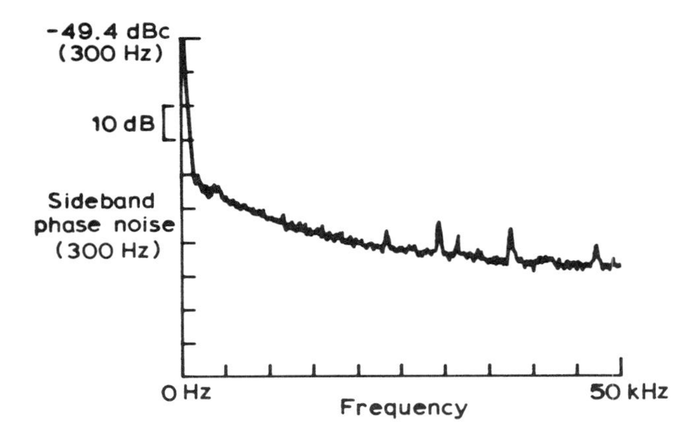

(a) Noise performance (Carrier power level = 23.4 dBm, sideband noise at 25KHz offset = –106.4 dBm (1 Hz) – 6dB for SSB = –135.8 dBc.Hz)

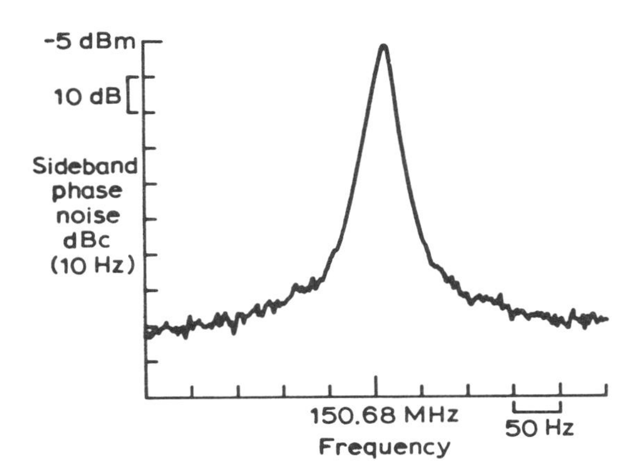

(b) Close to carrier noise performance (Sideband noise at 250 Hz offset = –92.5 dBm(1Hz) = –87.5 dBc/Hz)

Figure 8.9 Oscillator phase noise against frequency

measurement, meaning that the actual noise performance was possibly better than the measured performance. The oscillator amplifier was however very inefficient (2%) as the DC input power was 120 mW.

The close to carrier noise performance of the oscillator is shown in Figure 8.9.b. and demonstrates a sideband noise to carrier level in a 1 Hz bandwidth at 250 Hz offset of 87 dBc for a carrier frequency of 150 MHz. This was approximately 9 dB higher than would be expected if the oscillator noise sidebands increased at 20 dB per decade from the measurement at 25 kHz and was due to Flicker noise. The comparison of Flicker noise in silicon and GaAs oscillators will be shown in Section 8.4.2.3. Note Figure 8.9.b. shows the noise in a 10 Hz bandwidth. The noise in a 1 Hz bandwidth is 10 dB lower.

This design can be scaled up to microwave frequencies and built on GaAs MMIC's as inductors and capacitors can be put directly onto the substrate. An equivalent circuit using a balanced transmission line was built at UHF. At resonance a balanced line as shown in Figure 8.10, can

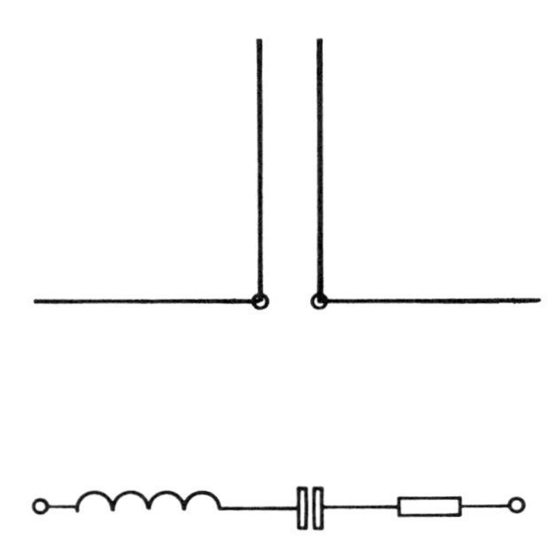

Figure 8.10 Equivalent circuit for balanced transmission line

be modelled as a series resonant circuit. If standard 300 ohm feeder is used an unloaded Q of 200 is possible. The noise performance of an oscillator using a balanced transmission line [20], close to the predicted value, is shown in Figure 8.11.

This type of transmission line is adequate for demonstration purposes but is not feasible for practical purposes because of the radiation fields, size and lack of screening. A variety of transmission line resonators have

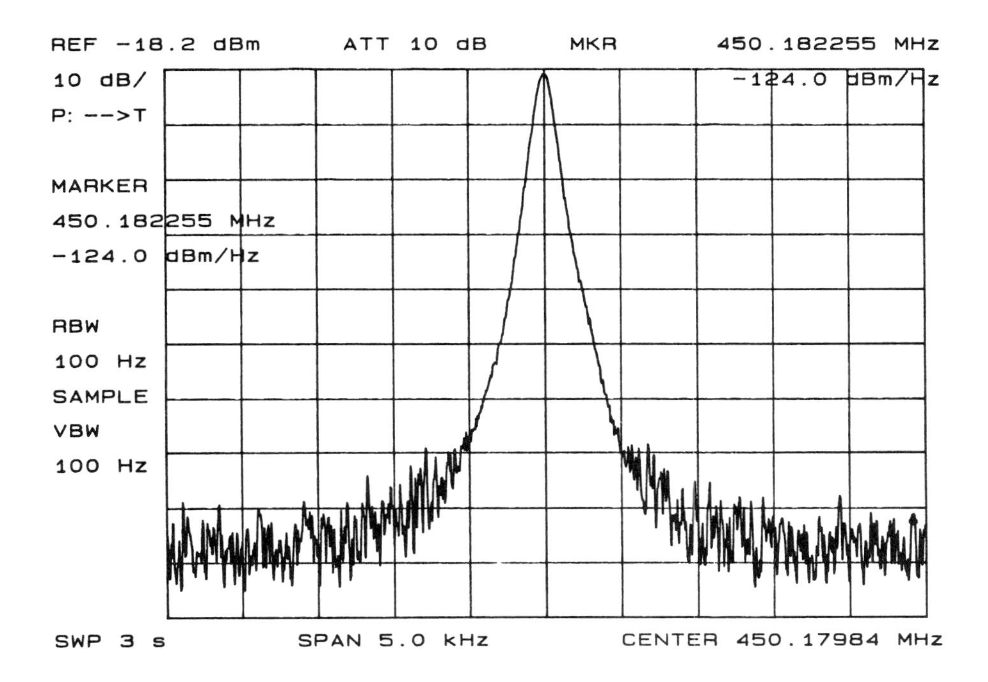

Figure 8.11 Noise performance of UHF oscillator

therefore been developed, some for direct use on Microstrip in hybrid or MMIC circuits with unloaded Q's up to 80 and some compact helical resonators with Q's over 600. These are described in the next section.

8.4.2. Transmission Line Resonators And Oscillators

At higher frequencies from above L band for hybrid circuits and 5 GHz for GaAs MMIC's, transmission lines offer a very suitable form of resonator which can be incorporated as part of the circuit board or substrate. A novel type of resonator has been developed at King's College London [21].

The resonator consists of a low–loss transmission line (length L) and two shunt reactances of normalised susceptance jX and is shown in Figure 8.12. This is very similar to the Fabry Perot resonator in optics where the reactances are equivalent to mirrors. If the shunt element is a capacitor of value C then $X = 2\pi fCZ_o$. The value of X should be the effective susceptance of the capacitor as the parasitic series inductance is usually significant. These reactances can also be inductors, an inductor and capacitor, or shunt stubs. If $Z_{oT} = Z_o$, where Z_{oT} is the resonator line

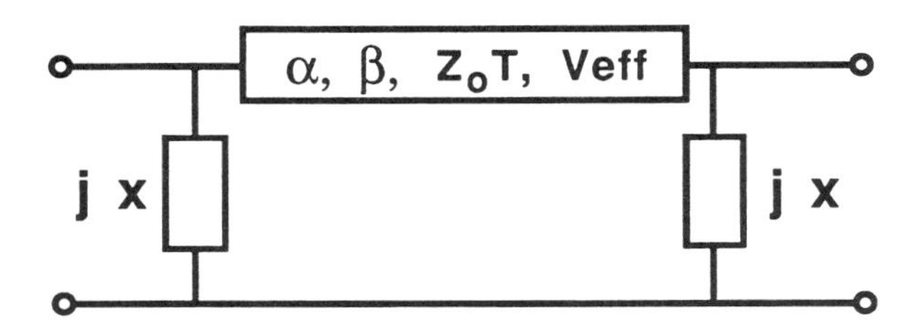

Figure 8.12 Transmission line resonator

impedance and Z_o is the terminating impedance, then S21 is given by the following equation,

$$S_{21} = 4\,\Gamma/\{(1+jX) - X^2(1-\Gamma^2)\} \tag{8.26}$$

where $\Gamma = \exp\{-(\alpha + j\beta)L\}$ (8.27)

α is the attenuation coefficient of the line

β is the phase constant of the line

For small αL (< 0.05) and $\Delta f/f_o \ll 1$, the following properties can be derived for the first resonant peak (f_o) of the resonator where $\Delta f = (f - f_o)$,

$$f_o = (v_{eff}/2L)\{1 + (1/\pi)\tan^{-1}(2/X)\}, \tag{8.28}$$

$$S_{21}(\Delta f) = S_{21}(0)/\{1 + j2Q_L(\Delta f/f_o)\} \tag{8.29}$$

$$S_{21}(0) = (1-Q_L/Q_o) = 1/\{1 + (\alpha L/2)X^2\} \tag{8.30}$$

$$Q_L = \pi S_{21}(0)X^2/4 \tag{8.31}$$

$$Q_o = \pi/2\alpha L \tag{8.32}$$

From equation 8.30 it can be seen that the insertion loss and the loaded Q factor of the resonator are interrelated. In fact as the shunt capacitors (assumed to be lossless) are increased the insertion loss approaches infinity and Q_L increases to a limiting value of $\pi/2\alpha L$ which we have defined as Q_o. It is interesting to note that when S_{21} = 1/2, $Q_L = Q_o/2$.

8.4.2.2 Slow Wave Helical Resonators

Mobile radios and other small systems often have very tight specifications on adjacent channel noise performance and as channel spacings get closer together further improvements are required. Conventional LC resonators often have low Q's which the screening can degrade. They are also more difficult to make above 1 GHz. Conventional transmission lines also take up considerable space at low microwave frequencies especially when using air as the dielectric. A form of helical transmission line resonator has been developed at King's College London which is both compact and as the screening is part of the resonator the Q is not degraded. It differs from conventional resonators in the form of the coupling. A typical 900 MHz resonator is shown in Figure 8.13 and consists of a helical transmission

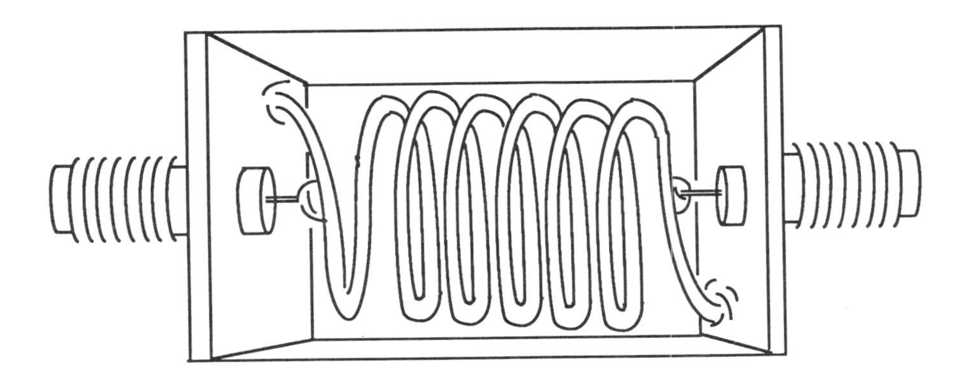

Figure 8.13 Helical Resonator

line wound between two SMA connectors with a portion of the coil connected to ground, This resonator has the same equivalent circuit as the transmission line resonator show in Figure 8.12 where the shunt capacitors are now inductors. The total length of a resonator is 30 mm, where the box is 12 mm wide and deep. The helix is 21 mm long close wound with an inside diameter of 4.5 mm and wire of 0.71 mm diameter. The total length of the helix unwound is 250 mm. This type of resonator can therefore be used from UHF frequencies to beyond X band and can be used to enhance the performance of GaAs devices where this is required. Varactors can be used to tune the resonator and this is currently being investigated.

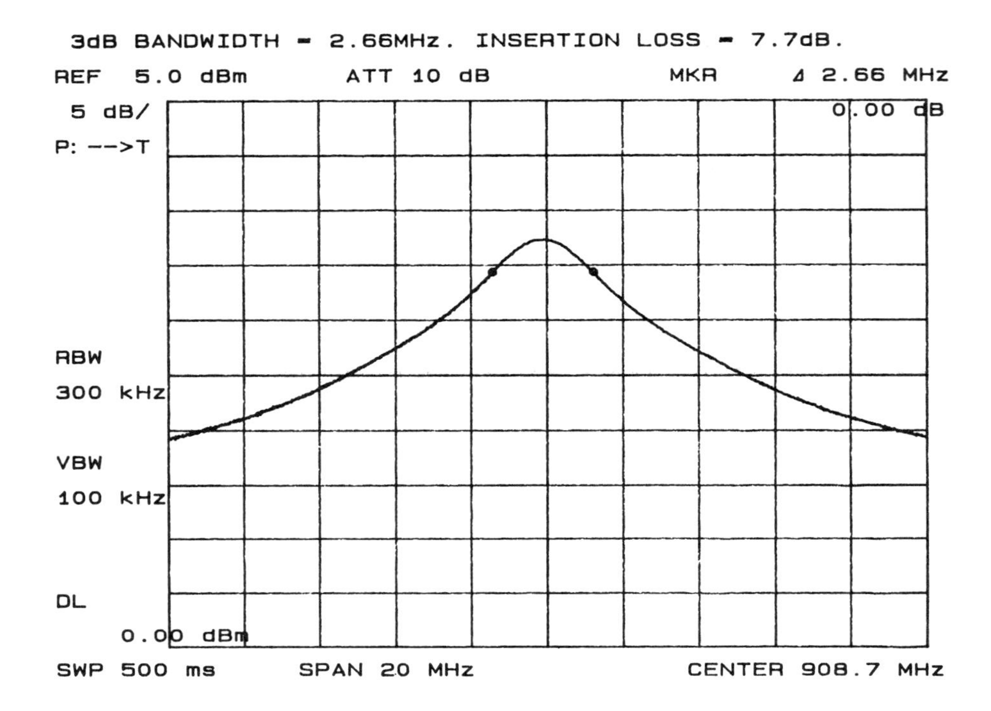

Figure 8.14 Frequency response of Helical Resonator

A typical response for a 900 MHz resonator is shown in Figure 8.14. The loaded Q is over 300 and the unloaded Q is over 600 where the insertion loss and the Q are related by the expression $S_{21}(0) = 1/(1-Q_L/Q_o)$.

The equations which describe this resonator are identical to those used for the 'Fabry Perot' resonator described in Section 8.4.2 except for the fact that X now becomes $-Zo/2\pi fl$ where l is the shunt inductance . As the Q becomes larger the value of the shunt l becomes smaller eventually becoming rather difficult to realise. This can be improved by using impedance transformers where these transformers can consist of a combination of series L's or C's and shunt L's or C's. Note the series L's would be part of the helix. This will often allow more convenient component values. The helical resonators can be built using high T_c superconductors which are currently under investigation at King's College London. It would also be possible to put down thin film high T_c superconductors on GaAs substrates to improve the noise performance although the losses in the substrate would become important.

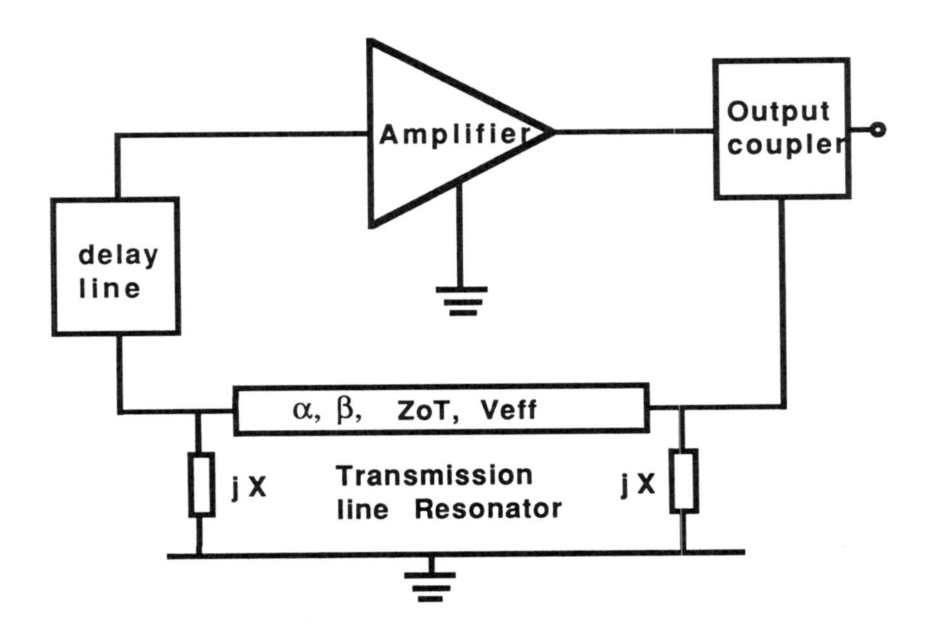

Figure 8.15 Block diagram of transmission line oscillator

8.4.2.3 L Band Oscillators

Two oscillators using microstrip transmission line resonators have been designed and built, one using a GaAs device and one using a silicon active device. The block schematic is shown in Figure 8.15 and the oscillator is shown in Figure 8.16. They have both been fabricated using microstrip techniques on RT Duriod (ϵ_r = 10.2).

The GaAs oscillator uses an AT 12535 MESFET (Id = 16 mA, Vds = 3 Volts) and the silicon based oscillator uses a bipolar NE 68135 (Ic = 30 mA, Vce = 7.5V).

Both oscillators use a 3 dB Wilkinson power splitter to deliver power to the external load. Phase compensation is accomplished by means of a short length of transmission line and is finely tuned using a trimming capacitor. The oscillation frequency is approximately 1.5 GHz and a L is found to be 0.019. The output power spectrum of the GaAs oscillator is given in Figure 8.17 and the silicon oscillator Figure 8.18.

Immediately it can be seen from Figures 8.17 and 8.18 that the GaAs oscillator has considerably higher noise sidebands due to the effect of Flicker noise. However the efficiency of the GaAs oscillator was found to

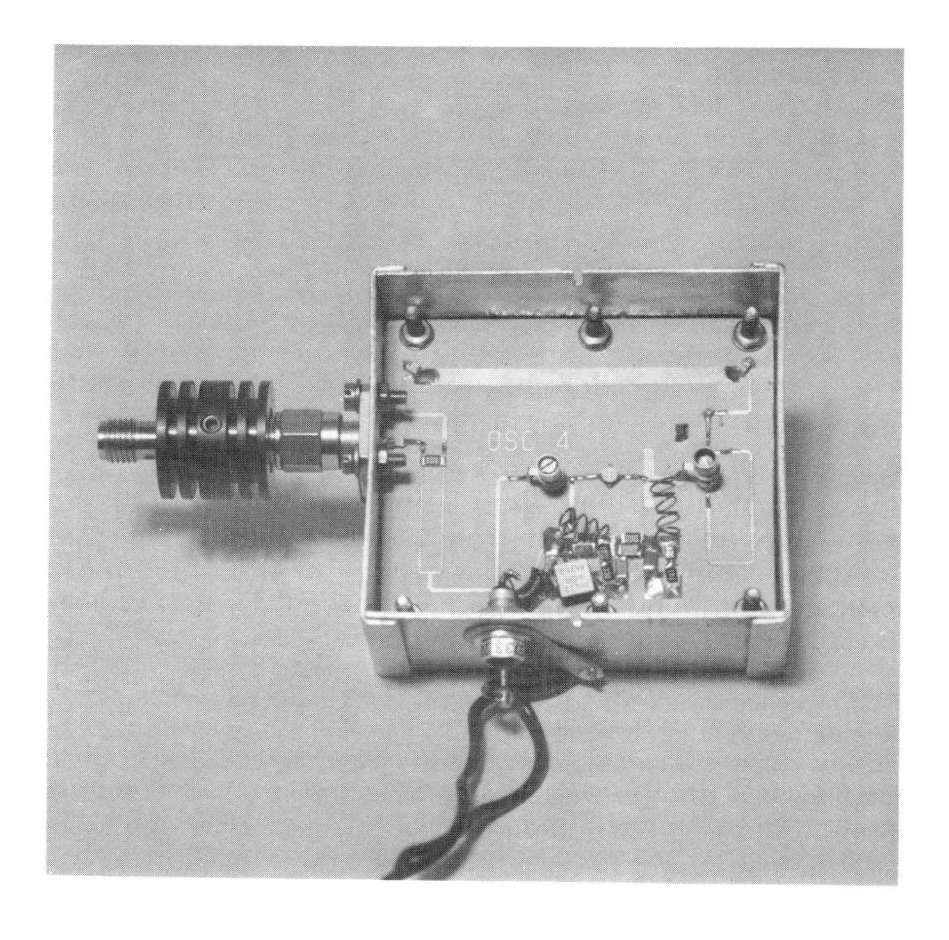

Figure 8.16 Photograph of transmission line oscillator

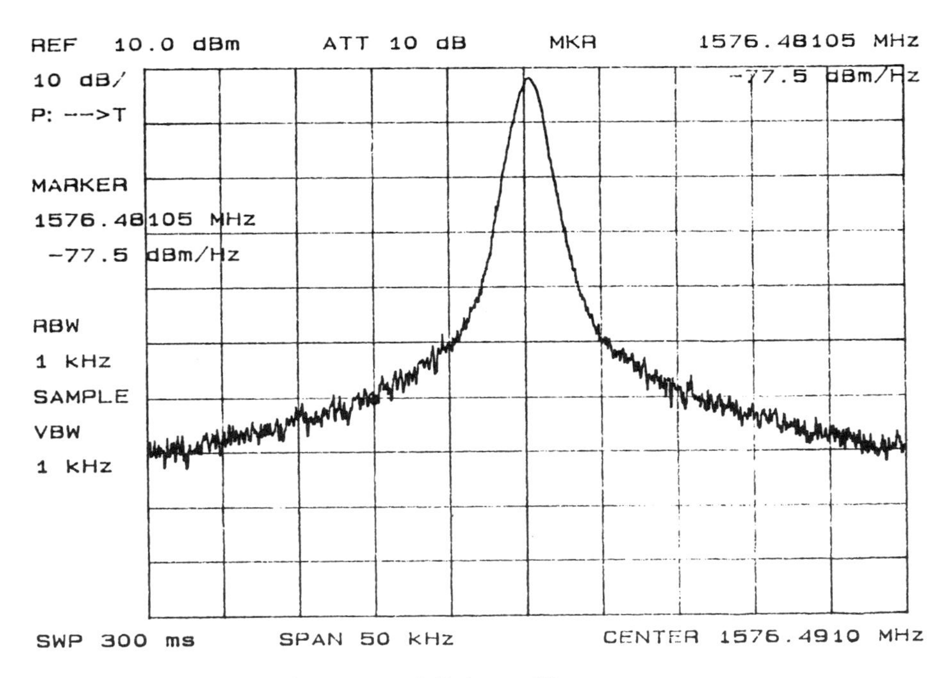

Figure 8.17 Noise performance of GaAs oscillator

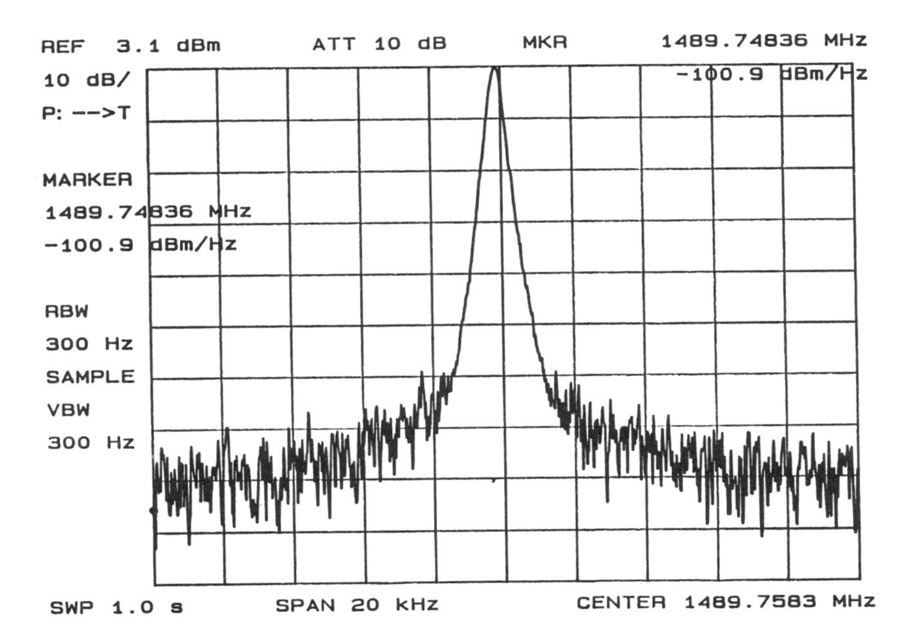

Figure 8.18 Noise performance of silicon oscillator

be over 14% whereas the silicon oscillator was about a percent. The absolute value of the noise power of the GaAs oscillator at 10 kHz offset is −77 dBm/Hz and at 20 kHz is 85.5 dBm/Hz which is almost 9 dB lower thereby showing a $1/f^3$ characteristic. Thus the oscillator is operating within the transposed Flicker noise corner. The silicon oscillator at 10 KHz offset is −100.9 dBm/Hz, some 24 dB lower than the GaAs oscillator and if it is measured at 5 kHz offset the noise power is 95.3 dBm/Hz which is 5.6 dB (nearly 6dB) higher which shows a $1/f^2$ characteristic therefore the Flicker noise corner is less than 5 kHz.,

The noise level of the silicon oscillator represents state of the art performance for an oscillator with a Q_o of 83. Better performance could be obtained with higher Q resonators (such as helical resonators) as the noise power at the output is proportional to $1/Q_o^2$. In these theories if the Flicker noise corner is ignored the output noise spectrum is independent of the oscillator power level. This means that the ratio of sideband noise to total output power improves linearly with power level. However if in practice the power level is increased by running the amplifier into saturation then non linear effects which change the effective noise figure need to be included. The results can therefore be quoted in terms of absolute noise power/Hz, $S_n(\Delta f)$. If the noise figure of the linear amplifier is assumed to be 3 dB then from equation 8.24, the theoretical limit of S_n(10 kHz) is approximately −102.6 dBm/Hz. The measured noise level at an offset frequency of 10 kHz is −100.9 dBm/Hz ± 1 dB.

It is therefore shown how important it is to reduce the effect of Flicker noise both in the device, bias circuit and resonator.

8.4.3 *Dielectric Resonator Oscillators*

The dielectric resonator (typically Barium Titanate) offers an excellent method for improving the phase noise of GaAs based transmission line oscillators due to their high Q, 40,000 at L band and over 1,000 at 20 GHz [1,3,7,9,10,12,22,23,28,32,41,43,47–50,53–55,80–84,121]. It consists of a sintered disc ceramic with a relative dielectric constant ϵ_r =30. There have been extensive publications in this area so only the salient points will be mentioned. GaAs dielectric resonator oscillators can be used to stabilise the local oscillator in satellite receivers where a 10 GHz GaAs Dielectric resonator oscillator with a mechanical tuning slug for centre frequency adjust can be used as the first local oscillator.

To be able to couple into this resonator the field pattern is required. This is shown in Figure 8.19.a,b,c where for the lowest order cylindrical $TE_{01\delta}$ there exists a circularly symmetrical E field and a solenoidal magnetic field. Coupling into these resonators initially appears simple, for example by placing the puck near a microstrip line as shown in Figure 8.19.c. However, it is found that this is a very lossy means of coupling due to the

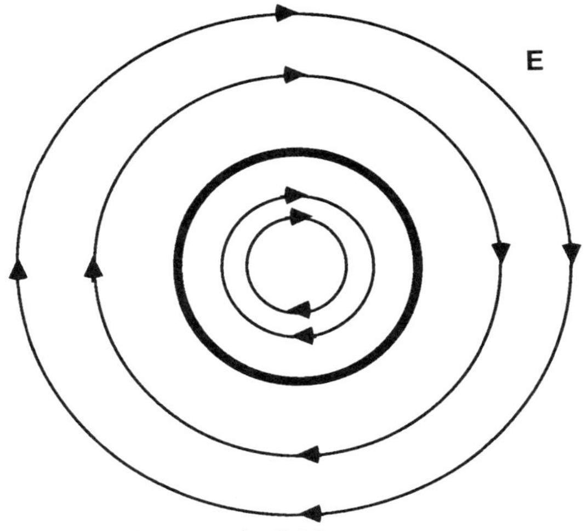

a) Electric field

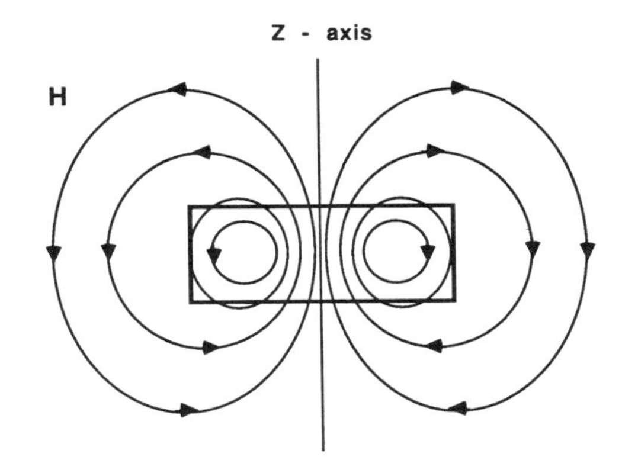

b) Magnetic field

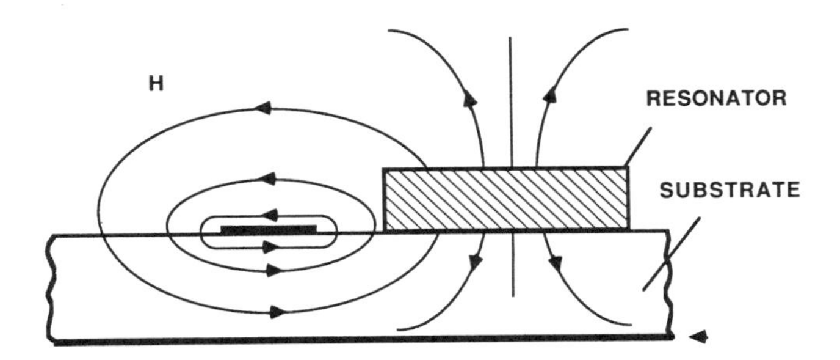

c) Magnetic field of dielectric resonator on microstrip

Figure 8.19 Field pattern of dielectric resonator

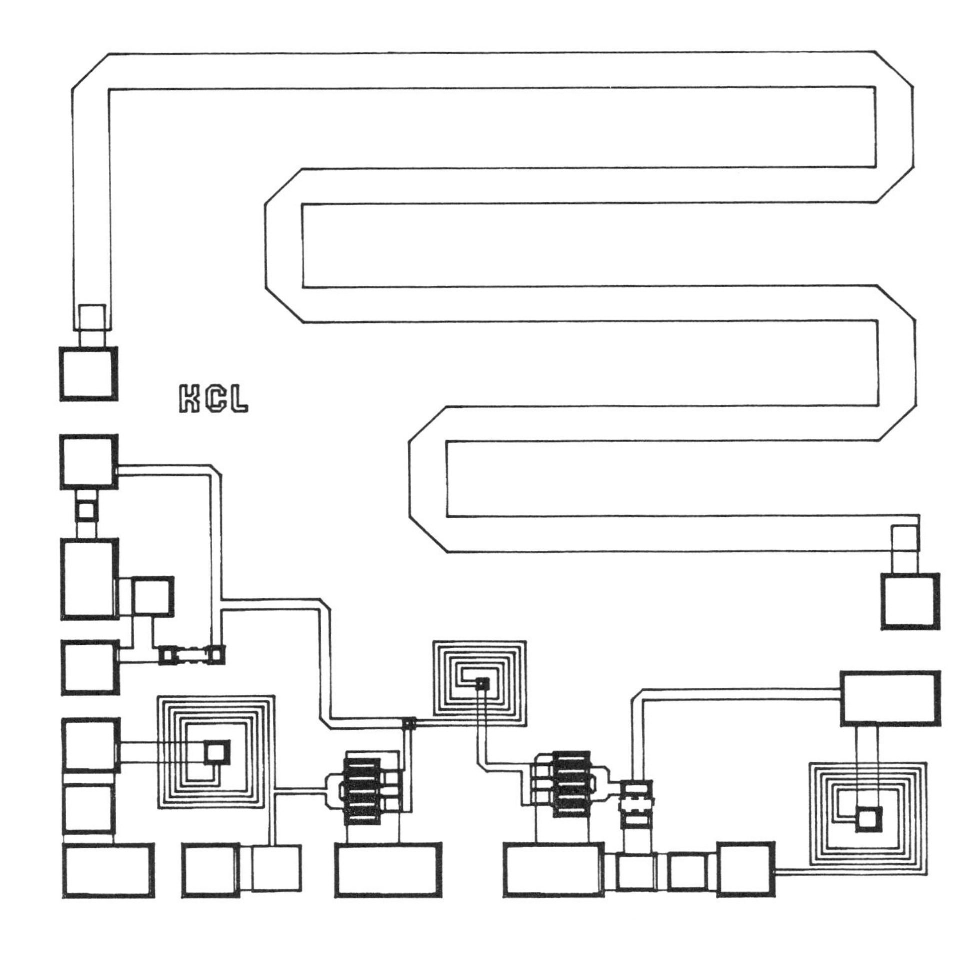

Figure 8.20 GaAs MMIC

radiation losses and losses in the microstrip under the transmission lines causing the available loaded and unloaded Q to be greatly reduced. For the optimum noise performance, the puck is usually mounted in a metal cavity of approximately twice the size and suspended using low loss material [53]. Coupling is then achieved by using low loss microstrip lines or wires. If Flicker noise is not taken into account the optimum coupling conditions of $Q_L/Q_o = {}^1/_2 \rightarrow {}^2/_3$ should still apply for low noise.

(i) GaAs Microwave Integrated Circuits

Many workers have designed oscillators on monolithic microwave integrated circuits. Where the noise performance has been important

DRO's are often used or the device is phase locked to a reference oscillator.

A low noise oscillator circuit has been designed by Cheng and Everard at King's College London on a GaAs MMIC using the transmission line resonator and the circuit layout is shown in Figure 8.20. The design consists of a low noise amplifier with impedance matching circuits to produce a 50 ohm amplifier and a folded transmission line resonator with series capacitors as the coupling elements. Note that the amplifier is not directly connected to the resonator to allow measurements on each part and to allow the phase shift to be varied if necessary. In fact fine tuning of the oscillator can be achieved by incorporating a phase shifter in between the amplifier and the resonator where for a range of f_0/Q_L the excess amplifier gain would need to be 3 dB and a noise degradation approaching 9 dB would exist at the band edges. For half this tuning range the noise degradations would be less than 3 dB. This oscillator was designed to operate at 7 GHz where the resonator insertion loss is around 5 dB.

8.4.4 Surface Acoustic Wave Resonator (SAW) Oscillators

Surface acoustic wave resonators and oscillators can now be built at frequencies above two GHz [24]. These can be used as a reference oscillator for microwave GaAs oscillators in synthesiser applications. Further as GaAs is piezo-electric the resonator could be incorporated in a GaAs IC substrate.

A surface acoustic wave resonator consists of one or more interdigital transducers, which convert electrical signals to surface acoustic waves, positioned in a cavity between two efficient reflectors of surface waves as shown in Figure 8.21.

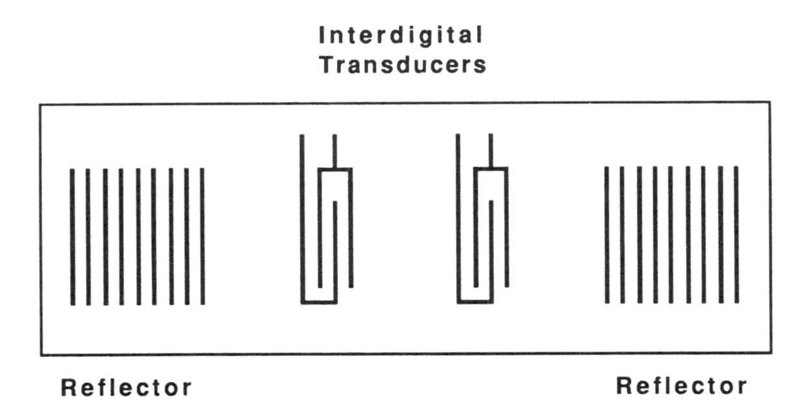

Figure 8.21 Surface Acoustic Wave resonator

A low loss surface wave is launched onto the substrate with a typical velocity of $3000ms^{-1}$ which is 10^5 times slower than the speed of light. The losses are proportional to the square of the frequency and are typically 0.1 dB/cm at 50 Mhz and 10 dB/cm at 500MHz. The usual substrates which exhibit piezoelectric properties are either quartz (SiO_2) or Lithium Niobate ($LiNbo_3$).

Unlike the more familiar bulk acoustic waves employed in bulk wave resonators, surface waves cannot be efficiently reflected by an abrupt discontinuity (e.g. substrate edge) because this would cause a significant proportion of the energy to be mode converted into bulk waves. SAW reflectors therefore consist of a large number of small impedance discontinuities in the form of metal strips or ion etched grooves spaced by half a wavelength. This results in bandpass reflectors with peak amplitude reflection coefficients of greater than 99%.

A typical resonator is shown in Figure 8.21. where two interdigital transducers are used to launch and detect the acoustic wave and the distributed reflectors are placed at either end. This form of resonator is a two port as it has two interdigital transducers. The resonator has an equivalent circuit which is the same as the complete resonator circuit shown for the inductor capacitor oscillator (Figure 8.8).

At King's College London Frazer Curley [14] and Everard have developed a 262 Mhz SAW oscillator using the design rules derived from the theories above. The two port SAW resonator (AT cut) with an unloaded Q of 15,000 was supplied as an engineering sample by STC. The oscillator consists of a resonator with an unloaded Q of 15,000, impedance transforming and phase shift networks and a hybrid amplifier as shown in Figure 8.22. The phase shift networks are designed to ensure that the circuit oscillates on the peak of the amplitude response of the resonator and hence at the maximum in the phase slope ($d\Phi/d\omega$). The oscillator will always oscillate at phase shifts of N*360 degrees where N is an integer but if this is not on the peak of the resonator characteristic, the noise performance will degrade with a Cos^6 relationship. The noise performance is shown in Figure 8.23 showing close correlation with the theory. Note the Change in slope ($1/f^2$ to $1/f^3$) at 1 kHz which is due to Flicker noise. The $1/f^2$ characteristic is caused by the Q multiplication filtering action of the oscillator on the flat (white) noise spectral density at the input and the $1/f^3$ output is caused by the same Q multiplication filtering, however now the low frequency 1/f noise is transposed up around carrier to give an input noise characteristic of $1/\Delta f$.

It is therefore advisable to use a transistor optimised for low Flicker noise. Audio transistors often have Flicker noise corners of less than 10 Hz and it is therefore believed that it is better to use a low Flicker noise transistor near the limit of its gain/frequency characteristic rather than a high frequency transistor which often has rather high Flicker noise

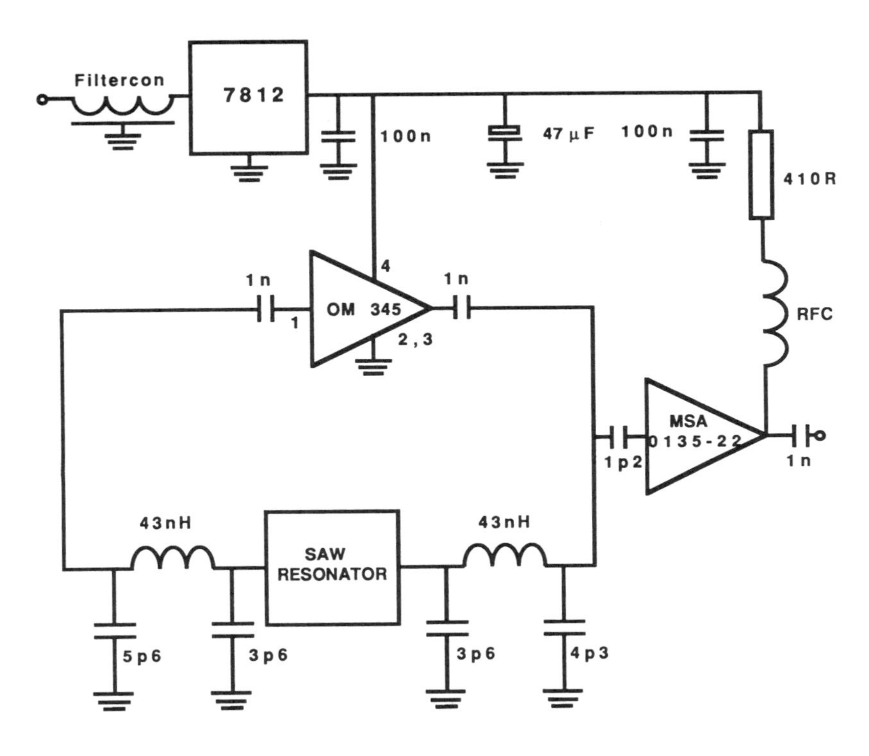

Figure 8.22 Low noise 262 MHz SAW oscillator

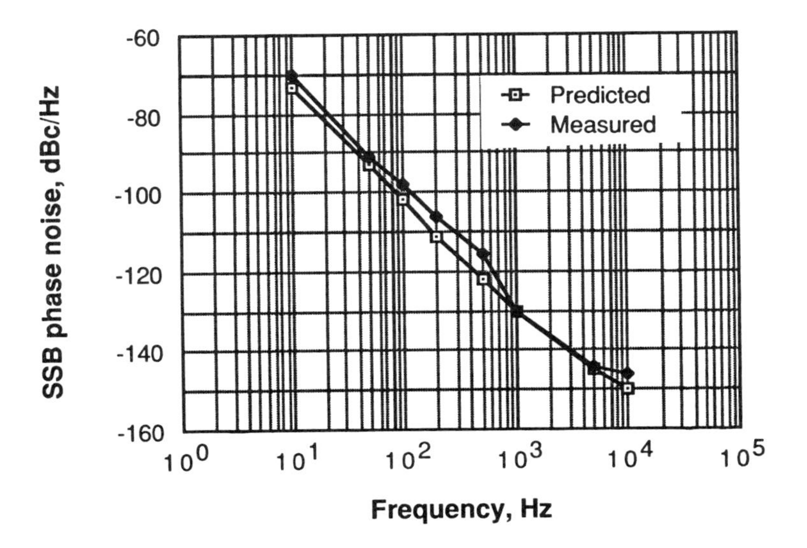

Figure 8.23 Phase noise performance of SAW oscillator

corners. The excess gain should, however, always be sufficient to ensure oscillation over the full temperature range. Parker [75,76,78,79] has also shown that SAW resonators generate Flicker noise which degrades the noise performance.

8.5 Frequency Tuning

Most modern systems require oscillators which can tune over large frequency ranges while still maintaining low noise performance. There is a number of limitations which should be considered. The first requirement is to achieve the correct value of Q_L/Q_o over the whole band where it can be seen that $S_{21}(0)$ should remain within 3 to 10 dB for optimum noise. The second requirement is to ensure that the phase error around the loop is kept low and constant over the whole tuning range as it can be shown that the phase noise degrades with a Cos^4 to Cos^6 relationship. A graph of the degradation of noise performance with phase error is shown in Figure 8.24. demonstrating a Cos^6 relationship. These are preliminary

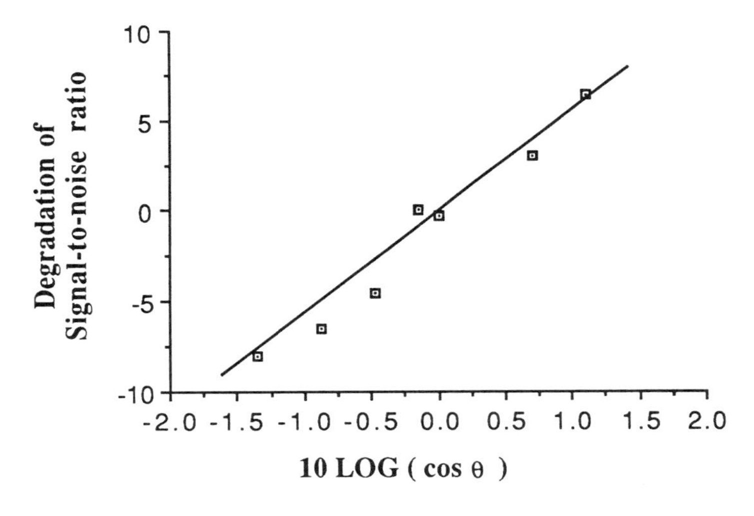

Figure 8.24 Noise degradation with phase error

results and are believed to be the first measurement of noise degradation with phase error (Cheng and Everard at King's College London) and was measured at 1.6 GHz in a GaAs oscillator. An isolator was placed at the input of the amplifier to ensure that a constant source impedance was always presented to the input of the amplifier, thus ensuring a constant amplifier noise figure. The details of this will be published in 1989. In oscillators the phase shift around the loop for resonance is 0 or N * 360 degrees. By phase error here we mean the difference in phase shift in the resonator between the point where the oscillator is operating at the peak in the amplitude and group delay response to the point at which it is operating where it is not at the peak.

8.5.1 *Power Limitations*

The noise performance of a broad tuning range oscillator is usually limited by the Q and the voltage handling capability of the varactor [122].

If it is assumed that the varactor diode limits the Q of the total circuit, then it is possible to obtain useful information from a simple power calculation. If the varactor is assumed to be a voltage controlled capacitor in series with a loss resistor (rs), The power dissipated in the varactor is

$$P = (V_{rs})^2/rs \tag{8.33}$$

The voltage across the capacitor V_c in a resonator is

$$V_c = QV_{rs} \tag{8.34}$$

Therefore the power dissipated in the varactor is

$$P_v = V_c{}^2/Q^2rs. \tag{8.35}$$

The noise power in oscillators is proportional to $1/PQ_o{}^2$. Therefore the figure of merit ($V_c{}^2$/rs) should be as high as possible and thus the varactor should have large voltage handling characteristics and small series resistances.

However the definition of P and the ratio of loaded to unloaded Q are important and these will alter the effect of the varactor on the noise performance. If it is assumed that the value of Q_L/Q_o is adjusted to the optimum value when the varactor is incorporated and that the varactor in an oscillator tuning over a large frequency range defines the unloaded Q of the resonator, the noise performance of such an oscillator can be calculated directly from the voltage handling and series resistance of the varactor.

If equation 8.19 is examined

$$L_{(fm)} = \frac{FkT}{8Q_o{}^2(Q_L/Q_o)^2(1-Q_L/Q_o)P_{rf}} \left(\frac{f_o}{\Delta f}\right)^2 \tag{8.36}$$

and the value of QL/Qo is put in as 2/3, this equation simplifies to

$$L_{(fm)} = \frac{27FkT}{32Q_o^2P_{rf}} \left(\frac{f_o}{\Delta f}\right)^2 \quad (8.37)$$

As $P_{rf} = 1.5.P_v$ and $P_v = Vc^2/Q_o^2rs$.

$$L_{(fm)} = \frac{9FkT.rs}{16V_c^2} \left(\frac{f_o}{\Delta f}\right)^2 \quad (8.38)$$

Similarly taking the equation for the general model.
If we take equation 8.24

$$L_{(fm)} = \frac{FkT}{8Q_o^2(Q_L/Q_o)^2(1-Q_L/Q_o)P_{avo}} \left(\frac{f_o}{\Delta f}\right)^2 \quad (8.39)$$

and substitute the value of $Q_L/Q_o = 1/2$ then this equation becomes

$$L_{(fm)} = \frac{2FkT}{Q_o^2.P_{avo}} \left(\frac{f_o}{\Delta f}\right)^2 \quad (8.40)$$

as $P_{avo} = 2P_v$ where $P_v = V_c^2/Q^2rs$

$$L_{(fm)} = \frac{FkT.rs}{V_c^2} \left(\frac{f_o}{\Delta f}\right)^2 \quad (8.41)$$

If we take a varactor with a series resistance of 1 ohm which can handle an rf voltage of 0.25 volt rms at a frequency of 1 GHz, then the noise performance at 25 kHz offset is approximately −97 dBc (F = 3 dB). This can only be improved by reducing the tuning range by coupling the varactor into the tuned circuit more lightly, or by switching in tuning capacitors using pin switches, or by using two back to back varactors.

8.5.2 Tunable Transmission Line Resonator

A new near octave tuning transmission line resonator (3.5–6 GHz) has been designed to achieve low noise operation over a broad band.

The design and experimental results of a novel varactor tuned transmission line resonator developed by Cheng and Everard at Kings College London are presented. Analysis based upon transmission line theory is used to model the resonator. A prototype with a tuning

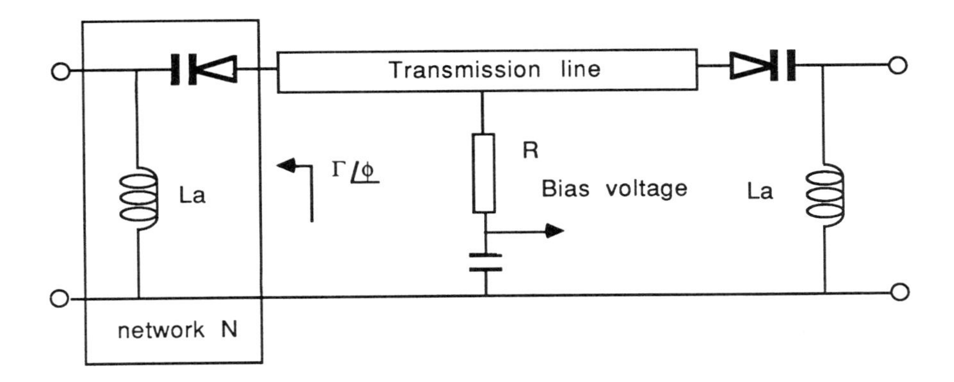

Figure 8.25 Tunable transmission line resonator

bandwidth of 53% (3.5–6 GHz) has been built. The tuned resonator could find applications in microwave oscillators and filters in both MMICs and hybrid circuits.

The resonator consists of a transmission line and a highly reflective network N at either end of the line and is shown in Figure 8.25. This is similar to the transmission line resonator described earlier where the network N was a fixed value shunt capacitor which was used in the design of a fixed frequency low noise oscillator. This is now extended to a wideband varactor tuned resonator. Frequency–stabilized circuits such as dielectric and ring resonators are generally difficult to tune electronically over a wide bandwidth [4,10,11]. The proposed resonator can be tuned over a wide frequency range and unlike dielectric resonators, the circuits are amenable to MMIC implementation. The design and performance of this tunable resonator and the effect of the finite Q of the tuning diode on the resonator is described.

The network N has a reflection coefficient Γ very close to one in magnitude. Its phase Φ varies with the diode junction capacitance C_{jv}, and hence the bias voltage. Resonances occur when $\beta_L - \Phi = n\pi$ where n is an integer, β is the phase constant and L is the length of the transmission line. The resonant frequency can therefore be varied by changing the bias voltage on the diodes. The lowest resonant frequency f_r (n=1), provided that $1/2\pi f_r C_{jv} \gg 2\pi f_r L_a$, is given by the solution of the equation,

$$2\pi\, f_r\, C_{jv}\, Z_o\, tan(\,\pi f_r/2f_o\,) \;=\; 1 \tag{8.42}$$

where $f_o = c_{eff}/2L$. Figure 8.26 shows the relationship between the frequency tuning ratio R (f_{rmax}/f_{rmin}) of the resonator and U ($2\pi f_o C_{j0} Z_o$) for a diode capacitance tuning ratio k of 6. This curve indicates that a high

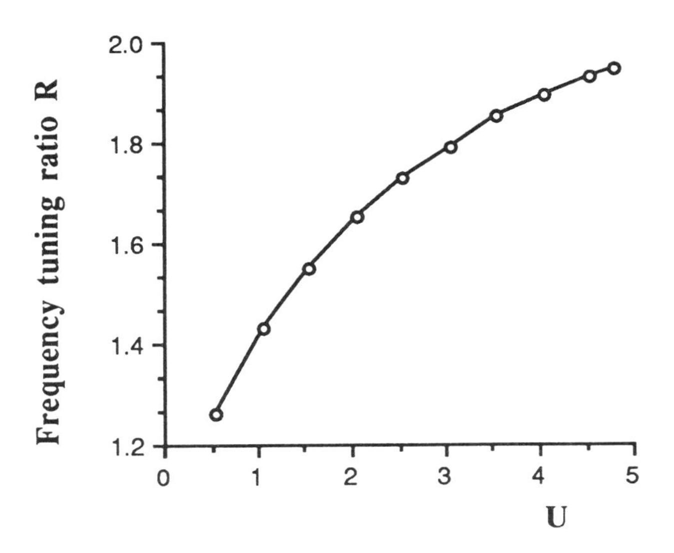

Figure 8.26 Frequency tuning ratio (f_{rmax}/f_{rmin}) vs $2\pi f_o C_{jo} Z_o$ (U)

value of U or C_{j0} is preferred when maximizing R. In theory, the maximum value of R is $\sqrt{k}$. However, the Q factor of the diodes degrades with higher values of C_{j0} and this affects the overall Q factor of the resonator (as will be discussed later). A tradeoff is therefore necessary. The variation of the insertion loss and the loaded Q factor of the resonator with W ($2\pi f_o L_a/Z_o$) are shown in Figures 8.27 and 8.28. The calculations are based upon the assumption that the varactor diode is lossless and $\alpha L=0.012$, where α is the attenuation coefficient of the line. Note that in Figure 8.28 the Q factor increases gradually with increasing value of f_r.

A prototype which can be tuned over the frequency range of 3.5 to 6 GHz has been built on 25 thou alumina ($\epsilon_r=9.8$) using hybrid circuit techniques. The varactor diode used is the Alpha CVE7900D GaAs tuning diode chip with C_{j0} = 1.5 pF, Q (−4V, 50 MHz) = 7000, breakdown voltage = 45V and k = 6. The microstrip line is 5 mm long and hence f_o is about 10 GHz. A resonator which consists of a 5 mm long microstrip line and two gaps (series capacitors) has been tested on alumina and αL is found to have a value of 0.012. The two shunt inductors are realized by gold wires about 0.4 mm long so that L_a is approximately 0.35nH and W is 0.44. Biasing of the diodes is achieved as shown in Figure 8.25. The impedance at the centre of the line is low for the fundamental resonance and high for the second resonance. A low value resistor can therefore be used to keep the bias circuit noise low and helps to suppress the second resonant peak. The variation of the resonant

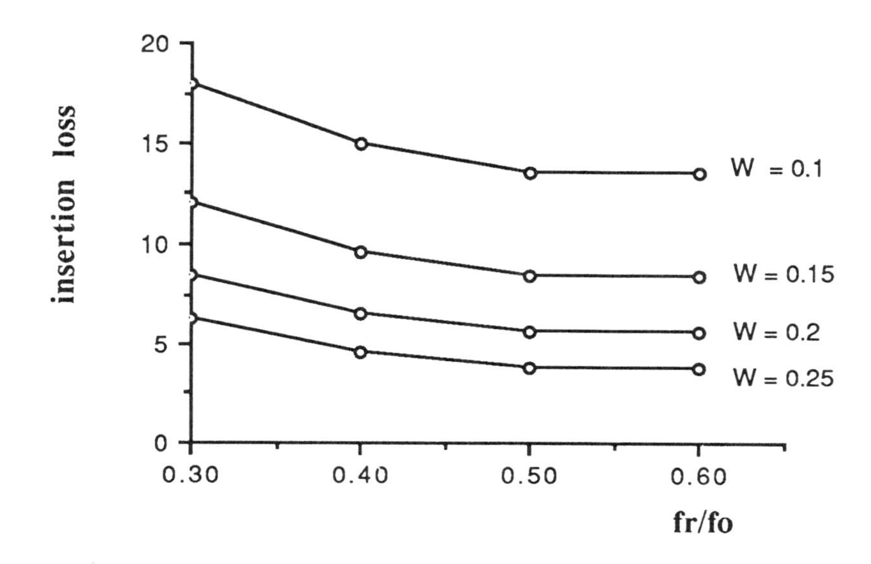

Figure 8.27 Insertion loss vs $2\pi f_o La/Z_o$ (W)

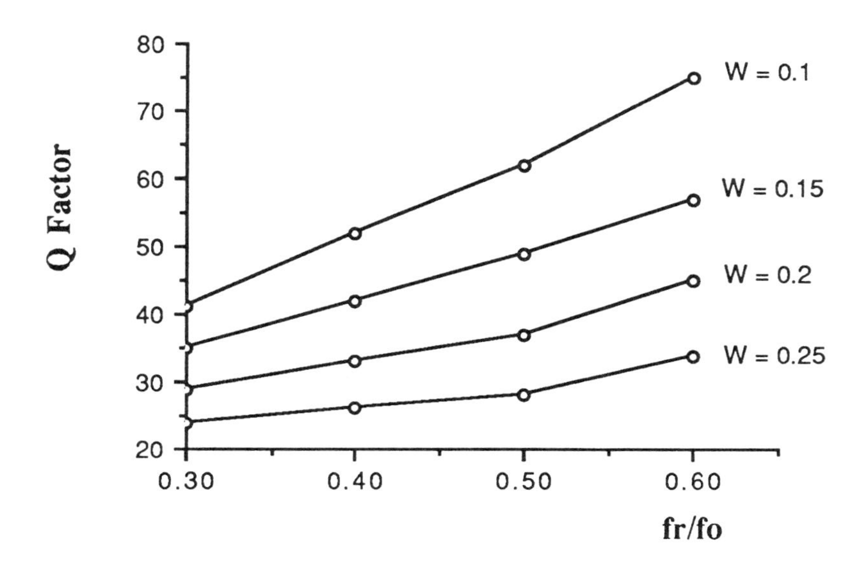

Figure 8.28 Q factor vs $2\pi f_o La/Z_o$ (W)

frequency against the reverse bias voltage v across the diode can be obtained from equation 8.42 together with the equation of the junction capacitance of an abrupt GaAs diode :-

$$C_{jv} = C_{j0}/(\ 1 + v/1.2\ )^{0.5} \tag{8.43}$$

Figure 8.29 shows that the theoretical calculations correlate well with the experimental results. The discrepancies at high resonant frequencies are due to parasitic capacitances which have not been included in the calculations. These parasitic capacitances need to be reduced in order to maximize R, such as by careful layout of the circuit. The measured insertion loss S21 (at the resonant peak) and the loaded Q factor of the resonator across the whole tuning range are shown in Figure 8.30. It has been observed that the effect of the series resistance (normalized value r_s with respect to Z_o) of the diode on the insertion loss and the loaded Q factor of the resonator is significant and cannot be ignored. Results which take into account these effects are given below:-

$$S21 = W' /\{\ W' + r_s + 0.5\ \alpha L\ sec^2(\pi f_r/2f_o)\ \} \tag{8.44}$$

$$Q\ = (\pi/4)\ (f_r/f_o)\ S_{21}\ sec^2(\pi f_r/2f_o)\ /\ W' \tag{8.45}$$

where $W' = (W\ f_r/f_o)^2$. For a constant value of W, the effect of r_s causes higher insertion loss and lower Q factor of the resonator. A higher value of W (larger inductance L_a) is therefore required for a given value of insertion loss. The resonator was designed to have a fairly constant insertion loss (5 ± 1 dB) over the frequency range from 3.5 to 6 GHz. It has been shown [19] that there is an optimum value of insertion loss and hence Q_L/Q_o of the resonator for minimum sideband noise in oscillators. For optimum noise performance, the insertion loss of the resonator should remain constant at the optimum value over the whole frequency tuning range. Such a tunable low noise oscillator is now under investigation.

8.6 Conclusions

A linear theory which describes the optimum operating conditions for minimum sideband noise in oscillators has been derived and experimentally verified using inductor capacitor, Surface Acoustic Wave and transmission line resonators. A number of designs are illustrated. The use of GaAs will enable the design of the complete front end including the oscillator and filtering components on a single integrated circuit. At the

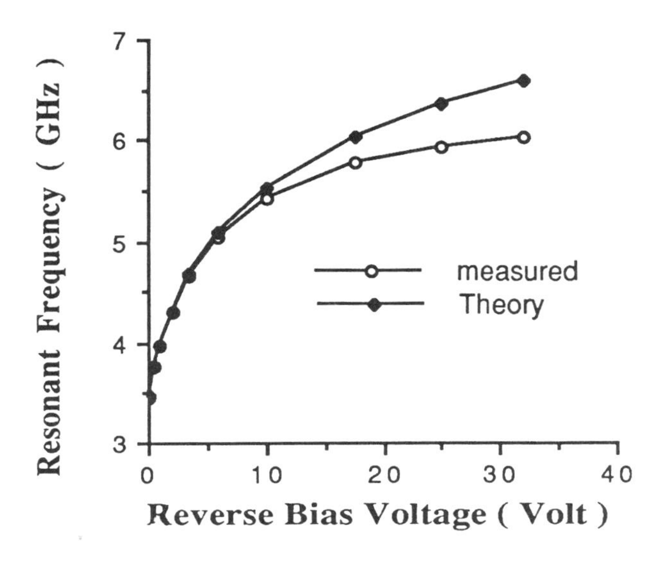

Figure 8.29 Resonant frequency vs Diode voltage

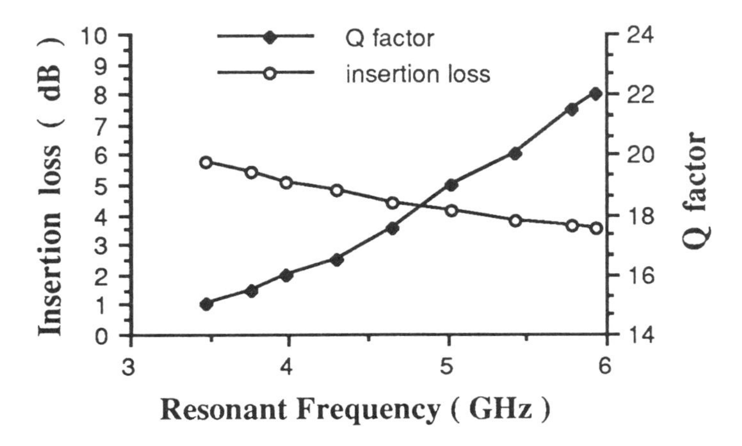

Figure 8.30 Insertion loss and Q vs frequency

moment the Flicker noise corner is considerably higher in GaAs than in silicon and this degrades the noise performance. Techniques for measurement and reduction of Flicker noise have been described. It is expected that with further advances in material processing and the development of new devices e.g. HBT's and HEMT's, the use of high Q resonators, DRO, YIG and transmission line, and the use of circuit techniques developed from accurate measurement and evaluation of Flicker noise, that these problems should be reduced to an acceptable level.

8.7 Acknowledgements:

I wish to thank the Science and Engineering Research Council for their generosity in supporting this work

I also wish to thank all of my students particularly Michael Cheng, Fraser Curley and Paul Dallas for their help in generating new ideas and results.

8.8 References

1 H. Abe et al: 'A highly stabilized low-noise GaAs FET integrated oscillator with a dielectric resonator in the C band', Transactions on Microwave Theory and Techniques, Vol.MTT-26, No.3, March 1978, pp.156-162.

2 K.K Agarwal C. Ho: 'Analysis of long-term frequency drift in FET oscillators', IEEE Transactions on microwave theory and techniques, Vol.MTT-35, No.12, Dec, 1987, pp.1328-1333.

3 K.K. Agarwal: 'Dielectric resonator oscillators using GaAs (GaAs/(Ga,Al)As as HBT's', 1986 IEEE MTT-S Digest, pp.95-98.

4 S.H. Al-Charchafchi and C.P. Dawson: 'Varactor tuned microstrip ring resonators' IEE Proc., Vol. 136, Pt. H, No. 2, April 1989, pp. 165-168

5 K.J. Anderson & A.M. Pavio: 'FET oscillators still require modelling but computer techniques simplify the task', Sept 1983, MSN, PP.60-68.

6 G.R. Basawapatna and R.B. Stancliff: 'A unified approach to the design of wide-band microwave solid-sate oscillators', IEEE Trans.Microwave Theory Tech., Vol.MTT-27, No.5, May 1979, pp.379-385.

7 M.J. Bianchini, J.B. Cole, R. Dibiase, Z. Galani, R.W. Laton, R.C. Waterman: 'A single-resonator GaAs FET oscillator with noise degeneration', IEEE MTT-S Digest 1984, pp.270-273.

8 T.J. Brazil & J.O. Scanlan: 'A non-linear design and optimisation for GaAs MESFET oscillators,IEEE MTT-S Digest pp.907-910.

9 M. Camiade, A. Bert, J. Graffenic, G. Pataut: 'Low noise design of dielectric resonator FET oscillators', pp.297-302.

10 O.Y. Chan and S. Kazeminejad : 'Voltage-controlled oscillator using dielectric resonator" Electronic letters, Vol. 24, No. 13, June 1988, pp. 776-777

11 K. Chang, S. Martin, F. Wang and J.L. Klein : 'On the study of Microstrip ring and varactor-tuned ring circuits" IEEE Trans., MTT-35, Dec., 1987 pp. 1288-1295

12 S.W. Chen, T.C. Chang, J.Y. Chin: 'A unified design of dielectric resonator oscillators for telecommunication systems', 1986, IEEE MTT-S Digest, pp.593-596.

13 H.C. Chien and J. Frey " 'Monolithically integrable active microwave inductor". Electronics letters, 7th July 1988, Vol. 24, No. 14.

14 F.M. Curley: 'The application of SAW resonators to the generation of low phase noise oscillators".MSc. Thesis 1987. King's College London.

15 B.T. Debney, & J.J. Joshi: 'A theory of noise in GaAs oscillators' IEE Transactions on Electron Devices, Vol.ED-30, No.7, July 1983, pp.769-776.

16 M.M. Driscoll: 'Low frequency noise quartz crystal oscillator', IEEE Trans, 1975, IM-24, (1), pp.21-26.

17 D.M. Drury & D.C. Zimmerman: 'A dual gate variable power amplifier', 1985, IEEE MTT-S Digest pp.219-222.

18 J.K.A. Everard: 'Low noise high efficiency oscillator'. Patent application 8502565, February 1985.

19 J.K.A. Everard: 'Low noise power-efficient oscillators: theory and design, IEE proceedings, Vol.133, pt.G, No.4, August 1986.

20 J.K.A. Everard: 'Low noise radio frequency oscillators". The IERE First International conference on Frequency Control and synthesis"Publication Number 75. The University of Surrey, Guildford. 8–10 April 1987.pp.75–78.

21 J.K.A. Everard and K.K.M. Cheng: 'Low noise transmission line oscillators".IEE Second International Conference on Frequency Control and synthesis, publication number 303,The University of Leicester 10–13 April 1989. pp58–61.

22 S.J. Fiedziuzsko: 'Practical aspects of and limitations of dual mode dielectric resonator filters', IEEE MTT-S Digest, 1985.

23 S.J. Fiedziuszko: 'Minature FET Osillator stabilizers by a dual mode dielectric resonator', IEE MTT-S Digest.

24 B. Fleischmann, W Ruile, G. Riha and A.R. Baghai-Wadji: 'Reproducible SAW Filters at 2.5 GHz". The IEE International Conference on Frequency Synthesis and Control., The University of Surrey, Guildford, Surrey 8–10 April 1987.

25 R.J. Gilmore and F.J. Rosenbaum: 'An analytic approach to optimum oscillator design using S-parameters', IEEE Trans.Microwave Theory Tech., Vol.MTT-31 No.8, August 1983, pp.633–639.

26 R.J. Gilmore,, and F.J. Rosenbaum: 'An analytical approach to optimum oscillator design using S-parameters', IEEE Trans, 1981, MTT-31, (8), pp.633–639.

27 J.A. Greer and T.E. Parker: 'Improved vibration sensitivity of the all quartz package surface acoustic wave resonator".42nd Annual Frequency Control Symposium 1988.pp.239–251.

28 P. Guillon, B. Byzery, M. Chaubet: 'Coupling parameters between a Dielectric Resonator and a Microstrip line', IEEE Transactions on Microwave Theory and Techniques, Vol.MTT-33, Resonator, No.3, March 1985, pp.222–226.

29 M.S. Gupta: 'Velocity Fluctuation Noise in Semiconductor Lasers and in MESFET Channels Under Hot Electron Conditions', IEEE Trans on Elect Dev. Vol.34, No.6, June 1987.

30 M.S. Gupta et al: 'Microwave noise characterisation of GaAs MESFETs; evaluation by on-wafer low-frequency output noise current measurement', IEEE 1987, pp.1208–1217.

31 M.S. Gupta, Greling: 'Microwave noise characterisation of GaAs MESFET's; determination of extrinsic noise parameters', IEÉE 1988, p.745–751.

32 Z. Galani et al: 'Analysis and Design of a single-resonator GaAs FET oscillator with Noise Degeneration, IEEE 1984 Trans. on Microwave Theory and Tech. MTT-32, No.12, Dielectric Resonator Oscillator, pp.1556–1565.

33 E. Hafner: 'The effect of noise in oscillators', Proc IEEE, 1966, 54, (2), 1966, pp. 179–198.

34 S. Hashigouchi, N. Aoki, H. Ohkubo: 'Distribution of 1/f noise in an expitaxial GaAs MESFET.Solid State Electronics Vol 29, No.7 pp745–749 1986.

35 B.H. Horine: 'SAW resonator filters; applications and capabilities', 1985, IEE MTT-S Digest, pp.247–250.

36 B. Hughes, N.G. Fernandez, J.M. Gladstone: 'GaAsFETs with a flicker-noise corner below 1 MHz', IEE Transactions on Electron Devices, Vol.ED-34, No.4, April 1987, pp.733–741.

37 R. Hans, Y.S. Chung and S.Charles: 'A study of the relation between device low-frequency noise and oscillator phase noise for GaAs MESFETs', IEEE MTT-S Digest 1984, pp.267–269.

38 W.H. Hayward: 'Introduction to radio frequency design, (Prentice Hall, 1982).

39 Hiroyuki: 'A GaAs self-bias Mode Oscillator, Microwave Theory and Techniques, Vol. MTT-34, No.1 Jan, 1986, pp.167–172.

40 Hurlimann & Hardy: 'Measurement of flicker phase noise of 1.4 GHz amplifier at temperatures between 300k and 1.26k', Elect. Letts., 12th March 1987, Vol.23, No.6.

41 O. Isihara et al: 'A highly stabilised GaAs FET dielectric resonator oscillator using a DRO feedback circuit form 9-14 GHz', Transactions on Microwave Theory and Techniques, Vol.MTT-28, No.8, Aug 1980, pp.817–824.

42 S. Iversen: 'The effect of feedback on noise figure, IEEE, Vol.63, pp.540–542, Mar 1975, pp.24–25.

43 A.P.S. Jhanna et al: 'A very high Q microwave transistor oscillators using dielectric resonators', pp.457–463.

44 K.M. Johnson: 'Large Signal GaAs MESFET oscillator design, Transactions on Microwave Theory and Techniques, Vol.MTT-27, No.3, March 1979.

45 K.M. Johnson: 'Large signal GaAs MESFET oscillator design', IEEE Trans.Microwave Theory Tech., Vol. MTT-27, No.3, March 1979, pp.217–227.

46 A.P.S. Khanna: 'Fast Settling, Low noise bipolar Ku band fundamental Bipolar VCO', 1987, IEEE MTT-S Digest.

47 A.P.S. Khanna, Y. Garault, M. Guediva: 'Dielectric resonator HE11d + 1 mode coupling to a shielded microstrip', 1983, IEEE MTT-S Digest, pp.527–529.

48 A.M. Khilla: 'Ring and disc resonator CAD model', Microwave Journal, Nov 1984.

49 Y. Komatsu and Y. Murakami: 'Coupling coefficient between microstrip line and dielectric resonator', Transactions on Microwave Theory and Techniques, Vol.31, No.1, Jan 1983, pp.-33-40.

50 P.S. Kooni, M.S. Leong Yeo: 'Circular microstrip disc resonator fo and Q calculations with HP41C programs.', Microwave Journal, Sept 1983, pp.165-169.

51 K. Kurokawa: 'Noise in synchronised oscillators, IEE Transactions on Microwave Theory and Techniques, Vol.MTT-16, No.4, April 1968.

52 K. Kurokawa: 'Some basic characteristics of broadband negative resistance oscillator circuits', Bell Syst. Tech. J., July-'Aug. 1969, pp.1973-1955.

53 M.J. Loboda,T.E. Parker G.Montress: 'Temperature sensitivity of dielectric resonators and dielectric resonator oscillators". 42nd Annual Frequency Control Symposium 1988. pp.263-271.

54 G. Lan, D. Kalokitis et al: 'Highly stabilised, ultra-low noise FET oscillator with dielectric resonator, 1986, IEEE MTT-S Digest, pp.83-95.

55 K.W. Lee, W.R. Day: 'Varactor tuned dielectric resonator GaAs FET oscillator in x-band', pp.274-276.

56 D.B. Leeson: 'A simple model of feedback oscillator noise spectrum', Proc. IEEE, 1966, 54, (2), p.329.

57 D.B. Leeson: 'Short term stable microwave sources', Microwave J. 1970, 13, (6), pp.59-69.

58 D.B. Leeson: 'A simple model of feedback oscillator noise spectrum', Proc. IEEE 54, February 1966, pp.329-330.

59 R.E. Lehman, D.H. Heston: 'X-band monolithic series feedback L.N.A., 1985, IEEE MTT-S Digest, pp.51-54.

60 M.F. Lewis: 'Some aspects of SAW oscillators', IEEE Conference on Sonics and Ultrasonics, Monterey, CA, USA, 1973, pp.344-347.

61 M.F. Lewis: 'The surface acoustic wave oscillator - a natural and timely development of the quartz oscillator', Proceedings of 28th Annual Symposium on Frequency Control, Atlantic City, NJ, USA. 29th-31st May 1974, pp.304-314.

62 M. Maeda, K. Kimura, and H. Kodera: 'Design and performance of X-band oscillators with GaAs Schottky-Gate Field-Effect Transistors', IEEE Trans. Microwave Theory Tech., vol. MTT-23, no.8, August 1975, pp.661-667.

63 Murray R.J. and White P.D: 'Surface Acoustic Wave devices, A practical guide to their use for engineers. Wireless world March 1981 pp.38-41, April 1981 pp.79-82.

64 S.B. Moghe and T.J. Holden: 'High-performance GaAs MMIC oscillators', IEEE Trans. Microwave Theory Tech.,Vol.MTT-35, No.12, December 1987, pp.1283-1287.

65 S. Moghe, T. Holden: 'High performance GaAs C-Band and Ku-Band MMIC Oscillators, 1987, IEEE MTT-S Digest, pp.911-914.

66 G.Montress, T.E. Parker,M.J. Loboda, J.A. Greer: 'Extremely low phase noise SAW resonators and Oscillators: Design and Performance. IEEE Transactions on Ultrasonics, Ferroelectrics and Frequency Control, Vol 35,No.6, November 1988. pp.657-667.

67 G.Montress, T.E. Parker,M.J. Loboda: 'Residual phase noise measurements of VHF, UHF, and microwave components".To be published in the Proceedings of the 43rd Annual Frequency Control Symposium 1989.

68 P.A. Moore., C.S. Barnes,. and P.D. White: 'A very high performance surface acoustic wave oscillator', IEE Colloquium Digest on Frequency Synthesis and Programmable Sources, 1981, pp.9/1-9/5.

69 P.A. Moore ., S.K. Salmon: 'Surface acoustic wave reference oscillators for UHF and microwave generators', IEE Proc. H., Microwave, Opt & Antenna, 1983, 130, (7), pp.477-482.

70 Moss: 'AM->AM and AM->PM measurements using the PM null technique', 1987, IEEE, Trans. on Mic. Theory and Tech., Vol.MIT-35, No.8, August 1987.

71 T. Otobe, Y. Komatsu, Y. Murakami: 'A low-drift oscillator stabilised by a highly sensitive discriminator'.

72 T.E. Parker: '1/f noise in quartz delay lines and resonators', Ultrasonic symposium, 1979, IEEE, pp.878-881.

73 Parker: '1/f noise in quartz S.A.W. Devices', Electronic Letts. 10th May 1979, Vol.15, No.10.

74 T.E. Parker: 'Current developments in SAW oscillator stability', Proceedings of the 31st Annual Symposium on Frequency Control, Atlantic City, NJ, USA, 1977, pp.359-364.

75 T.E. Parker: '1/f noise in quartz SAW devices', Electron. Lett., 1979, 15, (10), pp.296-298.

76 T.E.Parker: '1/f noise in quartz SAW devices'. IEEE Ultrasonics Symposium, 1979.

77 T.E. Parker: 'Characteristics and sources of phase noise in stable oscillators". 41st Annual frequency control symposium 1977.pp.99–110.

78 T.E. Parker G.Montress: 'Low noise SAW resonator oscillators" To be published in the Proceedings of the 43rd Annual Frequency Control Symposium 1989.

79 T.E. Parker G.Montress: 'Precision Surface Acoustic Wave SAW Oscillators". IEEE transactions on ultrasonics, Ferroelectrics and frequency control, Vol. 35 No.3, May 1988.

80 A.M. Pavio & M.A. Smith: 'Push-pull Dielectric Resonator Oscillator, 1985, IEEE MTT-S Digest.

81 J.K. Plourde: 'Application of Dielectric Resonators in Microwave components', Transactions on Microwave Theory and Techniques, Vol.MTT-29, No.8, Aug 1981, pp.754–771.

82 A Podcameni, L.F.M. Conrado: 'A new approach for the design of microwave oscillators and filters using dielectric resonators', 1985, IEEE MTT-S Digest, pp.169–172.

83 A. Podcameni, L.F.M. Conrado, M.M. Mosso: 'Unloaded quality factor measurement for MIC dielectric resonator applications', Elect. Letts., 3rd September, 1981, Vol.17, No.18, pp.656–658.

84 A. Podcameni, L. A Bermudez: 'Large signal design of GaAs FET oscillators using input dielectric resonators', 1983, Transactions on Microwave Theory and Techniques, Vol. MTT-31, No.4, April 1983, pp.358–361.

85 M. Pouysegur,. J. Craffenil,. J.F. Santereau,. J.P. Fortea: 'Comparative study of phase noise in HEMT and MESFET microwave oscillators 1987, IEEE, MTT-S Digest,pp.557–560.

86 P. Penfield: 'Circuit theory of periodically driven non-linear system, Proceedings of IEE, Feb 1966, Vol.54, No.2, pp.226–280.

87 Poole and Paul: 'Optimum noise measure termination for microwave transistor amplifiers'.

88 M. Pringent and J. Obregon: 'Phase noise reduction in FET oscillators by low-frequency loading and feedback circuitry optimization', IEEE Trans. Microwave Theory Tech., Vol.MTT-35, No.3, March 1987, pp.349–352.

89 Prinjent & Obregon: 'Phase noise reduction by low frequency loading and feedback'.

90 R.A. Pucel: 'The GaAs FET oscillator; its signal and noise performance'.

91 C. Rauscher: 'Large signal technique for designing single-frequency and voltages-controlled GaAs FET oscillators', IEE Transactions on Microwave Theory and Techniques, Vol. MTT-29, No.4, April 1981.

92 C. Rauscher & H.A. Willing: 'Simulation of a non-linear MESFET', simulation of non-linear microwave FET performance using a quasi-static model.IEEE transactions MTT, Vol MTT-27, No.10, Oct 1979, pp.834–840.

93 U.L. Rhode: 'Mathematical analysis and design of ultra stable low noise 100 MHz crystal oscillator with differential limiter and its possibilities in frequency standards'. Proceedings of 32nd Annual Symposium on Frequency Control, Atlantic City, NJ, USA, 31st May–2nd June 1978, pp.400–425.

94 A.N. Riddle and R.J. Trew: 'A new method of reducing phase noise in GaAs FET oscillators', 1984, IEEE MTT-S.

95 A.N. Riddle & R.J. Trew: 'Low frequency noise measurements of GaAs FETs', 1986, IEEE MTT-S Digest, pp.79–82.

96 A.N. Riddle, and R.J. Trew: 'A new method of reducing phase noise in GaAs FET oscillators', IEEE MTT-S Digest 1984, pp.274–276.

97 A.N. Riddle and R.J. Trew: 'A novel GaAs FET oscillator with low phase noise', IEEE MTT-S Digest 1985, pp.257–260.

98 V. Rizzoli, C. Cecchet, A. Lippanini: 'Frequency conversion in general non-linear multi-port devices', 1986 IEEE MTT-S Digest, pp.483–486.

99 W.P. Robins: 'Phase noise in signal sources (Theory and applications)', (IEE Telecommunications series 9, 1984).

100 F.N.H. Robinson: 'Noise in oscillators', Int. J. Electron, 1984, 56, (1), pp.63–70.

101 Rohdin, Su, Stolfe: 'A study of the relation between devices low frequency and oscillator phase noise for GaAs MESFETs', 1984 IEEE MTT-S Digest, pp.267–269.

102 J. Rutman: 'Relations between spectral purity and frequency stability', Proceedings of 28th Annual Symposium on Frequency Control, Atlantic City, NJ, USA, 29th–31st May 1974, pp.160–165.

103 S.K. Salmon: 'Practical aspects of surface acoustic-wave oscillators', IEEE Trans, 1979, MTT-27, (12), pp.1012–1018.

104 K.H. Sann: 'The measurement of near-carrier noise in microwave amplifiers', IEEE Transactions on Microwave Theory and Techniques, Vol.MTT-16, No.9, September 1968, pp.761–766.

105 J.F. Sautereau et al: 'Large signal design and realisation of a low noise X band GaAs FET oscillator', pp.464–468.

106 G. Sauvage: 'Phase noise in oscillators: A mathematical analysis of Leesons's model, ibid, 1977, IM-26(4), pp.408-410.

107 D. Scherer: 'Design principles and test methods for low phase noise RF and microwave sources, RF and Microwave Measurement Symposium and Exhibition, Hewlett Packard.

108 D. Scherer: 'Today's lesson- learn about low noise design', Part I, Microwaves, April 1979, 18, pp.72-77.

109 D. Schere: 'Today's lesson- learn about low noise design', Part II,ibid, May 1979, 18, pp.72-77.

110 H. Statz, H.A. Haws & R.A. Pucel: 'Noise characteristics of GaAs field-effect transistors'.

111 H.J. Siweris and B. Schiek: 'Analysis of noise upconversion in microwave FET oscillators', IEEE Trans.Microwave Theory Tech., Vol. MTT-33, No.3, March 1985, pp.233-241.

112 P.R. Shepherd. and E. Vilar: 'Optimisation and synthesis of low phase-noise microwave-feedback oscillators, IEE Proc. H., Microwave, Opt & Antenna, 1983, 130, (7), pp.437-444.

113 S.R. Stein, C.M. Maney., F.L Walls, J.E. Gray, and R.J. Besson: 'A systems approach to high performance oscillators', Proceedings of 32nd Annual Symposium on Frequency Control, Atlantic City, NJ, USA, 31st May-2nd June 1978, pp.527-530.

114 R.C. Smythe: 'Phase noise in crystal filters', Piezo Technology, Inc., Orlando, Florida.

115 H.J. Siweris & B. Schick: 'Noise analysis of up conversion in microwave FET oscillators', IEEE Transactions on Electron Devices, pp.233-24

116 A. Takoaka and K. Ura: 'Noise analysis of nonlinear feedback oscillator with AM-PM conversion coefficient', IEEE Trans.Microwave Theory Tech., Vol. MTT-28, No.6, June 1980, pp.654-661.

117 C.F. Schiebold: 'An approach to realizing multi-octave performance in GaAs-FET YIG-tuned oscillators', IEEE MTT-S Digest, 1985, pp.261-263.

118 T. Takano et al: 'Novel GaAs FET phase detector operable to Ka band', 1984, IEEE MTT-S Digest, pp.381-386.

119 R.J. Trew: 'Design Theory for broad-band, YIG-tuned FET oscillators', IEEE Trans. Microwave Theory Tech., Vol. MTT-27, No.1, January 1979, pp.8-14.

120 H. Thaler, G. Ulrich and G. Weidmann: 'Noise in IMPATT diode amplifiers and oscillators', IEEE Trans. Microwave Theory Tech.,Vol. MTT-19, No.8, August, 1971, pp.692-705.

121 C. Tsironis, P. Lesarte: 'X and Ku-band dual gate MISFET oscillator stabilized using dielectric resonators, pp.469-474.

122 M.J. Underhill: 'Oscillator noise limitations', IERE Conf. Proc. 39, 1979, pp.109-118.

123 M.J. Underhill: 'Comparison of the noise performance of some oscillators for tunable receivers', IERE Conf. Proc. 40, 1978, pp.237-252.

124 Van der Ziel: 'Noise resistance of FET's in the hot electron regime'.

125 G.D. Vendelin: 'Design of amplifiers and oscillators by the S parameter method', (Wiley Interscience, 1982).

126 G.D. Vendelin: 'Feedback effects on the noise performance of GaAs MESFETs, IEEE MTT-S int. Microwave Symp, 1975, pp.324-326, pp.294-296.

127 R.F. Voss: '1/f (flicker) noise, A brief review', Proceedings of the 33rd Annual Symposium on Frequency Control, Atlantic City, NJ, USA, 30th May 1st June 1979, pp.40-46.

128 F.L. Walls, A. De Marchi: 'RF Spectrum of a signal after frequency multiplication and measurement comparison with a simple calculation', IEE Trans. on Inst. and Measurement, Vol.IM-24, No.3, Sept 1975, pp.210-217.

129 A.E. Wainwright, E.L. Wallis, and M.D. McCaa: 'Direct measurements of the inherent frequency stability of crystal resonators', 28th Frequency Control Symposium, 1974, pp.177-180.

Chapter 9

Mixer Designs for GaAs Technology

I. D. Robertson

9.1 Introduction

GaAs technology has already had a tremendous impact on conventional diode mixers. GaAs Schottky–barrier mixer diodes have achieved cut–off frequencies of over 3000 GHz, and give lower conversion loss than silicon diodes at high frequencies. GaAs diode mixers have made it possible to build reasonably low–noise receiving systems at frequencies where low–noise amplifiers do not yet exist. In these applications GaAs technology has had another, less obvious, effect: The IF amplifiers operate in the GHz range and by reducing their noise figure, GaAs FETs have improved system sensitivity.

In addition to this, GaAs technology now makes it possible to use transistors at millimetre–wavelengths. HEMT devices with useable gain at 94 GHz have already been reported [1]. These new devices have two broad areas of application: Firstly they can be used to make state–of–the–art low noise amplifiers, replacing simple diode receivers and making existing systems more sensitive. Secondly the application of GaAs integrated circuits in millimetre–wave systems will reduce costs drastically, thus creating many new techniques and applications. For example it makes the widespread commercial use of millimetre–waves for communications economical.

This Chapter cannot hope to cover all aspects of microwave mixers in detail. Instead it concentrates on the practical demands on mixer performance in real systems, and how new GaAs circuit techniques can or can't be used to meet those demands. Extensive references are given for those who need information on detailed mixer theory and device characteristics.

9.2 Applications

Mixers are used in virtually all microwave and millimetre–wave transmission/reception systems. Mixing the received signal with a local oscillator to an intermediate frequency, or directly to baseband, allows a vast improvement in selectivity because the bandwidth of the filters is greatly reduced. Less bandwidth means less noise power, and increased sensitivity to low level signals. Simple envelope detectors can be as much as 60 dB less sensitive than homodyne and heterodyne receivers. Thus mixers of various types are found in communications transmitters and receivers, radar systems, a whole host of military systems, radio telescopes, and some types of medical scanners.

9.3 Mixer diodes

The Schottky barrier diode is used for both silicon and GaAs mixer diodes because it is a majority carrier device. PN–junction diodes are limited in frequency range by the minority carrier recombination time. The major advantage of GaAs over silicon is that the low–field electron mobility is higher, as discussed in Chapter 5. This means that diodes can be made with lower resistance if GaAs is used. It is not possible to make higher frequency silicon diodes by making them smaller to reduce their capacitance, because the resistance increases such that the cut–off frequency is roughly constant. The epitaxial layer thickness in a silicon diode is made very thin to reduce the series resistance, but this reduces the reverse breakdown to only a couple of volts; excess LO power can easily destroy a silicon mixer diode. In contrast, GaAs Schottky barrier diodes can be made with cut–off frequencies over 3000 GHz, and reverse breakdown voltages over 6 V. GaAs mixer diodes are thus more rugged and have lower conversion loss at very high frequencies. The disadvantages of GaAs diodes are cost and the larger knee voltage (typically 0.8 V compared with 0.5 V) which means that either DC bias or higher LO power is required. DC bias may be very difficult to apply for some balanced mixers.

9.3.1 Diode Large–Signal Modelling

The Schottky barrier is formed because of the difference in work functions between the metal contact and the semiconductor. Figure 9.1 shows the band diagram of the structure with zero bias before and after making contact. Before the metal and semiconductor are put together, electrons in the semiconductor have more energy (the Fermi level is higher). When the metal and semiconductor come into contact, electrons flow from the semiconductor to the metal until the Fermi levels are in equilibrium. The electrons leave behind positively–charged donors, and an electric field builds up between those and the electrons that collect at the metal surface, until electron flow is stopped. As shown in the band diagram there is now a built–in potential barrier, equal to the difference in work functions. Thus

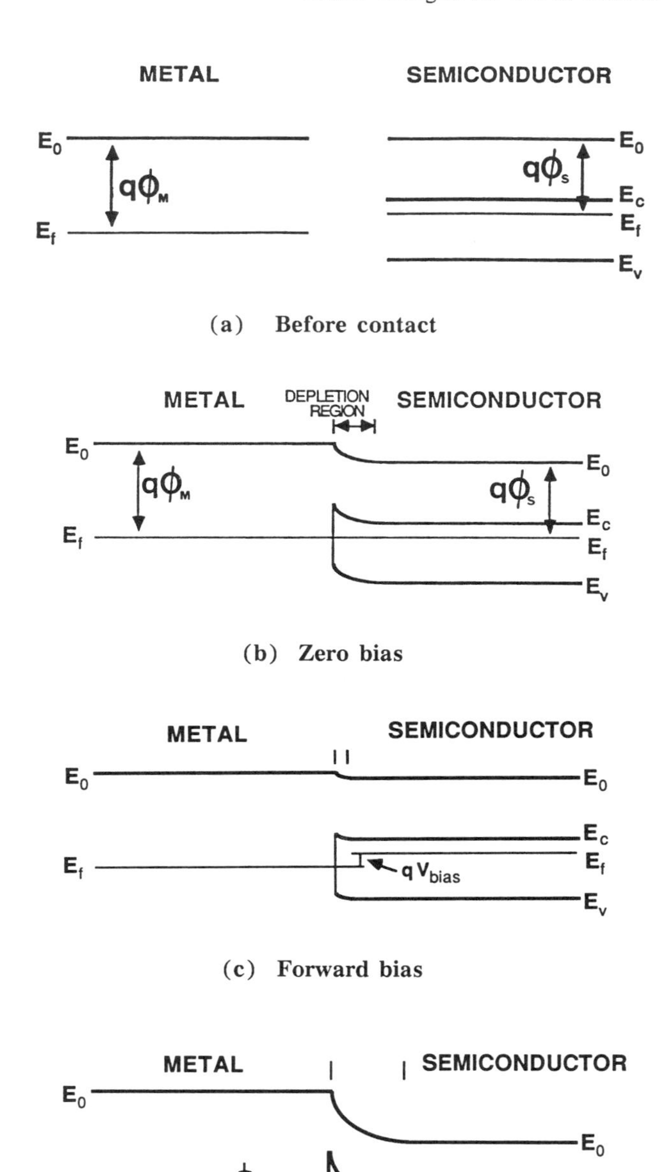

(a) Before contact

(b) Zero bias

(c) Forward bias

(d) Reverse bias

Figure. 9.1 Band diagrams showing the formation of the Schottky Barrier

in equilibrium there is a depletion region between the metal and semiconductor. When bias is applied the Fermi level is no longer constant. Thus the barrier potential drops for forward bias, so that electrons start to flow over the barrier by thermionic emission; the depletion region gets narrower, resulting in increased capacitance. For reverse bias the barrier potential increases, cutting off the electron flow and making the depletion region wider. Figure 9.2 shows the I–V and C–V characteristics for a typical GaAs Schottky diode. Figure 9.3 shows some of the most common practical diode structures.

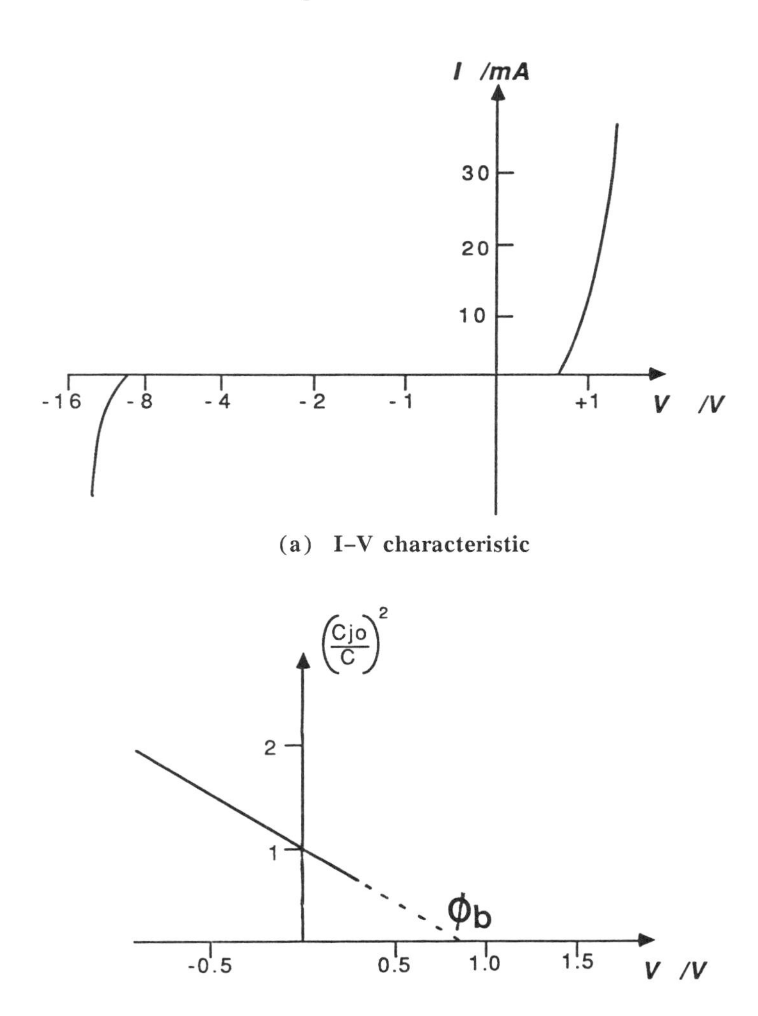

Figure 9.2 Typical I–V and C–V characteristics for a GaAs Schottky diode

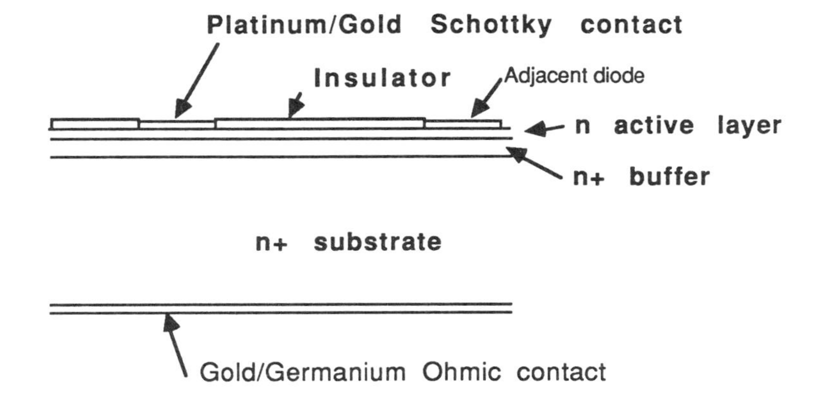

(a) The Dot–Matrix diode (bare chip)

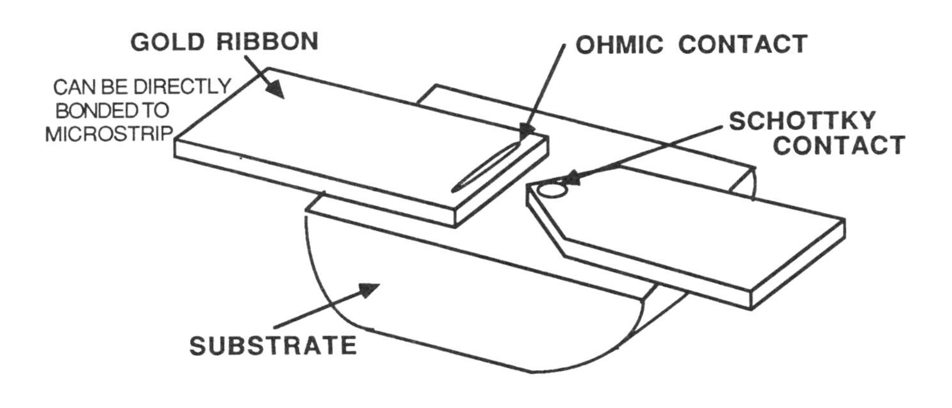

(b) Beam–lead diode

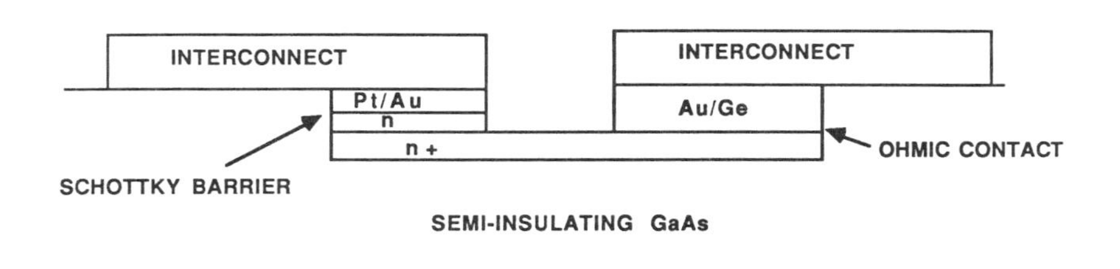

(c) Planar diode for MMICs

Figure 9.3 Practical diode structures

As the voltage across the diode varies, so the depletion region's width varies. Thus the intrinsic diode capacitance and resistance are both functions of the voltage across the intrinsic diode. Figure 9.4 shows the diode large-signal model. The series resistance comes from the contacts and the semiconductor bulk material; the shunt capacitance and series inductance represent the package parasitics. Modelling the diode involves measuring these component values at a variety of bias points, followed by extensive curve fitting. Alternatively a commercial CAD package's diode model can be used; then the basic parameters (such as zero-bias values and barrier voltage) are specified and one must assume their model fits.

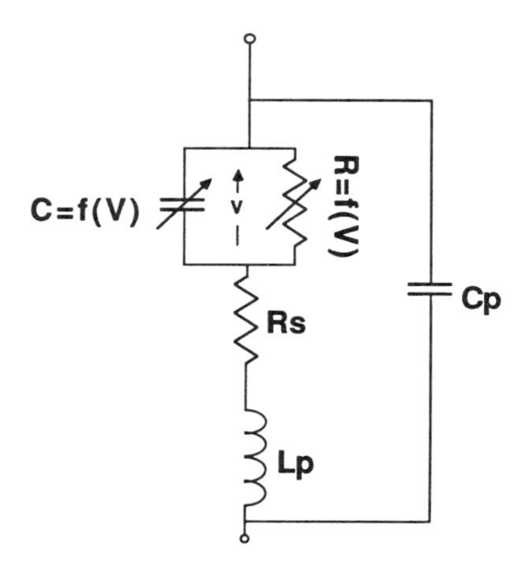

Figure 9.4 Diode large-signal model

9.3.2 Cooled Operation

GaAs is far better in cryogenically-cooled mixers where the lowest noise is required, because silicon suffers carrier freeze-out. At temperatures below around 70K, however, the Schottky diode's ideality factor increases so that the noise temperature improvement is not as good as expected. This is largely due to tunnelling. In addition, the capacitance variation with voltage in the Schottky can result in an increased LO power requirement: Firstly the time-varying diode impedance can prevent a good LO match, and secondly the capacitance variation can result in LO power being wasted in its harmonics. These problems have resulted in the development of the MOTT diode for cryogenically-cooled mixers operating at over 100 GHz. The major feature is that the active layer is lightly-doped but is so thin that the depletion region extends right through it. The light doping reduces tunnelling; having the active region fully depleted means that its capacitance does not vary much. Typically in active layer 0.1 μm thick, with a doping concentration of $1x10^{16}/cm^3$ is used. A true MOTT diode requires Molecular Beam Epitaxy to achieve the abrupt change in doping; because of this a compromise MOTTKY diode is often used.

9.4 Balanced Diode Mixers

A single diode can be used as a mixer by employing filters to isolate the RF, LO and IF signals, as shown in Figure 9.5. This technique gives the best conversion loss and lowest LO power requirement: Coupler losses are minimal, only one diode is being pumped, and it is easy to match the diode for optimum performance. This technique is thus used for millimetre-wave mixers, using low loss waveguide, because LO power is limited and low-noise amplifiers are not available which means that the mixer noise figure directly determines the receiver sensitivity. At frequencies above about 60 GHz this technique is used with GaAs diodes and gives the best sensitivity. HEMT low-noise amplifiers are possible at ever higher frequencies, however, and this reduces the mixer noise figure demands.

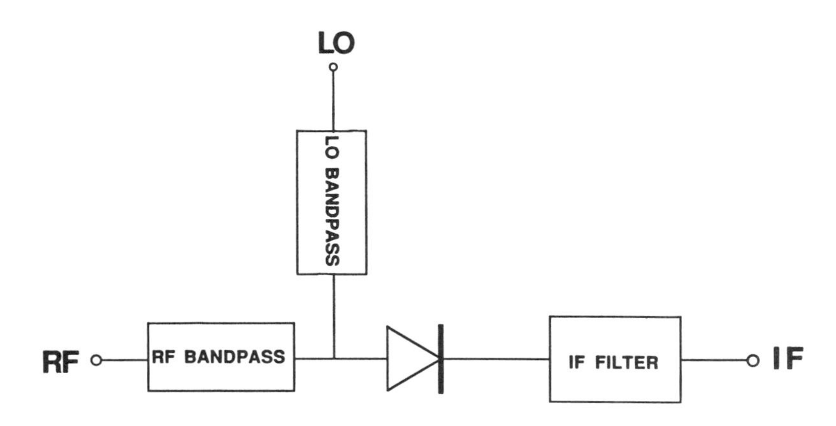

Figure 9.5 Single-diode mixer with filters for signal separation

In most applications, such as radar and communications, the need for the lowest possible mixer noise figure is lessened by the availability of excellent LNAs. Then the receiver selectivity and dynamic range are more important, because the spectrum is so crowded. In these applications balanced mixers are invariably used because they give good port-to-port isolations and cancel out a lot of spurious signals. This reduces the filtering problem and enables relatively simple receivers to be designed. Multiple diode mixers require more LO power, but this increases the signal compression level and is rarely a disadvantage.

9.4.1 Single-balanced Mixers

The single-balanced mixer consists of two diodes fed with RF and LO signals from a microwave hybrid. In a true single-balanced mixer one signal is fed in phase to the two diodes, the other is fed out of phase. The type of hybrid used and the orientation of the diodes must be correctly chosen so that the desired IF signals are in phase, whereas approximately

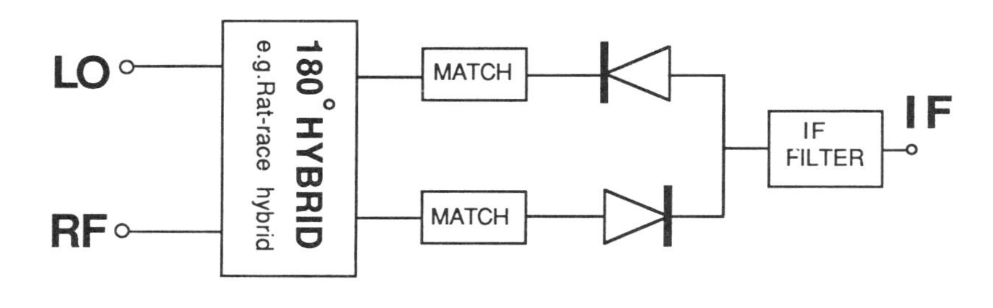

(a) 180 degree

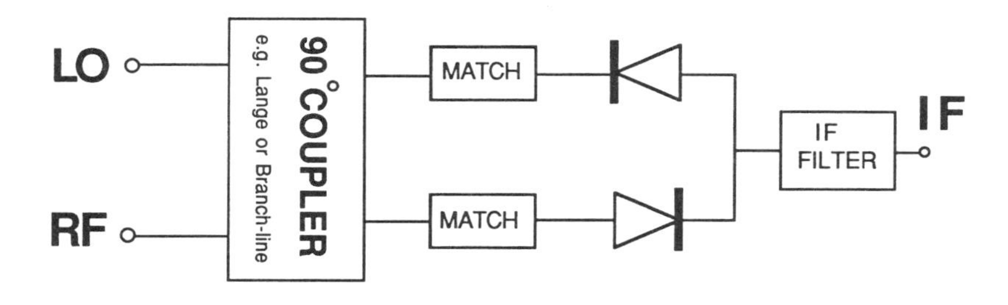

(b) 90 degree

Figure 9.6 Single-balanced diode mixers

half the unwanted signals cancel out. Exactly which ones cancel depends on the type of hybrid used and which ports the RF and LO are applied to. RF and LO signals are usually kept separate by the isolation of the hybrid. The hybrid can be realised as a waveguide magic–tee, a ferrite–cored transformer, or a microstripline coupler, depending on the frequency range and the cost/loss requirements. The two most common types of single–balanced mixer are shown in Figure 9.6. A detailed comparison of the properties of the various configurations can be found in Ref. 2. For applications with a low–noise oscillator is not available these are used because LO AM noise is rejected. More often though, the single–balanced mixer is used because of its simplicity.

9.4.2 *Double–balanced Mixers*

In double–balanced mixers four diodes are used and both the LO and RF are fed in anti–phase to them. The simplest arrangement is the ring mixer shown in Figure 9.7. The RF and LO now have separate 'baluns' to transform the unbalanced input signals into balanced signals for driving

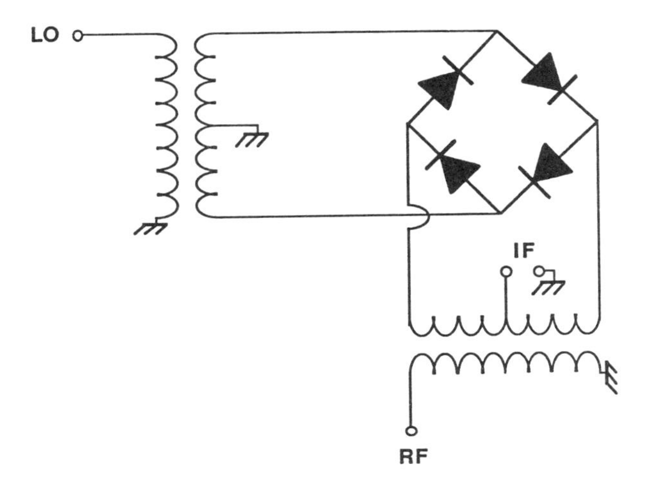

Figure 9.7 The double–balanced diode ring mixer

the diodes. The baluns can be realised as ferrite–cored centre–tapped transformers, or as transmission–line transformers. The double–balanced arrangement gives isolation between all three ports, and excellent rejection of spurious responses. Odd order intermodulation products are now

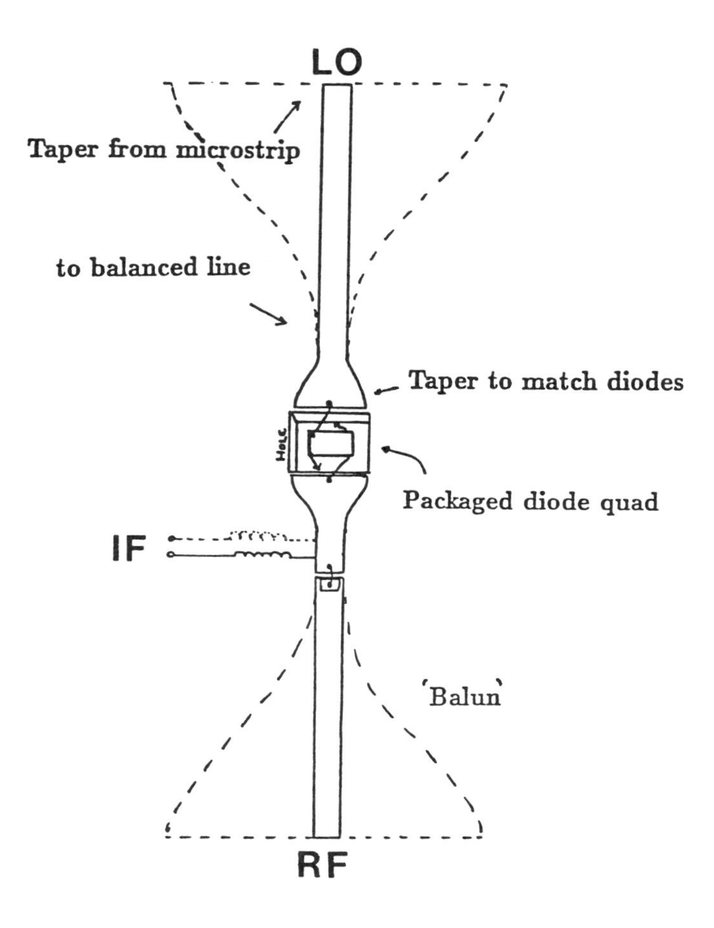

Figure 9.8 Schematic of a typical very wideband double-balanced diode mixer

rejected. The compression level is again increased at the expense of required LO power. To achieve good rejection the baluns must give accurate amplitude and phase balance and the four diodes must be well matched. Very well matched diodes are readily available in a special 4-terminal package (a diode 'quad') especially for this application: Figure 9.8 shows a typical wideband double-balanced mixer with transmission-line baluns and a diode quad. Such circuits can operate over 2 to 26 GHz with 25 dB port-to-port isolations. Their design is a bit of a black art however, and it is common to buy commercial packaged versions rather than attempting to design one from scratch.

There are some applications which require even more complicated balancing structures. Double–double–balanced mixers are one example, and are used where the IF band and RF bands must overlap [3].

9.4.3 *Image–rejection and Image–enhancement Mixers*

A superheterodyne receiver with a fixed LO and a certain IF bandwidth is sensitive to two RF bands, one above the LO, the other below the LO, as shown in Figure 9.9. If the desired RF signal is only in one of these bands, then the signal–to–noise ratio at the mixer output is degraded by 3 dB because noise in the other sideband is also downconverted. The value of 3 dB comes from the doubling of the downconverted noise bandwidth. The unwanted band is called the image band. In most fixed LO systems this degradation is avoided by using a filter to filter out the image noise. The IF frequency must be high enough to keep the two bands apart for filtering.

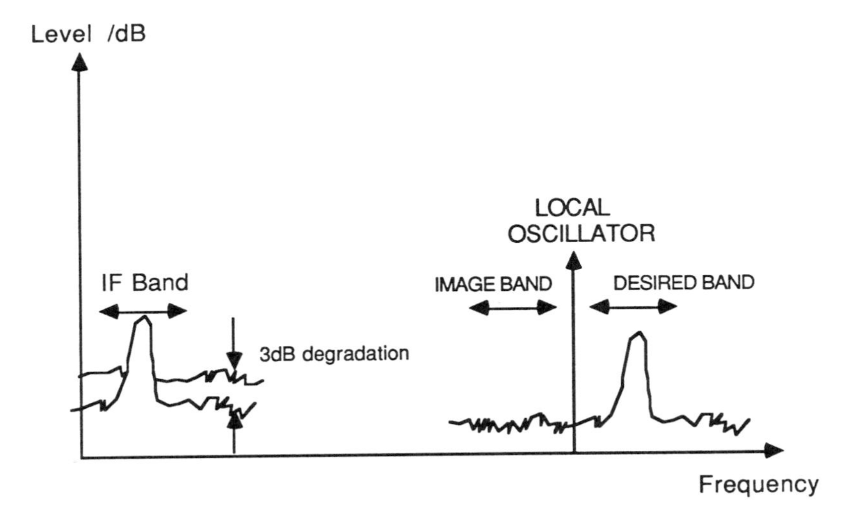

Figure 9.9 The image noise problem

There are applications where filtering the image may not be possible: For example multi–octave swept LO receivers, receivers which must have a low IF and MMIC receivers which cannot have high–Q filters. In these applications the image–rejection mixer can be used. Figure 9.10 shows the block diagram of an image–rejection mixer. To understand its operation both positive and negative frequency components must be considered, as shown in Figure 9.11. The crucial point is that a 90° delay

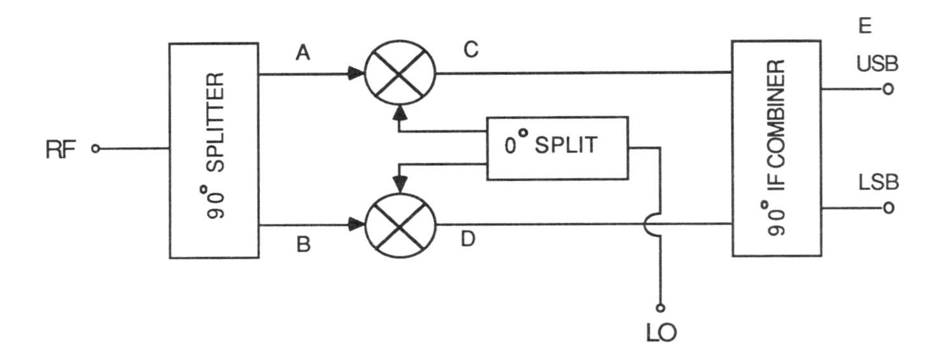

Figure 9.10 Image rejection mixer block diagram

means –90° for positive frequency components, but +90° for negative frequency components. After downconversion the LSB positive frequency component becomes negative and vice versa; the USB components remain as they were. Thus the USB and LSB components undergo different phase–shifts in the IF hybrid: For one sideband the 90° delay in the RF hybrid is cancelled; for the other sideband the two 90° delays add to make 180°, and the components from each mixer cancel out. In practice an image–rejection value of around 25 dB can be expected. This prevents image noise being a problem, but if there is a strong unwanted signal at the image frequency then a filter is still required.

Image–enhancement (also known as image–recovery) is different: In the normal mixing process frequencies of $F_{rf} - F_{lo}$ (the desired IF) and $2F_{lo} - F_{rf}$ are both generated. $2F_{lo} - F_{rf}$ is the image frequency but it represents an internally generated signal, not an input. The image–enhancement technique involves reconverting this wasted signal power to the IF frequency in order to improve the conversion loss a little [4]. This can be done with a filter at the RF port which is matched at F_{rf}, but reflects $2F_{lo} - F_{rf}$ back into the mixer. If using a filter is not possible, as in multi–octave receivers for example, it is also possible to construct an image–enhancement mixer using a phasing technique similar to that used for image–rejection [5].

In most applications, the RF input filter can be arranged to filter out input noise and unwanted signals at the image frequency, as well as reducing the conversion loss by image–enhancement. It is important to

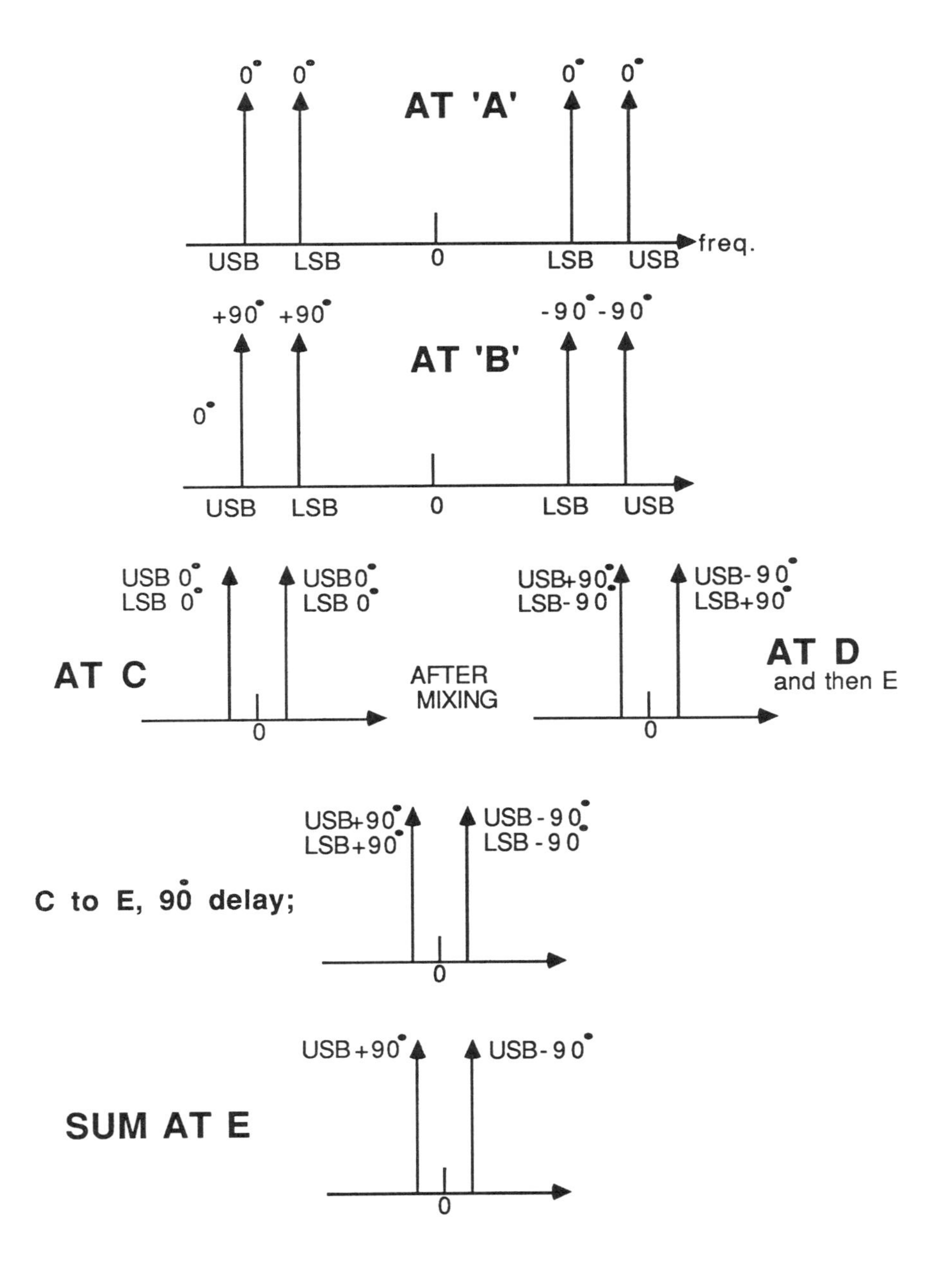

Figure 9.11 Spectra showing cancellation of the image signal

realise that a receiver, even with a noise–less LNA, suffers a 3 dB increase in output noise power if the image noise is not filtered out.

9.5 FET Mixers

The MESFET mixer was first investigated many years ago [6]. Unfortunately MESFET mixers have not achieved the excellent noise figures that MESFET amplifiers have, and so MESFET mixers are rarely encountered outside research. This situation is changing however because device technology is rapidly improving and because of the great interest in monolithic circuits for high volume applications. Already MESFET mixers can give better noise figures than diode mixers up to around 12 GHz if designed properly. The arrival of HEMTs promises even better noise figures, and millimetre–wave operation [7]. Thus FET mixers can offer better noise figure, power–handling and conversion gain than diode mixers and use less LO power. They are however harder to design because of feedback considerations, which affect the conversion gain and stability.

One problem that has not been overcome is that of l/f flicker noise, which is readily upconverted to the signal frequency in a non–linear device. This is attributed to effects at the active layer interfaces. It is hoped that HEMTs will be better than MESFETs in this respect, because of the different active layer structure.

9.5.1 Types of FET Mixer

The single–gate FET can be used in a variety of mixing modes. The LO can be fed into the source, drain or gate. It has been found that the 'gate mixer' gives the best performance, although the other techniques can have significant advantages – particularly in MMICs where simplicity is important. In the gate mixing mode the LO is combined with the RF and then fed into the gate. If the FET is biased near pinch–off then its transconductance can be made to vary significantly for a fairly small LO signal, as shown in Figure 9.12. With a time–varying transconductance the FET becomes a mixer. In practice most of the other FET components will change as well, and this makes theoretical, and even computational, modelling of the mixer behaviour very complicated. Reference 8 includes the most rigorous analysis to date, and detailed information on the design procedure: If this procedure is followed then a low noise figure should be possible, as well as conversion gain and a high compression level.

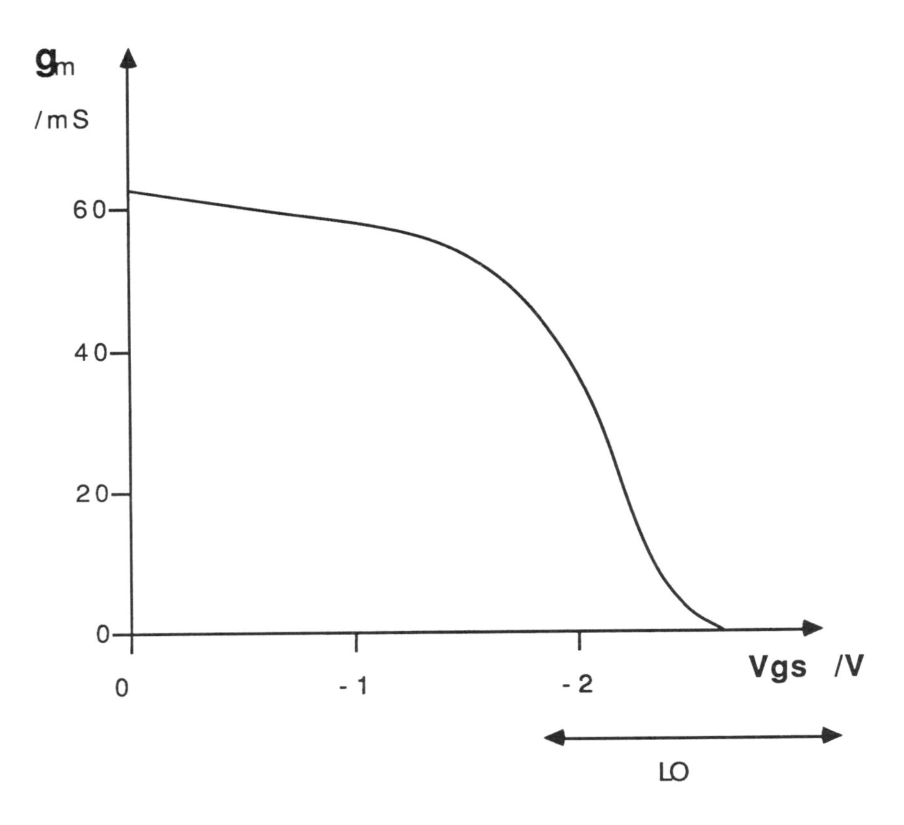

Figure 9.12 A graph of Gm vs Vgs for a FET, showing gate mixer operation

Dual–gate FETs can also be used as mixers, with the RF and LO applied to separate gates, to overcome the need for a coupler/combiner. The basic dual–gate FET cross–section and the model generally used are shown in Figure 9.13. Modelling the device as two series–connected FETs would seem to be applicable only if the two gates are kept far enough apart to keep the two depletion regions separate. In practice, if the gates are very close to reduce the parasitic resistance of the material between them, the model of Figure 9.13 is unlikely to be valid for all DC bias conditions. Sometimes a low resistance n+ layer is put between the two gate stripes, and sometimes a third, unconnected, gate stripe is put between the two signal gates. These techniques generally improve the noise figure, because the inter–FET resistance is reduced, but can result in a greatly increased inter–gate capacitance which affects stability and signal isolation.

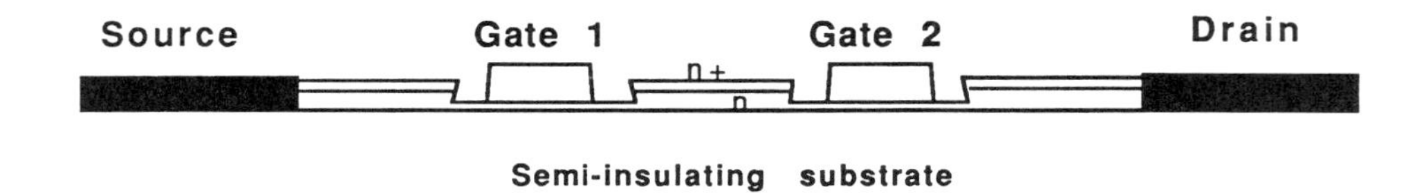

(a) Typical cross-section

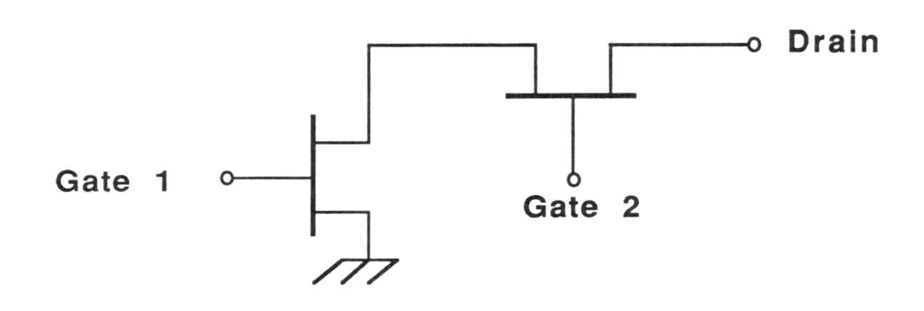

(b) Model with separate series connected FETs

Figure 9.13 The dual-gate FET

The dual-gate FET can be operated at a variety of bias points: Figure 9.14 shows the 'bi-dimensional' DC characteristic commonly used to describe dual-gate FET behaviour. The horizontal axis represents the fact that the total drain voltage of 5V is split between the two FETs. Because the source of FET2 is at potential Vds1, Vg2 can be positive voltage. The LO can be applied to either gate1 (nearest the source) or gate2. The various operating modes have been described in great detail by Ashoka [9,10]. The lowest noise figure is said to be achieved by applying the LO to gate2, and operating the device in region A of Figure 9.14. In this mode FET1 is driven between its linear and saturated region because the LO changes Vds1; FET2 is driven between saturation and pinch-off. Thus both FETs in the model are non-linear, and the mixing occurs in both: FET1 is a drain-mixer, FET2 acts as a gate-mixer and IF amplifier. The impedances presented by the 3 ports at each frequency must be carefully chosen to ensure stability, conversion gain and low noise figure. It is not usually possible to match all three ports at their respective frequencies without causing instability, and a rough-and-ready design approach will thus often cause problems.

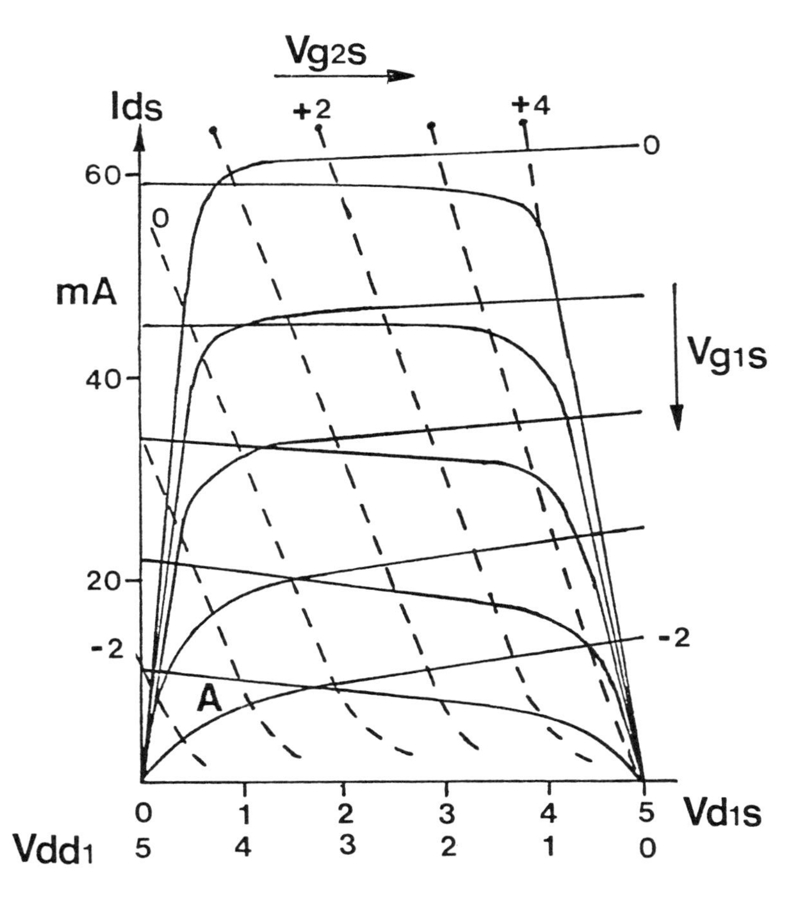

Figure 9.14 Bi-dimensional DC characteristic of a typical dual-gate FET

9.5.2 Device Large-Signal Modelling

The large-signal FET model is described in great detail in Chapter 3. The complexity of this model is a major reason for the limited use of FET mixers outside research. Only recently has this type of model been incorporated into commercial microwave CAD packages. Before these become available, theoretical and practical work were used to define the basic design principles, but actually optimising the circuit was not possible. Theoretical analysis of noise figure for single-gate FETs has been attempted [11] but, as is usual for microwave circuits, computer modelling techniques are really required because of the need to model so many details such as parasitics. This is not yet possible and so designing a low-noise FET mixer is by no means straightforward, although at least the large-signal operation can now be predicted accurately, using packages such as Libra(TM), Harmonica(TM), and Hewlett-Packard's Microwave Design System(TM). Details of all these packages can be found in Chapter 4. The dual-gate FET model is not yet included because it is much more complicated: At least 8 model elements have a significant dependence on the three separate bias voltages.

9.5.3 Balanced FET Mixers

Balancing techniques can be employed with FET mixers to give the same advantages of port-to-port isolation, spurious signal cancellation and LO AM noise rejection as for diode mixers. Balancing in diode mixers can be arranged by simply using back-to-back diodes fed with balanced signals. FETs cannot be used in this way and so the IF signals from individual mixers must be combined in an IF hybrid. This is not a serious problem for simple balanced arrangements with narrowband IFs. However, using wideband IFs or complicated balancing arrangements is much more involved than for diode mixers.

Figure 9.15 shows the block diagram of a single-balanced mixer using a 180° hybrid and single-gate FET mixers. The rejection characteristics are basically the same as for diode mixers, but the additional hybrid may reduce bandwidth and signal rejection. If single-gate FET mixers are used in much more complicated arrangements the number of hybrids becomes too great, leading to large circuits with a lot of loss before the mixers, and signal routing problems. Using dual-gate FET mixers instead can reduce the number of hybrids required because of the inherent LO/RF isolation. This makes double-balanced and image-rejection FET mixers very attractive. Figure 9.16 shows a double-balanced dual-gate FET mixer,

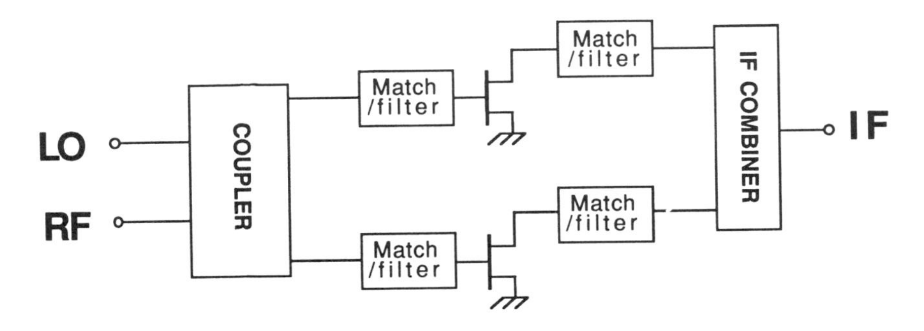

Figure 9.15 Single-balanced FET mixer

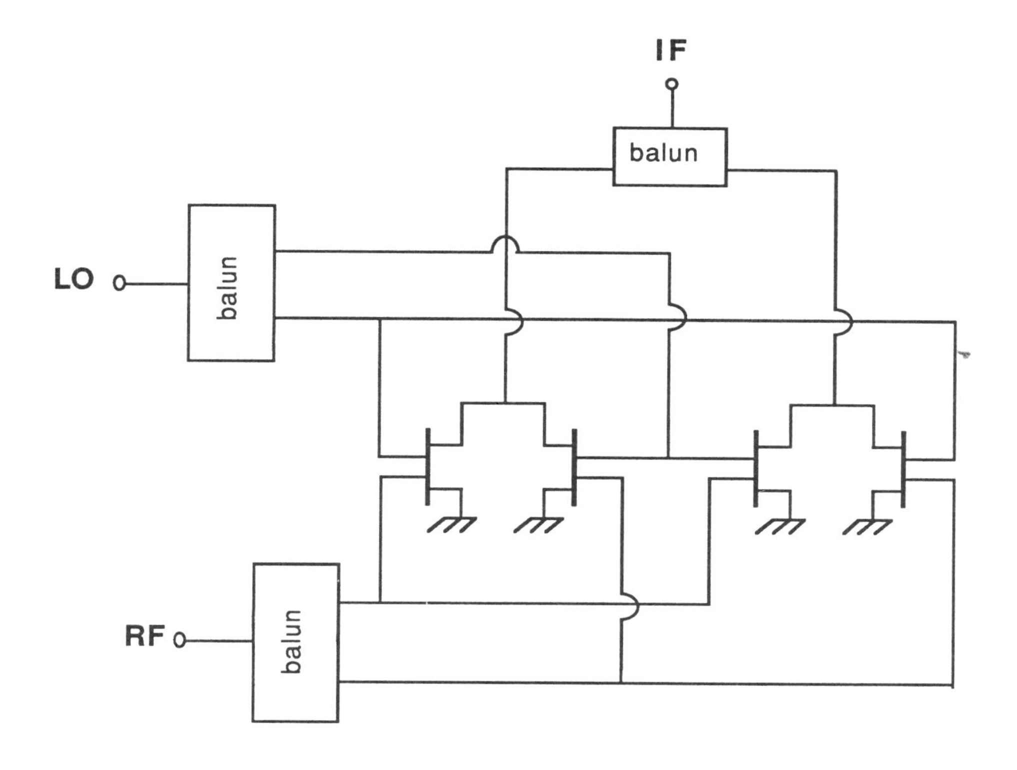

Figure 9.16 Double-balanced dual-gate FET mixer

for example. For high frequencies this type of circuit is better suited to monolithic implementation [12] because in hybrid form the connections between FETs would have too much inductance. An image-rejection mixer using dual-gate FETs is shown in Figure 9.17. This technique is useful for image noise rejection in MMIC receivers, where image filtering may be impossible, and in the very wideband receivers used in Electronic Warfare and measurement systems.

9.5.4 Distributed FET Mixers

The distributed MESFET mixer was first demonstrated by Tang and Aitchison [13] in hybrid form, using single-gate FETs with a passive coupler to combine the RF and LO signals (Figure 9.18). The principle is similar to the distributed amplifier with the gate and drain

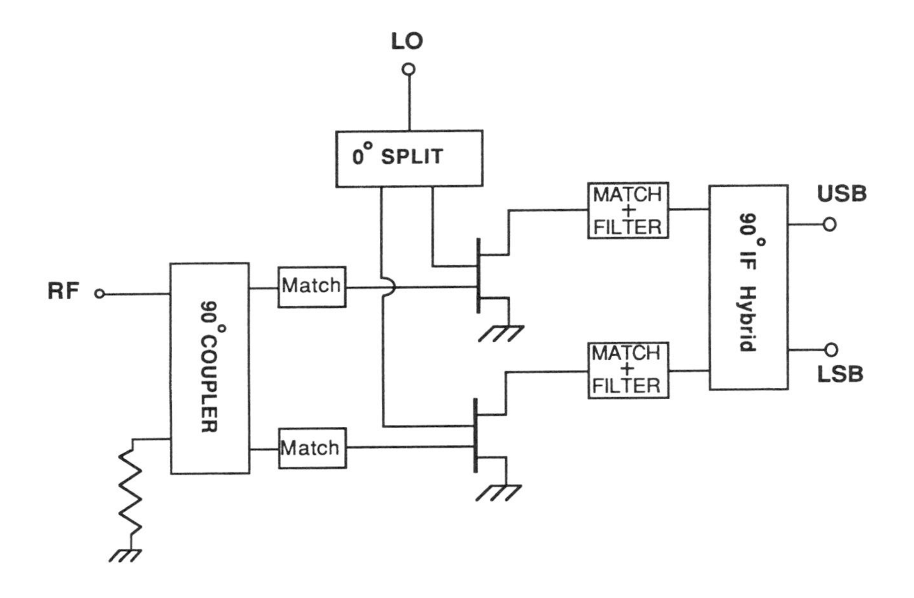

Figure 9.17 Image–rejection mixer with DGFETs

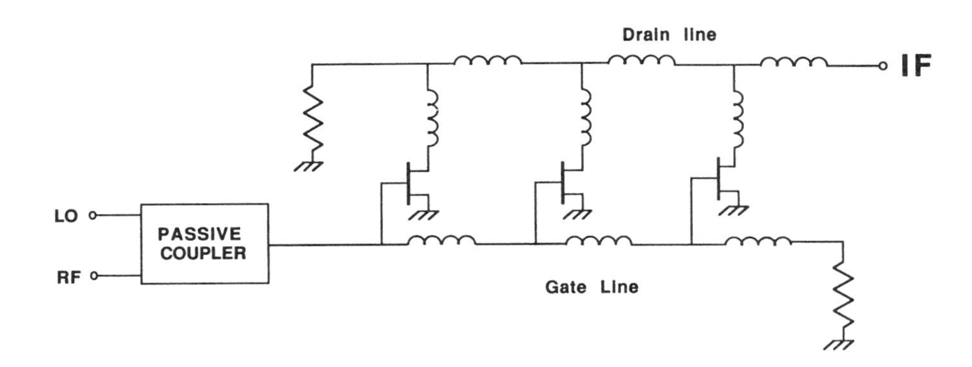

Fig. 9.18 The original distributed mixer configuration

capacitances incorporated into artificial transmission–lines, so that multi–octave bandwidth is achieved. Dual–gate FETs can also be used so that the coupler is not needed [14]. The mixer then has a gate–line for the RF, a gate–line for the LO and a drain–line for the IF, as shown in

Figure 9.19. By having a distributed drain–line very high IFs are in principle possible, although a way of obtaining LO and RF to IF isolation is needed. This can be done with a double–balanced arrangement.

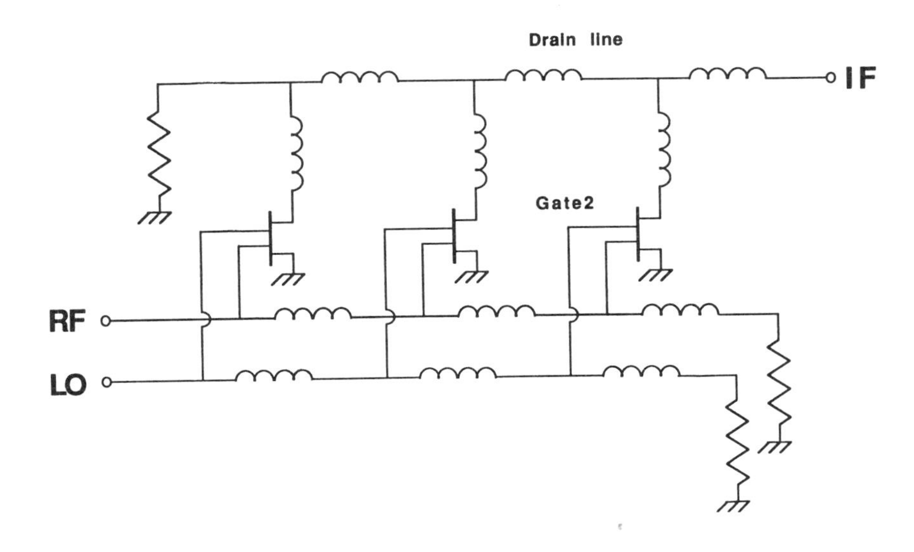

Figure 9.19 A distributed mixer using dual–gate FETs

A few dB gain with a noise figure of around 10 dB single–sideband is achievable over 1 to 12 GHz using 0.5 μm gate–length MESFETs. Shorter gate–length MESFET and HEMTs could be used to give state–of–the–art noise/gain/bandwidth performance. In practice it is just as important to find equally wideband techniques for using distributed mixers in balanced and double–balanced arrangements. Work in this area is largely confined to MMIC circuits as will be discussed. Wideband power splitters, combiners and baluns can be made using the distributed amplification technique [15]. MMICs give considerable scope for a range of different distributed mixer ideas to be tried [16,17].

9.6 MMIC Mixers

GaAs MMIC mixers have so far had a limited impact on systems, particularly compared with the new generation of low cost MMIC amplifier gain blocks. In this section the reasons for this will become clear, and the new modelling and circuit techniques which are expected to change the situation will be described. One of the most promising applications for MMICs is in millimetre–wave communications, where the reduced costs and high–volume capability of the transmitter–receiver circuits makes the use of new frequency bands economical. The development of advanced MMIC mixers is crucial to this, and other applications.

9.6.1 Difficulties in MMIC Mixer Design

Fortunately FETs and diodes can be used in MMIC mixers, just as for hybrids, although the accuracy of the models becomes more important. The major problem in realising mixers is that the transmission–line circuits, couplers and filters used in conventional microwave mixers are at best wasteful of space and therefore expensive, and at worst they are non–planar and hence not realisable. At frequencies above about 20 GHz, couplers can be used with great success, but even at these frequencies they are a major obstacle to keeping integration levels high. One approach is to use folded and spiral couplers instead, but ultimately stray coupling limits how compact the couplers can be made. Another approach is to replace them with active circuits, by extending MOSFET design techniques to GaAs circuits. This requires non–linear analysis software which can accurately model the transistors, as well as lumped and distributed passive components. Standard SPICE™ does not cope well with transmission lines, and at microwave frequencies this is a big problem. Software which can cope with both time–domain and frequency–domain type components is required. The recent introduction of commercial harmonic–balance analysis software is a milestone for MMIC mixers because complex actively–matched mixers can now be modelled properly. In the near future, high–density active MMIC mixers will probably replace techniques based on hybrid technology.

The major remaining problem for mixer modelling is that device and passive component models must be improved significantly. The mixer includes many harmonics, and the impedances presented to them do affect the performance. Thus a spiral inductor model that is accurate to 20 GHz may be adequate for an 18 GHz amplifier design, but to predict the gain and stability of an 18 GHz FET mixer requires a much better inductor model because of the signals at 36 GHz etc. Field–based modelling techniques are better in this respect, and can model spirals well beyond their resonant frequency [18].

9.6.2 Choice of Device

Diodes, MESFETs and HEMTs can all be used to realise a GaAs MMIC mixer. Whether the mixer can be integrated with other circuits on the same chip depends on the active layer technology used and the mixer device chosen. Each device type requires a different active layer profile, as described in Chapter 2, and it may not be possible to include different devices on the same wafer. So, for example, to integrate a 30 GHz LNA and mixer on the same chip it is best to use a HEMT mixer. It is possible that new processing techniques will overcome this limitation, but extra processing means more costs and so MESFET and HEMT mixers are very attractive for cheap single–chip receivers. In lower frequency circuits selective ion–implantation is quite common and this allows MESFETs and diodes to be used on the same wafer. Diode mixers are generally more compact because balancing and biasing are simpler. For very wideband applications MESFETs and HEMTs have the advantage of being suitable for distributed matching techniques.

9.6.3 Balancing Techniques

Filters cannot be used in most MMIC applications. Thus if the mixer has poor port–to–port isolations or generates a lot of spurious output signals then it cannot be integrated on a receiver chip without causing problems of intermodulation, IF amplifier nonlinearities etc. Thus good balanced mixers are essential for realising integrated receivers, unless system performance degradation is allowed.

At frequencies above around 20 GHz the same techniques can be used as for hybrid FET mixers, because the couplers become relatively small. Thus mixers can be realised easily at frequencies like 30 GHz [19], or even 94 GHz [20], using diodes and couplers in conventional circuits. Figure 9.20 shows a typical mm–wave balanced MMIC mixer layout. The advantages of using MMICs are potential low cost and easy manufacture, and also the device parasitics are reduced and accurately predictable. The disadvantage is that state–of–the–art noise figures are not usually achievable because MMICs are quite lossy. The degradation is not serious for most applications, however.

At lower frequencies (below about 10 GHz) the couplers are far too large and so either lumped–element splitters and combiners are used [21], or active circuits can be employed. Figure 9.21 shows the circuit diagram of a lumped–element 90° phase splitter, for example. Figure 9.22 shows an actively–matched double–balanced FET mixer circuit [12]. As the frequency goes up, lumped–element and active techniques become problematical because of component parasitics, such as the feedback capacitance of spiral inductors. In this frequency range a combination of lumped–elements, active circuits, and folded couplers can be used. Folded couplers help integration because their aspect ratio is more convenient [22]. Figure 9.23 shows the layout and coupling characteristics of a folded MMIC coupler which uses a novel overlayed structure [23].

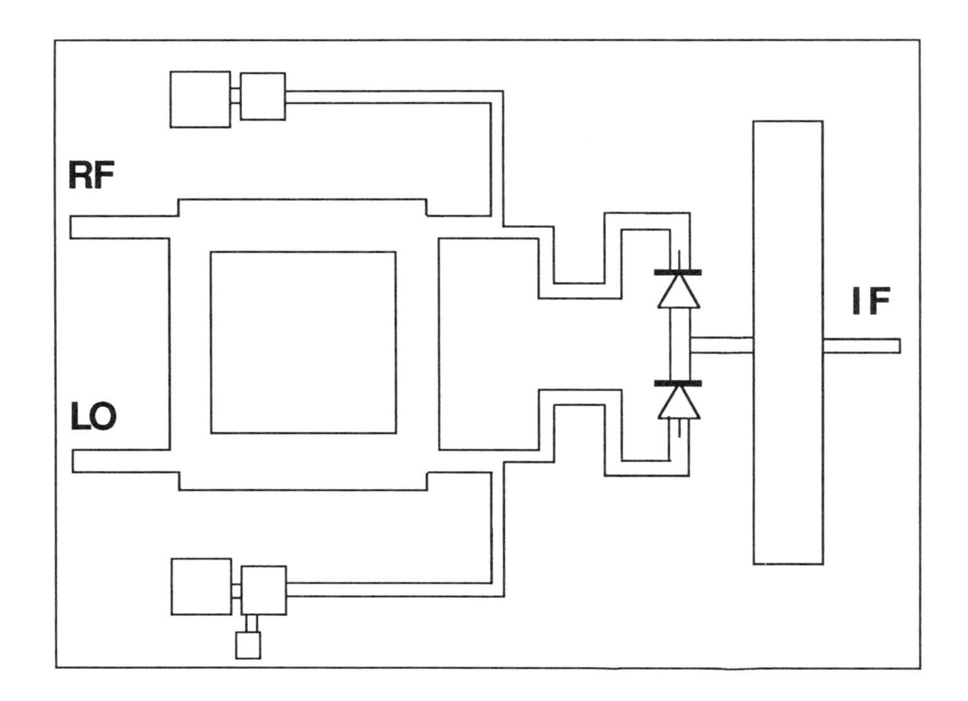

Figure 9.20 Schematic of a typical mm-wave balanced GaAs MMIC mixer

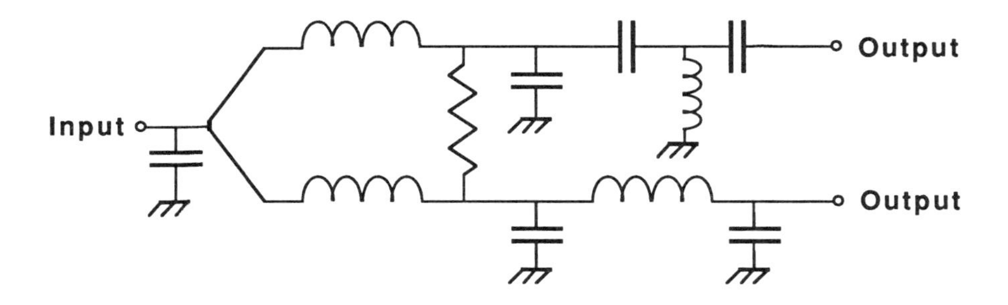

Figure 9.21 Lumped element 90° phase splitter, suitable for MMICs

Field modelling programs like LINMIC+ are needed for such techniques [24].

Spiral transformers, consisting of inter-wound spiral inductors, are a form of folded coupler, and can be used in diode mixers in place of conventional bi-filar wound transformers. They can perform bias

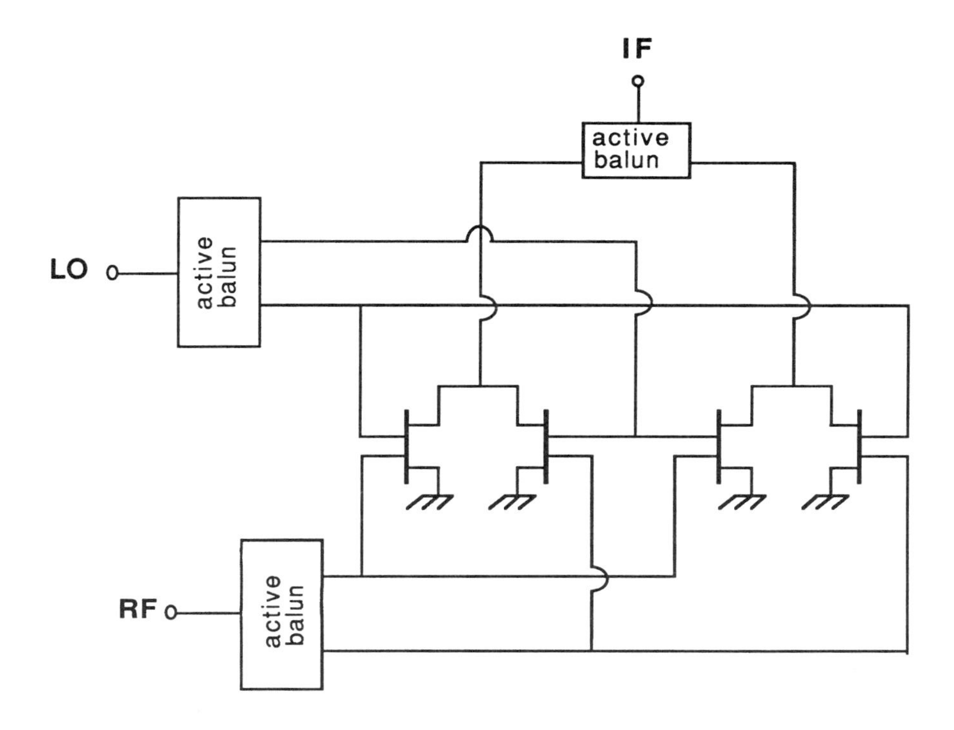

Figure 9.22 A double–balanced dual–gate FET mixer suitable for MMIC integration

injection, matching, DC blocking and phase–shifting all at once. This technique can result in very high packing density, with even the space in the centre being used by a transistor [25]. The frequencies that this technique can be used at are limited by the size and DC resistance at low frequency and the inter–spiral capacitances at high frequencies.

Figure 9.24 shows a circuit which can be used as an active balun [26]. This has been demonstrated at X–band with good results. With the new non–linear analysis programs it is possible to model the effect of parasitics easily, and so active techniques like this can be used at higher and higher frequencies.

Dual–gate FETs provide an attractive solution for MMIC mixers because the RF and LO can be input to separate gates, and so the number of couplers can be reduced. This approach is commonly used for DBS receivers and excellent results have been obtained [27,28]. The advantages and disadvantages of dual–gate FETs have been discussed already.

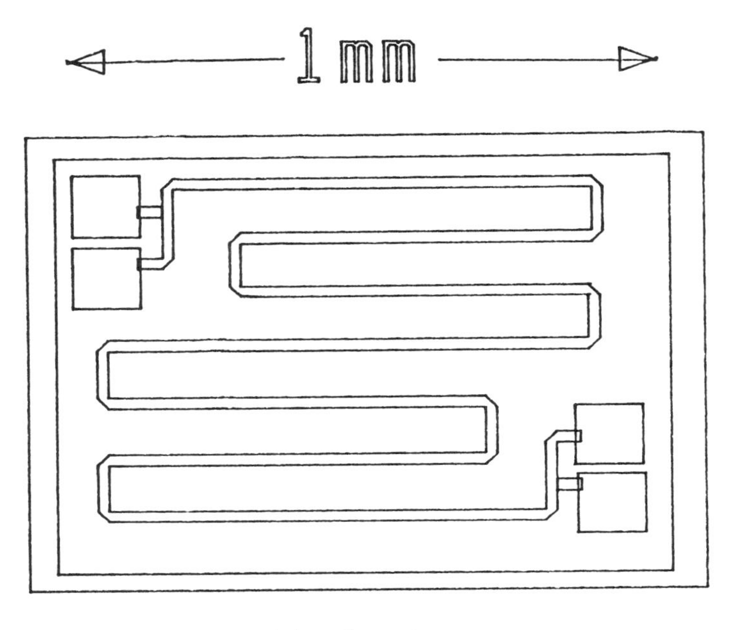

(a) Layout

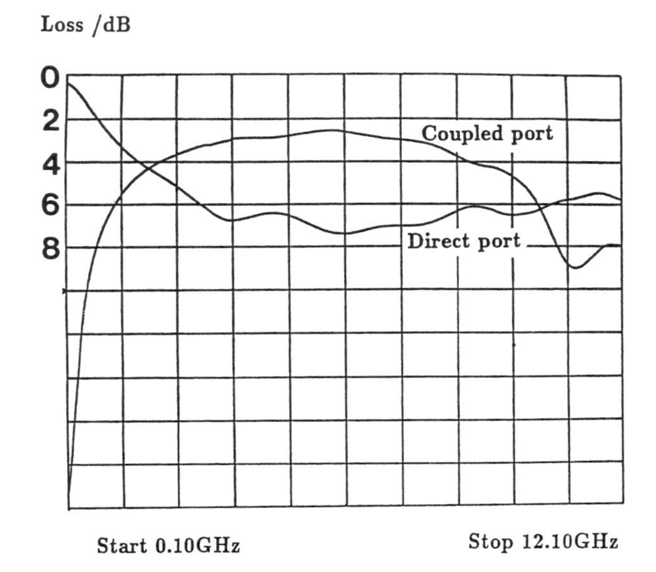

(b) Measured coupling characteristic

Fig. 9.23 An experimental overlayed MMIC coupler

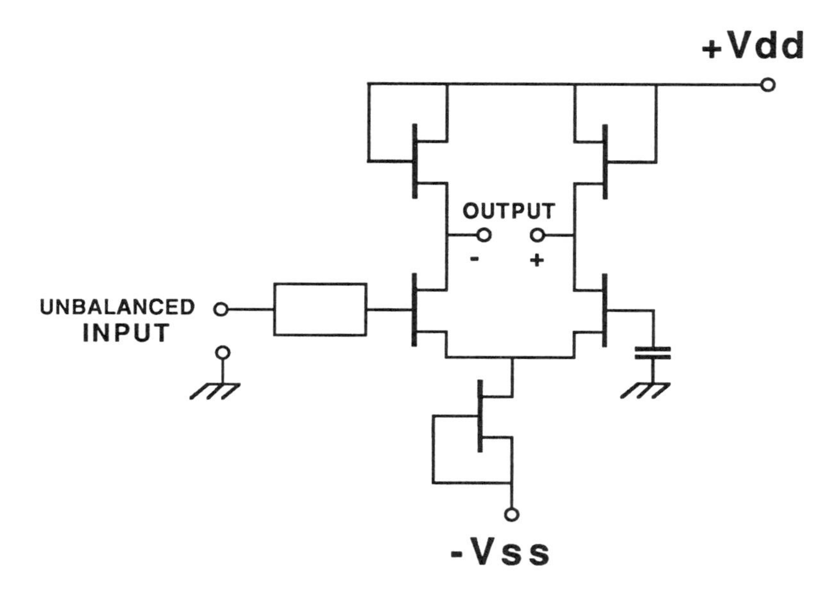

Fig. 9.24 An active balun circuit, to overcome the need for large couplers

Virtually all GaAs Foundries now offer dual–gate FETs, but accurate large–signal models are not commonly available.

9.6.4 Novel Circuit Designs

For the circuit designer, MMICs have a major advantage over hybrid techniques: Devices of any shape and size can be incorporated in a circuit with virtually zero parasitic bond–pad capacitance and bond–wire inductance. This makes it possible to devise novel circuits which are impossible in hybrids because the extra parasitics cause gain roll–off, phase shifts, instabilities and other problems. The ubiquitous MMIC feedback amplifier is an excellent example of how the reduced parasitics lead to better performance (particularly bandwidth) and easier design [29].

In distributed mixers this advantage is considerable. Distributed matching demands a simple FET equivalent circuit, and package parasitics degrade the achievable bandwidth considerably. For example, a 1 to 18 GHz active balun has been demonstrated in MMIC form [30]. This uses the natural signal inversion of the FET to generate anti–phase output signals, and excellent amplitude and phase balance is achievable. Double–balanced MMIC mixers are now possible using this type of circuit. At present these are slightly inferior to conventional mixers, especially for noise figure. However, MMIC mixer techniques are still near the bottom of the learning curve, whereas the double–balanced hybrid MIC mixer has been optimised over many years. Thus it can be confidently predicted that a distributed MMIC mixer using HEMTs will fairly soon offer state–of–the–art noise figure and bandwidth performance.

Many other novel circuit techniques will be developed by using the advantages of MMICs to the full. For a plethora of possible circuits it is useful to consult books on MOSFET analogue design (e.g. Ref. 31), and even old high–frequency valve circuit design. The distributed MMIC amplifier is in fact an extension of a valve circuit technique [32].

9.7 Future Developments

The importance of the GaAs mixer diode as the lowest noise receiver is being eroded at both ends of its useable frequency range. At frequencies below around 60 GHz, HEMTs can already provide lower noise temperatures than diode receivers. Another new device which may have a big impact on mixers is the Heterojunction Bipolar Transistor. This may be particularly useful for mixers if the l/f noise problems encountered with FETs are overcome. At frequencies over 100 GHz the SIS (semiconductor–insulator– semiconductor) tunnel junction mixer is now capable of bettering Schottky diode receiver noise temperatures [33]. GaAs diode mixers will continue to be used in most applications for some years however, because of their simplicity and the wealth of design techniques available.

GaAs MMIC mixers are expected to become very much more important over the next few years. New systems are being developed which rely on the possibilities for mass producing low–cost microwave and millimetre–wave circuits. For example, a system to broadcast TV and other services over the 'last mile' to the subscriber via short mm–wave line–of–sight links has been proposed [34]. The system, shown in Figure 9.25, is known as M^3VDS (Millimetre–wave Multichannel Multi–point Video Distribution System), and relies on the cost advantages of using a GaAs MMIC for the subscribers' downconverter, which will have an input frequency around 40 GHz. Also the possibility of using 60 GHz links for micro–cellular mobile communications is being investigated, and this again required GaAs MMIC technology to reduce the cost of the millimetre–wave transmitters and receivers.

In radar and many other applications there is tremendous interest in the active phased–array antenna system. Circuit reproducibility and cost is of paramount importance in making such systems viable. Hence most work on the individual Transmit/Receive modules involves using GaAs MMICs. IC mixers must be improved, or else they will be a major weak

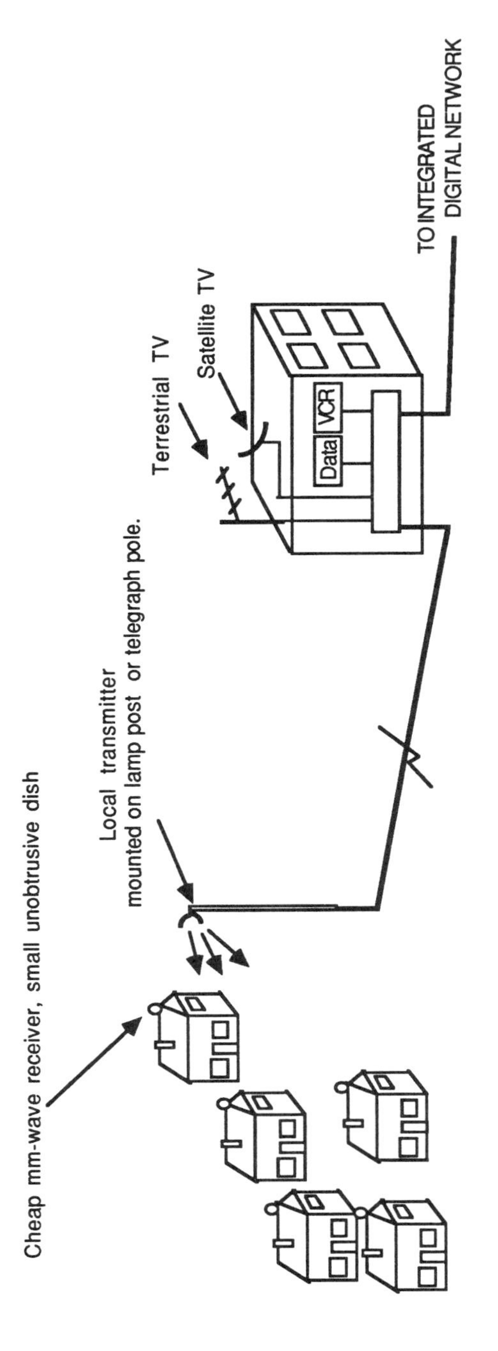

Figure 9.25 Schematic of a proposed system to distribute TV by short mm-wave links

link in realising the single-chip TX/RX module that is so often talked about.

In the low microwave frequency range, however, the future for GaAs technology is quite bleak: Low cost silicon MMIC mixers operating at 5 GHz are already commercially available. This suggests that for many such applications GaAs will only be required for special functions such as low noise amplifiers.

9.8 Conclusions

GaAs has already had a big impact on conventional systems, thanks to the Schottky barrier mixer diode. The future for GaAs technology in millimetre-wave mixers is very promising, particularly in the form of integrated circuits. There are many systems being developed and proposed which rely on the low cost and reproducibility of GaAs integrated circuit transmitter/receivers. There is no reason why GaAs MMICs should not be used at frequencies as high as 200 GHz or more; the devices are becoming available very rapidly. Digital electronics has proved that the capital costs of IC manufacture are easily paid for by the ability to mass produce circuits, and the ability to make circuits more and more complex. In millimetre-wave applications the monolithic circuit has the additional advantage of reducing the parasitics between components and devices.

However, because filters are not practical, the IC mixer must have excellent performance. It is not possible to build a complete receiver around a mixer that generates a wide range of spurious outputs, and is sensitive to a wide range of spurious inputs. In communications the result would be regular outages due to interference, and a generally poor service. In military systems the result would be false alarms, and susceptibility to jamming. There are very few applications where the system performance can be degraded simply to enable MMICs to be used. Thus considerable improvements in IC mixers are required in order to take full advantage of the capabilities of GaAs. Some researchers have made excellent progress in this field, and now that non-linear microwave circuit analysis is possible the performance of IC mixers should improve significantly, using novel techniques as described.

9.9 References

1 K.H.G. Duh, et al: 'Millimeter-Wave Low-Noise HEMT Amplifiers', IEEE MTT International Microwave Symposium Digest, 1988, pp923–926

2 S. Maas: 'Microwave Mixers', Artech House, 1986, Chapter 7

3 B.C. Henderson: 'Mixers: Part 1 Characteristics and Performance', The Watkins Johnson Signal Processing Handbook 1988

4 B.R. Hallford: 'An Experimental Study of Image Termination Methods for Low Noise Mixers', IEEE MTT International Microwave Symposium, 1976

5 T.H. Oxley, et al:'Phasing Type Image Recovery Mixers', IEEE MTT International Microwave Symposium, Philadelphia 1980.

6 J.E. Sitch and P.N. Robson: 'the Performance of GaAs MESFETs as Microwave Mixers', Proc. IEEE, Vol.61, p399

7 P.D. Chow, et al: 'Design and performance of a 94 GHz HEMT Mixer', IEEE MTT International Microwave Symposium, June 1989

8 S. Maas: 'Microwave Mixers', Artech House, 1986, pp281–293

9 H. Ashoka and R.S. Tucker: 'Modes of operation in dual–gate MESFET mixers, Electron. Lett., Vol.19, pp428–429, 1983

10 H. Ashoka: 'Modelling and Analysis of Dual–Gate MESFET Mixers', PhD thesis, University of Queensland, Australia, June 1985

11 G.K. Tie and C.S. Aitchison: 'Noise Figure and Associated Gain of a Microwave MESFET Mixer', proc. 13th Annual European Microwave Conference, 1983, pp579

12 K. Kanazawa, et al: 'A GaAs Double–Balanced Dual–Gate FET Mixer IC for UHF Receiver Front–end Applications', IEEE MTT International Microwave Symposium Digest, 1985, pp60–62

13 O.S.A. Tang and C.S. Aitchison: 'A practical Microwave Travelling Wave MESFET Gate Mixer', IEEE MTT International Microwave Symposium Digest, 1985, pp605–608

14 T.S. Howard and A.M. Pavio: 'A Distributed 1–12 GHz Dual–Gate FET Mixer', IEEE MTT International Microwave Symposium Digest, 1986, pp329–332

15 I. D. Robertson and A.H. Aghvami: 'Novel Techniques for Multi–Octave GaAs MMIC Receivers', IEE Colloquium Digest, 'Multi–Octave Active and Passive Components', Savoy place, May 1989

16 M. Fairburn, B. Minnis and J. Neal: 'A Novel Monolithic Distributed Mixer Design', IEE Colloquium Digest, Microwave and Millimeter–wave Monolithic Circuits, November 1988

17 I. D. Robertson and A.H. Aghvami:'' 'A Practical Distributed FET Mixer for MMIC Applications', IEEE MTT International Microwave Symposium Digest, 1989

18 R. H. Jansen, R.G. Arnold and I.G. Eddison: 'A Comprehensive CAD Approach to the Design of MMICs up to Millimetre–wave Frequencies', IEEE Trans. MTT–36, February 1988, pp208–219

19 L.C. Liu, et al: 'A 30 GHz Monolithic Receiver', IEEE Trans., MTT–34, December 1986, pp1548–1552

20 P. Bauhahn, et al: 'A 94 GHz Planar GaAs Monolithic Balanced Mixer', IEEE Microwave and Millimeter–wave Monolithic Circuits Symposium Digest, 1984, pp70–73

21 G.B. Beech, C.W. Suckling and J.R. Suffolk: 'An S–band Image Rejection Receiver Front–end Incorporating GaAs MMICs', proc. European Microwave Conference, 1985, pp1019–1024

22 R.C. Waterman et al: 'GaAs Monolithic Lange and Wilkinson Couplers', IEEE Trans, 1981, ED–29, pp212–216

23 I.D. Robertson and A.H. Aghvami: 'Novel Coupler for GaAs MMIC Applications', Electronics Letters, 8th December 1988, pp1577–1578

24 R.H. Jansen: 'LINMIC, a CAD Package...', Microwave Journal, February 1986, pp151–161

25 D. Ferguson et al: 'Transformer Coupled High–Density Circuit Technique for MMICs', IEEE Microwave and Millimetre–wave Monolithic Circuits Symposium Digest, 1984, pp34–36

26 R. Van Tuyl: 'A Monolithic GaAs IC for Heterodyne Generation of RF Signals', IEEE Trans, ED–28, February 1981, pp166–170

27 C. Kermarrec, et al: 'The First GaAs Fully Integrated Microwave Receiver for DBS applications at 12 GHz', Proc. 14th European Microwave Conference, Liege, 1984

28 S. Cripps, O. Nielsen and J. Cockrill: 'An X–band Dual–Gate MESFET Image–Rejection Mixer', IEEE MTT International Microwave Symposium Digest, 1978, p300

29 P.N. Rigby, J.R. Suffolk and R.S. Pengelly: 'Broadband Monolithic Low Noise Feedback Amplifiers', IEEE Microwave and Millimetre–wave Monolithic Circuits Symposium Digest, June 1983, pp71–75

30 A.M. Pavio, et al: 'Double Balanced Mixers using Active and Passive Techniques', IEEE Trans. MTT–36, pp1948–1957, December 1988

31 R. Gregorian and G. Temes: 'Analog MOS Integrated Circuits for Signal Processing', John Wiley and Sons, 1986

32 W.S. Percival: 'Thermionic Valve Circuits', British Patent No. 460562, January 1937

33 T.H. Buttgenback, et al: 'A Broadband Low Noise SIS Receiver for Submillimeter Astronomy', IEEE MTT International Microwave Symposium Digest, 1988, pp469–472

34 M. Pilgrim and R.P. Searle: 'Millimetre–wave Direct–to–Home Multichannel TV Delivery System', IEEE MTT International Microwave Symposium Digest, June 1989

Chapter 10

Switched Capacitor Circuits and Operational Amplifiers

D. G. Haigh, C. Toumazou and A. K. Betts

10.1 Introduction

The switched capacitor circuit approach has very successfully allowed the realisation of high precision analogue functions, including filtering and analogue–to–digital and digital–to–analogue conversion, in integrated circuit form. This has brought enormous benefits in the form of high performance and low cost and also the possibility of mixed mode systems combining digital and analogue sub–systems on a single chip, at least for high volume requirements. Using CMOS technology, switched capacitor filters with switching frequencies of up to 30 MHz have been achieved [1]. Over the past 2–3 years we have seen a successful and rapidly growing use of GaAs MESFET technology in the area of sample data analogue signal processing circuits based upon the switched capacitor circuit approach [2]. Here, the higher electron peak velocity and resulting higher low field mobility of GaAs have been exploited to increase switching frequencies to 250 MHz [3,4,5] with the successful realisation of a bandpass filter response centered at 10 MHz in [4] and [5]. This chapter presents some concepts developed in recent work which has been aimed at extending this early work using two case study designs. Both of these designs are being fabricated at the time of publication and therefore some of the ideas have a degree of speculation. They are, however, firmly based on experience gained through measurements on the earlier filters and therefore, we believe, indicate the correct direction for further work. The work to be described is helping to realise the possibility of high precision filtering in the frequency range 10 MHz to 100 MHz, with important applications in communications and radar.

Despite the speed advantages of GaAs technology, there are a number of deficiencies and impracticalities which are most undesirable for analogue integrated circuit design [6]. These include electron trapping and surface effects, poor uniformity, backgating, substrate oscillations and hysterisis, together with a high 1/f noise. Furthermore, due to the low hole mobility of GaAs, P–channel devices are generally not fabricated and design is restricted to N–channel, depletion–mode GaAs Metal Semiconductor FET's (MESFET's). In addition the gate and channel

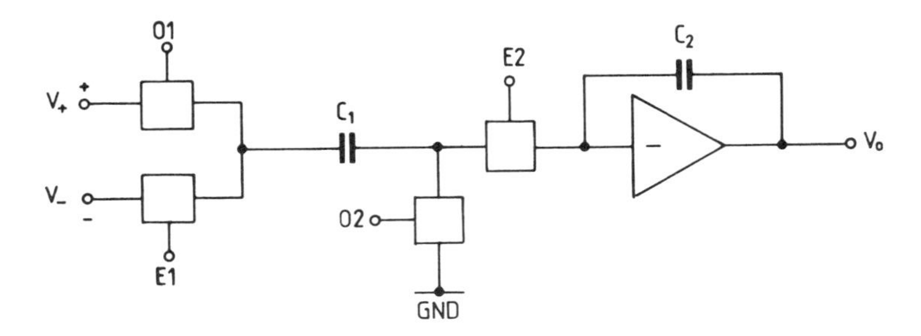

(a) Integrator circuit

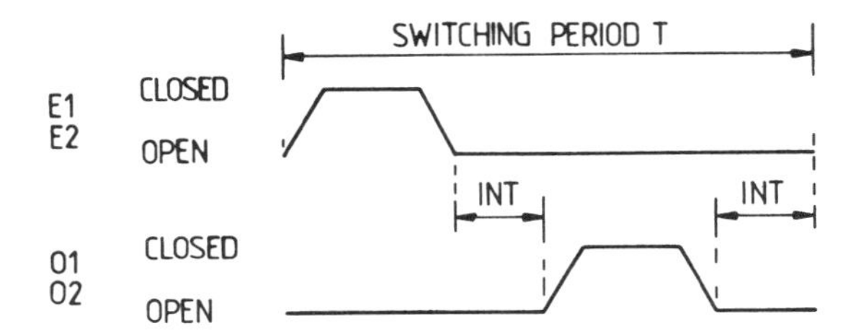

(b) Switching waveforms

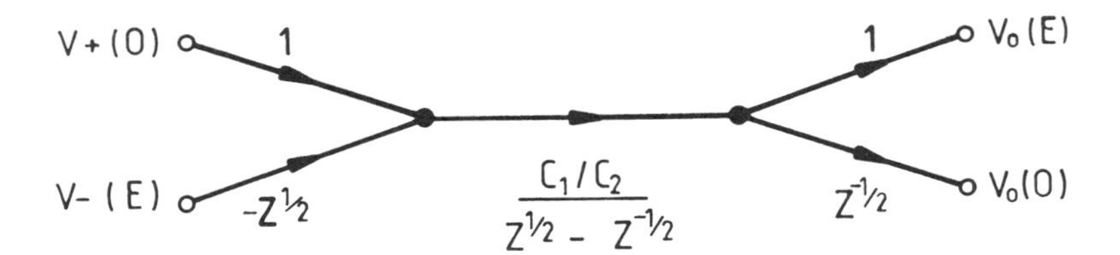

(c) Signal flow graph

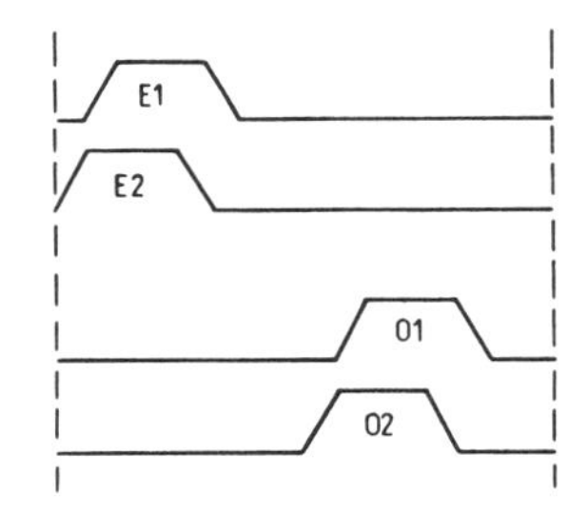

(d) Special switching waveforms

Figure 10.1: SC integrator, switching waveforms and signal flow graph

form a Schottky barrier diode limiting the maximum gate–source voltage to about 0.6 V causing particular problems for analogue switch design. The successful use of GaAs for high speed analogue integrated circuit applications is therefore placed in the hands of the analogue circuit designer who is faced with the challenge of developing novel circuit design techniques to alleviate some of these problems and exploit as far as possible the speed potential of GaAs. For practical implementation of complex systems, the designer has to accept the additional constraints of chip area and power consumption limitations. The successful realisation in GaAs of SC filters (and also analogue–to–digital and digital–to–analogue converters, which are covered in Chapter 7) will allow higher levels of system integration, optimisation of system performance and a greater degree of system flexibility.

10.2 Switched Capacitor Circuit Architectures

An excellent review of switched capacitor circuit design is given in [7] and only a very brief summary of a few relevant concepts will be quoted here. Switched capacitor (SC) circuits use, as basic components, operational amplifiers, capacitors and periodically actuated switches. When these components are realised in integrated circuit form, parasitic capacitances appear associated with them. The values of these capacitances can be as high as 10% of the wanted circuit capacitor values and therefore cause unacceptable deviations in response. This problem was alleviated by the introduction of parasitic insensitive SC circuits, whose transfer function is relatively unaffected by grounded parasitic capacitance at any node of the circuit.

For the realisation of high order systems, the differential input integrator has been recognised as a powerful universal building block and a parasitic insensitive differential input integrator in SC form is shown in Figure 10.1a. Like most SC circuits, the integrator operates from 2–phase switching waveforms which are shown in Figure 10.1b. For correct operation of the circuit, it is important that these waveforms do not overlap and, in order to guarentee this, there are guard intervals (denoted INT) when all switches are open. Analysis of the circuit in Figure 10.1a, applying the principle of charge conservation, leads to a signal flowgraph representation shown in Figure 10.1c. Since, SC circuits are sampled data networks, they are described by the discrete time complex frquency variable z. The branch gain:

$$G = \frac{C1/C2}{z^{1/2} - z^{-1/2}} \tag{10.1}$$

in Figure 10.1c corresponds to a sampled data integration of the LDI (linear discrete integrator) type. The circuit capacitor values C1 and C2

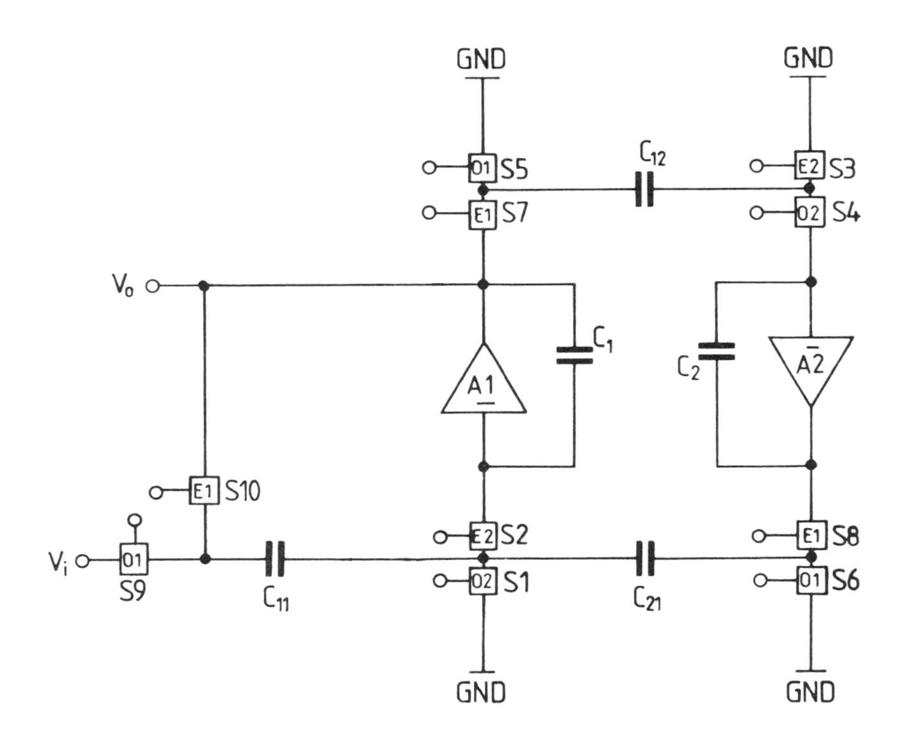

Figure 10.2: 2nd order SC filter using high gain amplifiers; normalised capacitor values C11=1, C12=C21=16, C1=C2=63.5

appear in the transfer function as a ratio. This is a property shared by all SC circuits and is very important from the point of view of implementation and is, in fact, the reason for the prominance of SC circuits. It follows from this property that absolute capacitor values may be made small and hence occupy small chip area. Although absolute capacitance tolerances for an integrated circuit can be as large as +/- 10%, the ratio can usually be made accurate to within +/- 0.1 %. Thus high precision circuits can be realised in small chip areas.

Although the concept of parasitic insensitivity led to considerable improvements in SC circuit performance, performance is still not acceptable in terms of response precision and linearity for demanding applications. These residual deviations are due to non–grounded parasitic capacitances associated with the control terminals of the switches [8]. However, by slightly delaying the control signals to the pair of switches E1 and O1 at the inputs of the integrator in Figure 10.1a, as shown in Figure 10.1d, the effect of these capacitances can be dramatically reduced and response precisions of 0.02 dB and distortion levels at −80 dB relative to the fundamental achieved.

Higher order systems may be realised by interconnecting a number of integrators and Figure 10.2 shows a 2nd order filter which realises a bandpass response. The design of this filter is discussed in reference 4. It consists of two integrators, one of which (based on amplifier A2) is the same as in Figure 10.1a. The other (based on amplifier A1) requires additional inputs, which are realised by duplicating the capacitor C1 and pair of switches E1/O1 in Figure 10.1a. The capacitor values shown provide a Q–factor of 16, a midband gain of unity and a midband frequency equal to 1/25th of the switching frequency, which is a typical value taking into account the problems of aliassing and imaging [7].

The integrator circuits used in Figure 10.2 require a gain for the amplifiers of at least 60 dB. However, alternative circuits have been developed which allow lower amplifier gains of 40 dB, simplifying the problem of amplifier design [3,7,9]. These circuits are known as finite gain insensitive (FGI) SC circuits. A differential input integrator of the FGI type, which will be used in a filter design in Section 10.9.2, is shown in Figure 10.3 [9].

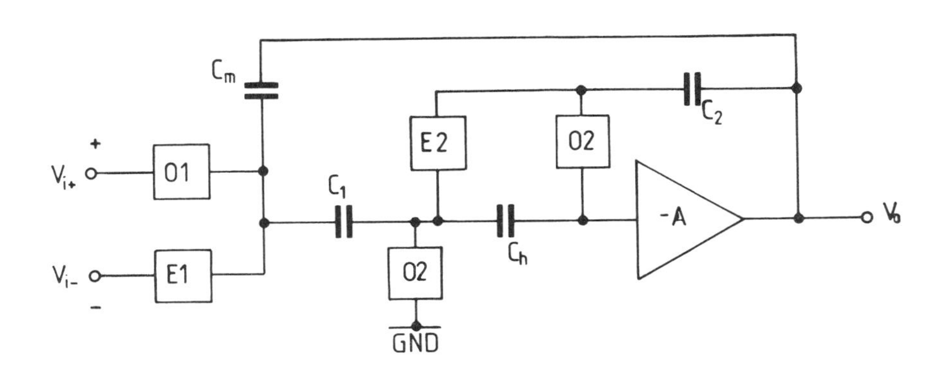

Figure 10.3: Finite gain insensitive differential input integrator

For filters with order greater than two, a well established design procedure has evolved [7]. Since LCR filters have the desirable property of exceptionally low sensitivity of their response to component value variations, the usual practice is to begin with an LCR filter design which meets the required specifications. The LCR filter is converted into a signal flow graph which preserves the required overall transfer function. The next step is to carry out a transformation from the continuous time frequency variables, appropriate to continuous time systems such as LCR filters, to the discrete time frequency variable z, which is appropriate for sampled data systems such as switched capacitor circuits. The transformations usually used are the bilinear transformation:

$$s \rightarrow 2(1-z^{-1})/[T(1+z^{-1})] \quad (10.2)$$

or the LDI transformation

$$s \rightarrow (z^{1/2} - z^{-1/2})/T \quad (10.3)$$

where T is the switching period. Application of such transformations to the frequency variable s in the signal flow graph leads to a signal flow graph in terms of z which may be rearranged using well-known transformations in such a way that parts of the signal flow graph have the form of the integrator building block (Figure 10.1c) and may be replaced accordingly. This design procedure correctly applied can lead to a high order SC filter structure with a very low sensitivity of its response to capacitor ratio variations.

10.3 GaAs Devices, Modelling and Simulation

10.3.1 Devices and the modelling problem

The most generally adopted active device in GaAs technology is the depletion mode N-channel Metal Semiconductor FET (MESFET) which can be produced with a relatively high degree of uniformity and which has a typical gate length in the range 0.5 to 1 micron. Gate widths vary between a minimum of about 10 microns to several hundred microns. It is possible to model such devices using the JFET model in SPICE in order to allow simulation of sampled data circuits. The general problem of MESFET modelling is covered in Chapter 3, and only a brief summary of some aspects relevant to sampled data system design will be presented here.

Measured characteristic curves for a MESFET produced by a typical 1 micron (gate length) GaAs process [10] are shown in Figure 10.4. The parameters given in Table 10.1 for a SPICE JFET model were obtained by optimisation (using TECAP) to give a best fit to the measured curves, as indicated by 'simulated' in Figure 10.4. The ideal JFET model assumes that saturation of Id occurs at Vds = Vgs – VT, whereas actual devices can exhibit 'early saturation' at a lower voltage than this. It can be seen from Figure 10.4 that, for Vgs = 0, the saturation region begins when Vds equals 1 V, which is the device threshold voltage. Thus these devices do not exhibit the 'early saturation effect' of some GaAs processes [11]. Since

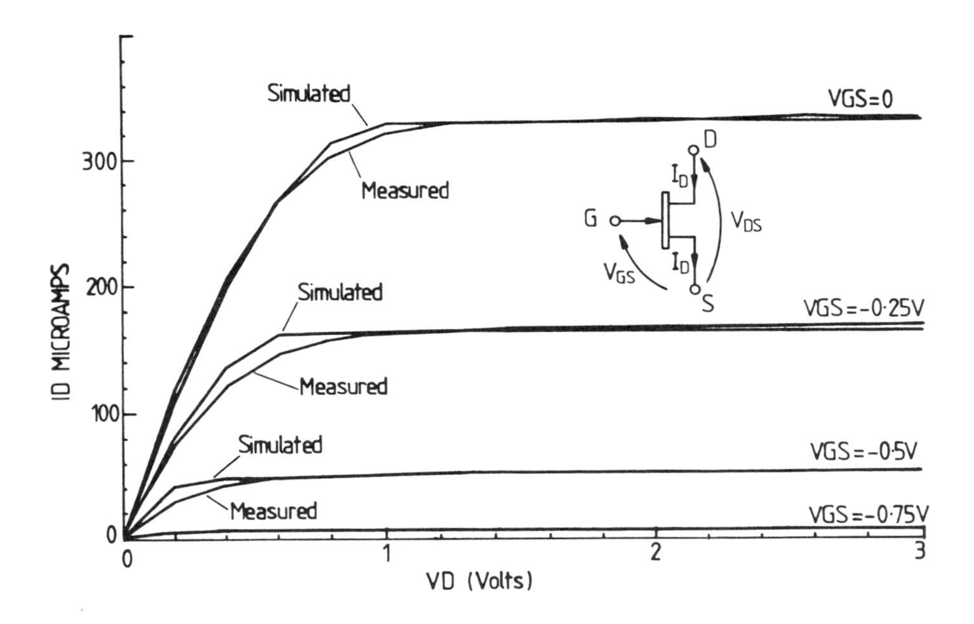

Figure 10.4: Measured and simulated characteristics for typical GaAs MESFET

Table 10.1 : Key parameters for JFET model of GaAs MESFET

Parameter Name	Value	Units
VTO	−1	V
BETA	0.067E−3	mAV^{-2}
LAMBDA (LF)	0.06	V^{-1}
LABDA (HF)	0.3	V^{-1}
Rd	2920	Ohms
RS	2920	Ohms
CGS	0.39E−15	F
CGD	0.39E−15	F
PB	0.79	V
IS	0.075E−15	A

the curves in Figure 10.4 are obtained by a curve tracer and are therefore low frequency measurements, the gradient in the saturation region does not reflect the high frequency output conductance (go) which can be many times higher due to dispersion effects, as indicated by the LF and HF values for LAMBDA in Table 10.1 [12].

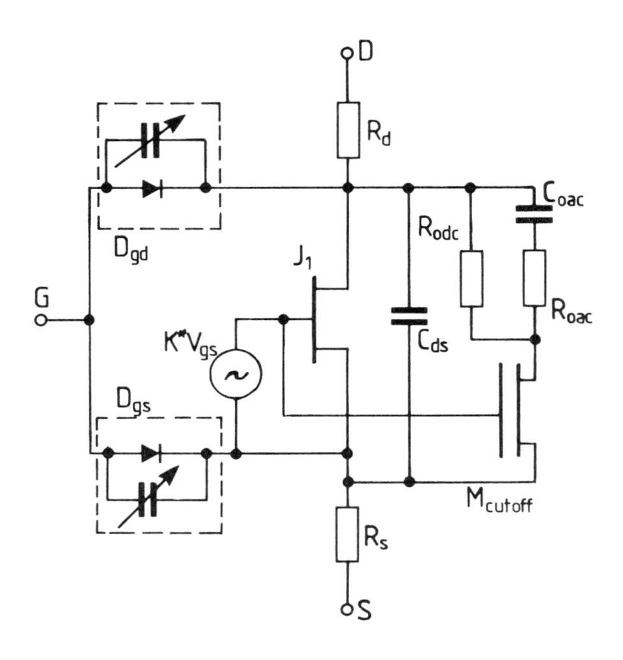

Figure 10.5: MESFET model subcircuit

High frequency MESFET characterisation is generally in the form of s–parameter measurements in the frequency range from 1 GHz to 14 GHz as required for microwave applications. Small signal parameters can be derived from this data by optimisation using the microwave simulation package TOUCHSTONE. Since SC circuits usually operate at or just below 1 GHz, the data at the minimum frequency of 1 GHz is usually used.

The modelling procedure suffers from a number of problems and inaccuracies. In the saturation region, actual devices often do not obey the square–law relationship between Id and Vgs and can exhibit early saturation. The linear dependence of output conductance, go, on Id in the saturation region is not always observed. Finally, as mentioned earlier, the high value at high frequencies of the output conductance leads to a very low figure for the intrinsic MESFET voltage gain (gm/go) of the order of 20 (gm is the transconductance). These problems are aggravated in the modelling of GaAs sampled data circuits because of the variety of device operating modes in such systems. A viable approach is to add additional components to the basic JFET model to form a model subcircuit [12,13] and to develop a different model for each operating mode.

10.3.2 Digital mode model

In the digital parts of the system which provide the clock signals, the devices go through the extreme range of operating points. For this mode, it is important to have an accurate large-signal model, and a suitable model subcircuit is shown in Figure 10.5. The go dispersion effect is modelled by devices Coac, Roac and Rodc. The early saturation effect can be modelled using the voltage-controlled voltage source at the JFET gate [10] giving correct DC conditions. MOSFET Mcutoff gives zero conductance when in the off state as required. If the on-state output conductance is only weakly dependent on Id, the JFET output conductance parameter LAMBDA can be set to zero.

10.3.3 Analogue mode model

In the operational amplifiers and buffers, the devices are permanently biased in the saturation region at a relatively stable operating point. The model for these devices should be optimised to give correct values for transconductance, output conductance, capacitance values and DC operating point. A sub-circuit for this model can be obtained from that in Figure 10.5 by removing the MOSFET (Mcutoff) and adjusting the model parameters.

10.3.4 Switch mode model

The devices used to realise the switches in the SC circuit experience extreme operating points and must be symmetrical with respect to drain and source. The model for this mode must also give the correct values of on-resistance and cutoff voltage. A model subcircuit can be obtained from the analogue-mode model by removing Coac and Roac, setting Cgd = Cgs in the JFET model and adjusting the parameters appropriately.

10.3.5 Simulation

Computer simulation using a general non-linear circuit analysis program, such as SPICE, usually plays a major role in the design process at a number of levels [14]. At the component level, all of the major components, i.e. amplifiers, buffers and switching circuits, are verified and characterised using SPICE with appropriate device models. An SC circuit analysis program, such as SCNAP [15,16], can be used to check the response of basic SC architectures and to determine initial conditions for SPICE analysis of some of the SC circuits. Finally, SPICE can be used, as will be described, for verification of a complete system.

10.4 Switches

It can be seen from the SC architecture in Figure 10.2 that switches form one of the main components of SC filters. When such circuits are implemented in CMOS technology, the switches can be realised by MOSFET's, the gates of which may be driven from the negative to the positive power supply rails to open and close the switches. This is possible because in CMOS technolgy the gate is seperated from the channel by an insulating layer (oxide). The situation in GaAs technology is quite different in that the gate forms a Schottky diode with the channel, and therefore large gate currents will flow if the gate voltage exceeds the voltage at the source or drain by more than about 0.6V, resulting in a failure of normal FET operation. Thus, we cannot apply the simple switching arrangement used in CMOS SC circuits.

Consider the realisation of the switches in Figure 10.2. These switches may be divided into two categories. The switches adjacent to the amplifier inputs (e.g. S1 – S4) are switching signals close to ground and they may each be realised as a MESFET with gate voltage switched between 0V (on–state) and V1, where V1 < VT (off–state). The remaining switches in Figure 10.2 (S5 – S10) are associated with the integrator input terminals. Of these, four (S7 – S10) are switching large analogue signals and it is necessary to introduce a switch control circuit (SCC) for these switches which limits the high voltage on the switching MESFET gate to track the analogue voltage so that the switching MESFET's gate Schottky diode is not forward biassed [3,5,17]. The remaining two switches (S5 and S6) are switching signals at around 0V and therefore do not strictly need the introduction of SCC's, although they may be introduced if desired. A reason for introducing them is that the switching delay introduced by the SCC's then applies uniformly to all switches at integrator inputs (E1 and O1 switches) and this is precisely the condition illustrated in Figure 10.1d, which gives maximum response accuracy and minimum distortion.

A number of SCC circuits will be presented, all of which have the general form shown in Figure 10.6a. The digital input is fed to an inverter, the output of which drives the gate of the switch MESFET M1. The output of the inverter switches between the inverter negative supply voltage –Vs, which opens the switch, and the inverter positive supply voltage. The inverter positive supply is derived from the output of a voltage follower, the input voltage of which is the voltage on one of the switch terminals X. As a result the voltage on the gate of M1 is equal to the voltage at node X, as required, when the switch is in the closed state. It may appear that the gate diode of M1 might become forward biassed if the voltage at Y is

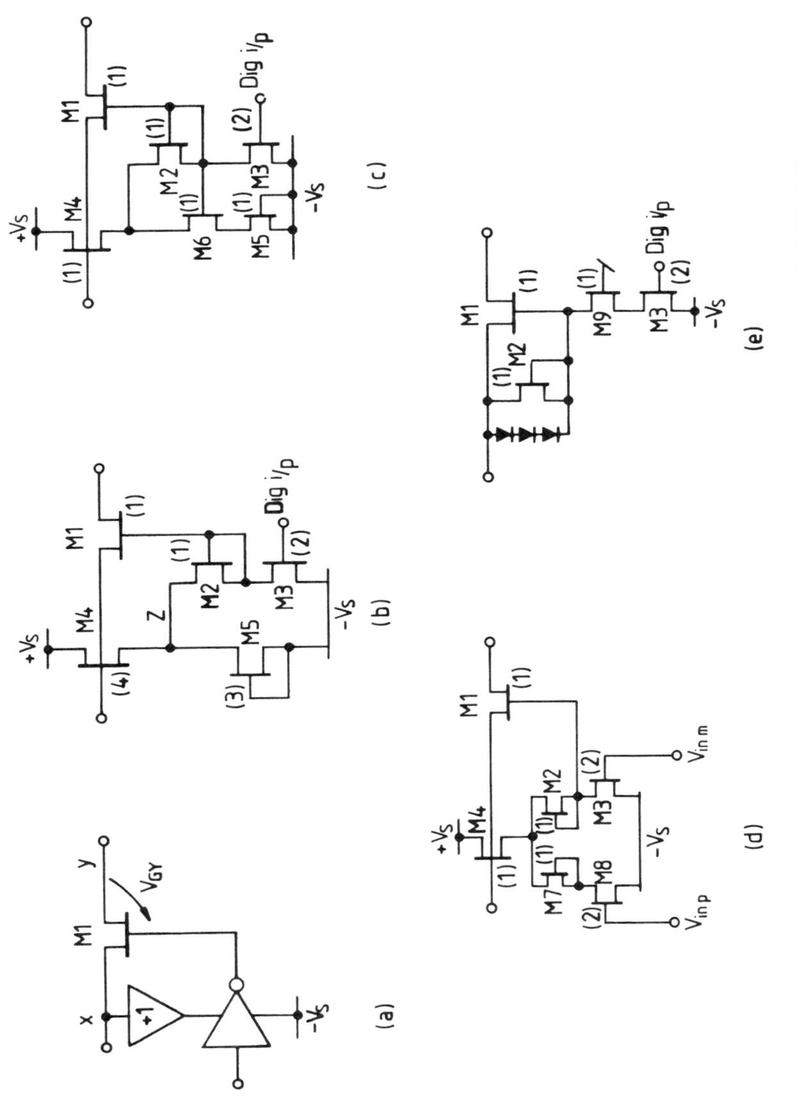

Figure 10.6: Switch control circuits for GaAs technology; typical Vs=5V
(a) Basic architecture, (b) Circuit used in early work, (c) Low power circuit, (d) Circuit using complentary switching, (e) Circuit omitting voltage follower

significantly less than that at X at the start of a closing transition. In fact this does not occur because, well before the voltage V_{GY} reaches the critical 0.5 V, the switch would start to go into its closed state tending to make the voltage at Y change towards that at X.

A switch driver circuit implementation used in early work [4] is shown in Figure 10.6b. Here, MESFET's M2 and M3 constitute the inverter and source follower M4 constitutes the voltage follower. Typical recommended gate widths are given in Figure 10.6 in the form of numbers in brackets (alongside each FET) which are factors multiplying the minimum realisable gate width for the process (typically 10 microns). M1 and M2 are minimum gate widths. M3 is double M2 so that the gate of M1 is pulled down properly in the open state. The current in the inverter devices changes from one switching state to another and flows through M4 affecting its source voltage, and causing an error on the gate voltage of M1. As a consequence, the constant current source connected device M5 is introduced and the gate width of M4 is increased to allow for the extra current. This reduces the relative variation in the drain current of M4 and hence reduces the voltage error, but the additional power consumption is undesirable since SCC's are required in large numbers in SC circuits.

A possible solution to this problem is shown in Figure 10.6c. The additional device M6 has been included and all device widths are equal to the minimum, apart from M3. The device M6 reduces the voltage across, and therefore the current in, M5 to zero when the switch is in the open state. Consequently, the current in M4 is always equal to its IDSS and therefore its gate–source voltage is always zero, as required, thus eliminating gate voltage error. The limitation on this circuit is that, during switch transitions, it is possible for M2 and M6 to be conducting simultaneously, and the consequent increase in current in M4 can cause M4's gate to become forward biassed presenting a very low transient impedance to the SC circuit which can affect proper operation.

Another approach is shown in Figure 10.6d [18,19]. Vinp and Vinm are complementary clock signals which turn M3 and M8 on and off in a complementary manner. As a result, the source–follower M4 has a drain current which is the same in both switching states and its gate–source voltage is therefore zero. As in the previous circuit, the source follower device M4 has the minimum gate width which is advantageous from the point of view of consuming minimum power. The performance of this complementary switched circuit is dependent on the balancing of the complentary signals, especially in the transitions between switching states.

An alternative SCC is shown in Figure 10.6e [17]. The main difference between this circuit and the previous circuits is the omission of the source follower device [3]. As a consequence, the circuit draws a finite DC current at its input when the switch is in the open state. This can be accepted when the operational amplifiers driving these nodes can supply the current and, in any case, the voltages at these nodes when the switch is in the open state are not sampled. The circuit features diodes, as in [3], to make clock feedthrough signal independent; but the signal dependence is further reduced by the introduction of the cascode device M9, which reduces the drain conductance of the digital driver device M3.

The complementary clock signals Vinp and Vinm required by the SCC in Figure 10.6d can be generated by a circuit having the structure of Figure 10.7 [18,19]. This circuit is essentially a hard–driven long–tailed–pair with diodes to define the low–level output signals.

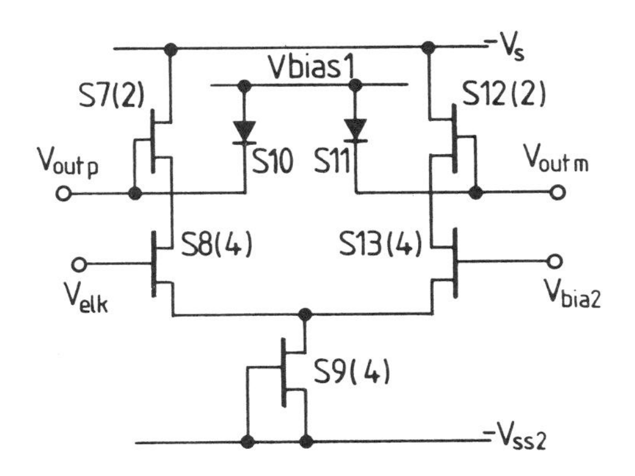

Figure 10.7: Clock generation circuit; typical Vs=5V, Vbias1=–7.5V, Vbias2=–10V, Vss2=–12V

10.5 Capacitors

For GaAs process's, three kinds of capacitor are available, namely overlay types using Silicon Nitride and Polyimide dielectrics and the interdigital type. Interdigital capacitors are best suited for use at microwave frequencies and are not really appropriate for sampled data applications. The decision as to whether to use Silicon Nitride or Polyimide can be based on Q–factor measurements, some examples of which are given in Table10.2 for the case of the Plessey III–V 0.7 micron foundry process (for details of this process, see Chapter 2).

The results in Table 10.2 show that, from Q–factor considerations, the use of Silicon Nitride capacitors is essential for high performance SC circuits. The large difference between the nominal and measured capacitor values for low–value capacitors shows the important role played by undercut (sideways etching) in giving the true capacitor value, since the

Table 10.2 : Measurements on capacitors at 1 MHz

Type	Nom(pF)	C(pF)	G(uS)	Q(= ωC/G)
Polyimide	1.89	2.1	0.2	66 +/- 13
	0.84	0.983	0.1	61 +/- 25
	0.84	0.98	0.06	102 +/- 68
Silicon	44	49.02	1.1	280 +/- 10
Nitride	2.7	3.29	0.07	295 +/- 169
	0.2	0.322	--	
	0.2	0.323	--	
	0.2	0.327	--	

nominal capacitor values are calculated from the drawn capacitor sizes. This makes it desirable that capacitors should be realised by the interconnection of unit-sized elements. The Plessey process, for example, has a minimum sized Silicon Nitride capacitance of 0.11 pF having a via size of 8 x 8 microns. Although this is less than the minimum capacitance allowed in continuous time circuits, owing to absolute capacitance tolerance, in the case of SC circuits, it is the capacitor ratio tolerance which determines circuit precision and such low absolute capacitor values are acceptable.

10.6 Operational Amplifiers

10.6.1 General approaches

The operational amplifier is a centrally key component for the realisation of all sampled data analogue circuits and therefore its realisation must now be considered. For application in sampled data circuits, such as SC filters and analogue-to-digital and digital-to-analogue converters, the requirements differ in key aspects from those for general purpose amplifiers [20-23]. In particular, a high output impedance is preferred (current output instead of voltage output) in view of the capacitative loads, and DC gain must be above specified limits which are typically 60 dB, or 40 dB using finite gain insensitive SC circuits. Furthermore, settling time is a very important parameter since it determines the maximum speed of circuit operation.

GaAs technology poses a number of difficulties for the amplifier designer. These include the lack of a P-channel device, in general the lack of enhancement mode devices (depletion mode only generally available) and low device gain (typical gm/go = 20). In this Section, we shall present techniques which have been developed specifically to overcome these problems. The lack of P–channel devices will be overcome by the development of a negative current mirror using N–channel MESFET's [24,25]. The lack of enhancement mode devices when required for DC level shifting will be overcome using diodes and specially developed double level shifting techniques relying on device width ratios [10]. Finally, to overcome the low device gain, double cascode techniques will be introduced [10]. Two–stage GaAs operational amplifiers have been proposed [20,23]; however, in order to fully exploit the inherent wide bandwidth capability of GaAs, we shall consider single stage designs only. For minimum complexity and maximum performance, designs with single ended input are initially considered where the quiescant input voltage is equal to the negative power supply voltage in order to avoid internal level shifting circuitry; in this case the ground connections for switches S1 and S3 in the filter of Figure 10.2 have to be grounded to the negative power supply [26]. Finally, a differential to single ended converter which provides a differential input facility for the single ended input amplifiers developed will be presented.

10.6.2 GaAs current mirror

In a non–complementary technology providing N–channel devices, the negative current mirror is a very useful building block for circuit design. However, in high speed bipolar npn technology, the realisation of the negative current mirror has proved difficult [27]. In the case of N–channel GaAs technology, on the other hand, an elegant realisation of the negative current mirror has been discovered [24,25].

Figure 10.8a shows a simple circuit consisting of a pair of N channel MESFET's with gates and sources cross–coupled. The operation may be explained using the simple equivalent circuit in Figure 10.8b. Here, the controlled sources of the small signal models for the two MESFET's are shown in a central block with the (parasitic) passive components outside. Small signal analysis of the central block gives:

$$I_1/I_2 = -gm_1/gm_2 \qquad (10.4)$$

corresponding to an inverting negative current mirror. In addition, the circuit has a voltage following property, and so may be shown to be a form of positive impedance converter. Hence, the input impedance at port 1 is approximately given by

$$Z_{i1} = (gm_2/gm_1)Z_{L2} \qquad (10.5)$$

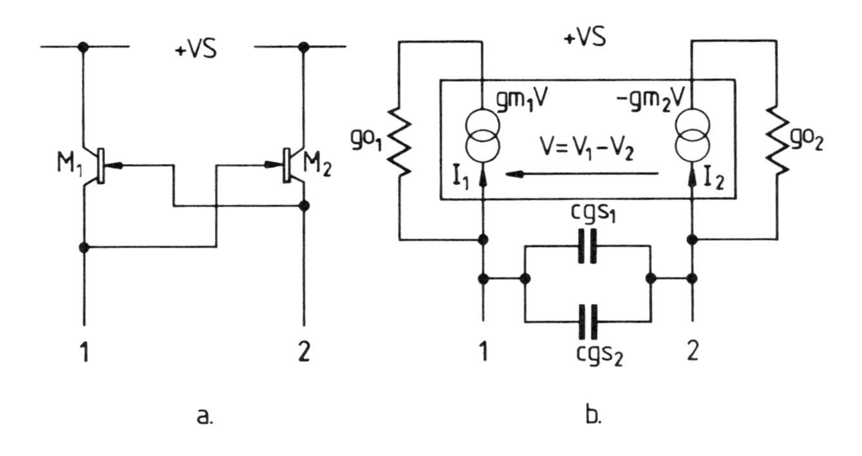

Figure 10.8: GaAs current mirror
(a) circuit, (b) equivalent model

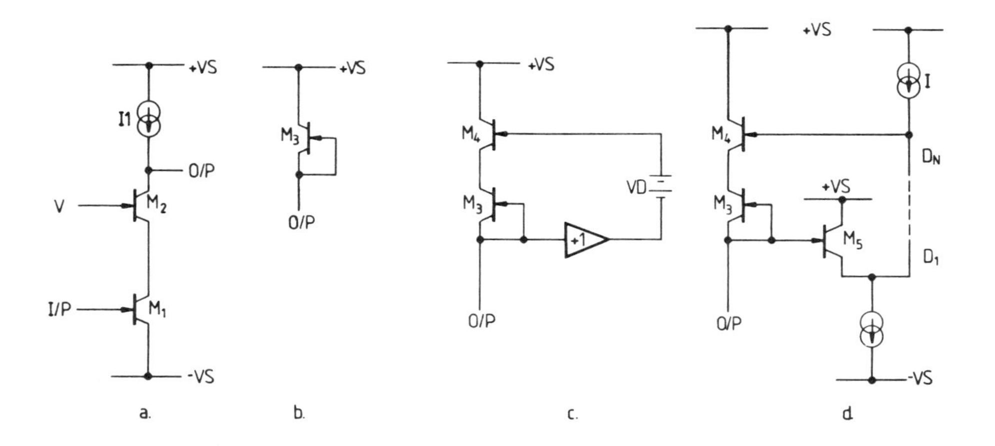

Figure 10.9: Basic amplifier techniques
(a) Basic inverter architecture, (b) Single MESFET current source,
(c) Cascode current source architecture, (d) Cascode current source

where Z_{L2} is the load impedance at port 2. Similarly, the input impedance at port 2 is

$$Z_{i2} = (gm_1/gm_2)Z_{L1} \tag{10.6}$$

This voltage following, inverting, negative current mirror provides the key, as we shall show, to the controlled current sources required for push–pull amplifier design, overcoming the lack of P–channel devices and forming an important component in differential to single–ended converters, thus providing a valuable addition to the available GaAs building blocks.

10.6.3 Circuits using diode level shifting

A basic structure for a moderate gain inverting amplifier is shown in Figure 10.9a; M1 is the driver transistor and M2 the cascode transistor. Assuming an ideal current source, the voltage gain is $gm_1gm_2/(go_1go_2)$ which would have a typical value of 400. For the realisation of the current source, a single MESFET connected as in Figure 10.9b may in principle be used. However, this would reduce amplifier voltage gain to a low value of approximately gm_1/go_3 and so it is necessary to develop a high impedance current source using cascode techniques. This is shown in Figures 10.9c and d. In Figure 10.9c, the drain–source voltage of M3 is kept relatively constant (independent of Vo) by means of the buffer amplifier, DC voltage source (VD) and source follower M4. A realisation of the buffer and DC source is as in Figure 10.9d, using the source follower M5, diode chain and associated current sources.

The realisation of a complete amplifier is shown in Figure 10.10. The devices M1 to M5 are as identified in Figure 10.9; M6 realises the upper current source in Figure 10.9d; the lower current source in Figure 10.9d is realised in two parts, namely M7 and M8; finally, M9, M10 and M11 are cascode transistors which improve source follower performance. DC gain for this amplifier is typically $(gm/go)^2/4$ (about 40dB). For the bias MESFET's in Figure 10.10, M5 – M11, the gate widths (indicated as factors multiplying the minimum gate width) may be equal to the minimum, i.e. unity. The gate widths of the input/output chain in Figure 10.10 comprising M1–M4 determine the amplifier transconductance and slew rate and the gate widths of these devices are defined as an amplifier scaling factor k multiplied by the minimum gate width. SPICE simulation of a unit scaled amplifier using a typical MESFET model, as in Table 10.1, indicates a DC gain of 41 dB, a gain–bandwidth product of 3.6GHz and a phase margin of 70 degrees, when used with an effective capacitative load of 80fF. Simulated performance figures for this amplifier, using the model parameters of Table 10.1, are given in Table 10.3.

Table 10.3 : Operational Amplifier Performance Parameters

Amplifier circuit	Min settling time (ps)	Load capaci–tance (pF)	DC gain (dB)	GB product (GHz)	Phase margin (deg)
Fig 10	200	1.0	41	3.6	70
Fig 13a	630	0.8	63	3.1	64
Fig 14	243	0.4	61	3.2	71
Fig 15 + 13a	480	1.0	68	2.0	56

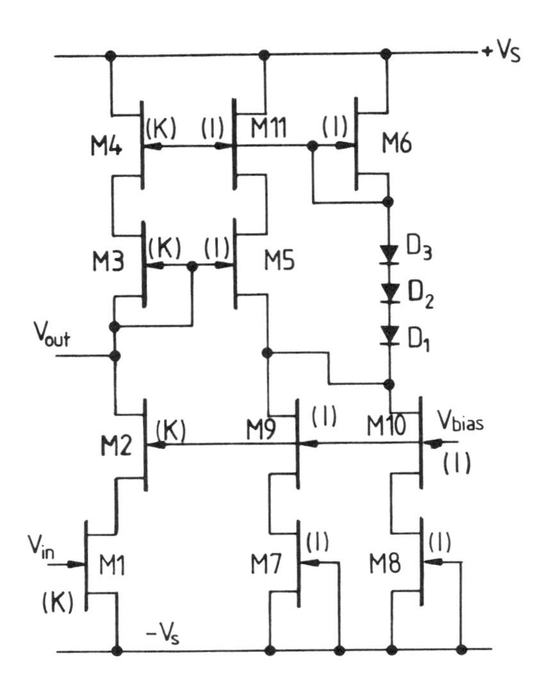

Figure 10.10: 40 dB op-amp using diodes; Vs=5V, Vbias=-3V

10.6.4 Circuits using device width ratioing

(i) High impedance current sources

A high impedance current source, as required in Figure 10.9a, can be obtained without the use of diodes. Figure 10.11a shows a self-bootstrapped cascode current source [11]. The drain source voltage of M4 is equal to the negative of the gate source voltage of M5, which is given by

$$Vds4 = -Vgs5 = |VTO|\ [1 - \sqrt{(W4/W5)}] \qquad (10.7)$$

W5 is made greater than W4 and so that Vds4 is sufficient to keep M4 saturated assuming that early saturation exists. This principle may be extended as shown in Figure 10.11b to realise a double cascode self-bootstrapped source [10]. However, many processes (e.g. see Figure 10.4) do not exhibit any or sufficient early saturation behaviour in

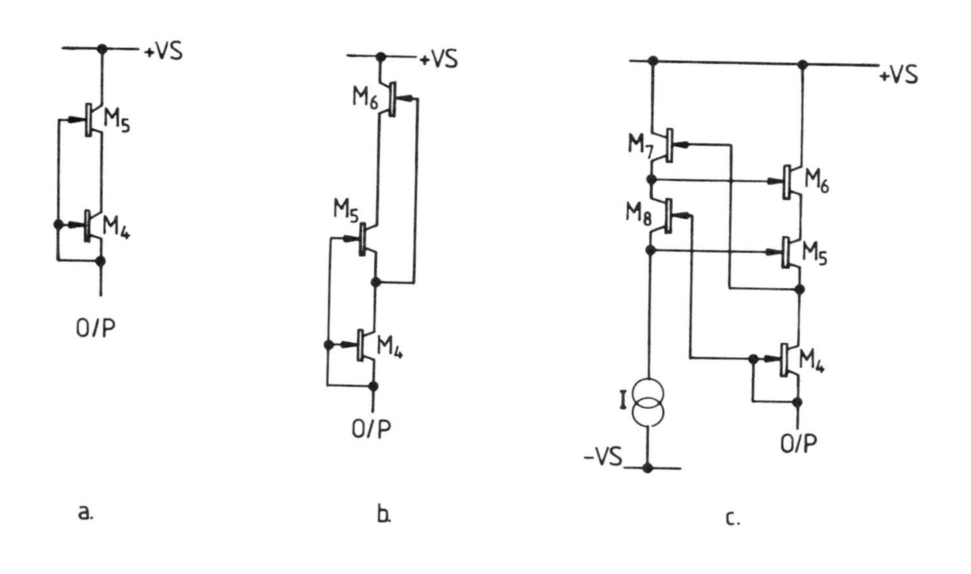

Figure 10.11: Further amplifier techniques
(a) Self bootsrapped cascode current source, (b) Self bootstrapped double cascode current source, (c) Double level shift double cascode current source

which case we require that Vds4 ≥ VTO and eqn (10.7) indicates that an infinite device ratio is required. This leads to the development of some alternative circuit techniques.

The double level shifting technique is shown applied to the realisation of a double cascode current source in Figure 10.11c [10]. We have introduced an auxilliary bias chain comprising M7 and M8. Now the level shifting source follower M8 assists M5 in keeping M4 saturated. The device M7 similarly helps to maintain M5 in saturation and also bootstraps the drain of M8 to maintain adequate source follower performance. The circuit in Figure 10.11c realises a high impedance double cascode current source without any reliance on early saturation of devices. This principle has been successfully implemented using a state of the art 0.2 micron GaAs process [21,28].

(ii) Push pull amplifier design

Although the current source of Figure 10.11c is a valuable component for the realisation of high performance operational amplifiers, a current source which can be controlled is advantageous because it allows the realisation of a push–pull output. A basic architecture for a push–pull

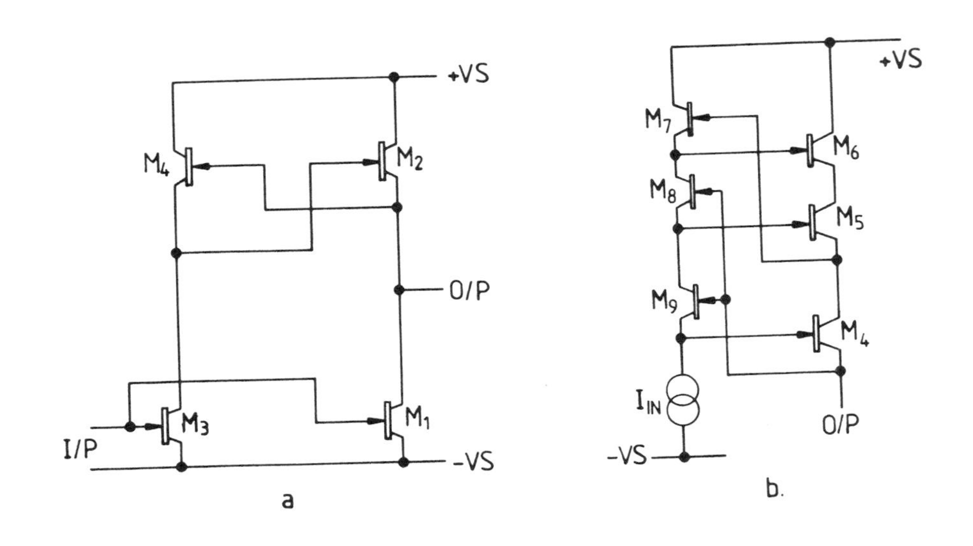

Figure 10.12: Pushpull amplifier design techniques
(a) Basic pushpull architecture, (b) Double cascode current mirror

amplifier is shown in Figure 10.12a [10]. M1 and M3 are driver MESFET's and cross-coupled MESFET's M2 and M4 form an inverting voltage following current mirror. The advantage of this push-pull arrangement is that, by reducing the widths of M3 and M4 below those of M1 and M2, it is possible to obtain double the transconductance and maximum output current for the same effective area and power consumption compared with a simple (non push-pull) inverter (Figure 10.9a). However, the gain of the amplifier in Figure 10.12a is only of the order of gm/(2go) and it is thus necessary to introduce double cascode techniques for each MESFET. A double cascode version of the current mirror (M2 and M4) is shown in Figure 10.12b. It is obtained from the current source in Figure 10.11c in a number of stages; we first break the gate source connection of M4, introduce the additional MESFET M9 and cross-couple it to M4. The final amplifier is shown in Figure 10.13a [10]. The upper part is identical to the high quality current mirror of Figure 10.12b. The same circuit is also used in the lower part with appropriate input connections to form a double cascode dual driver as required (see Figure 10.12a). The MESFET gate widths in Figure 10.13a are defined relative to the gate width x for the amplifier driver device M1. The simulated frequecy response of the amplifier in Figure 10.13a, for a gate width factor x of 100 microns, a 0.8 pF load capacitance and device parameters as in Table 10.1, is shown in Figure 10.13b. Key performance parameters are tabulated in Table 10.3. It can be seen that the higher gain

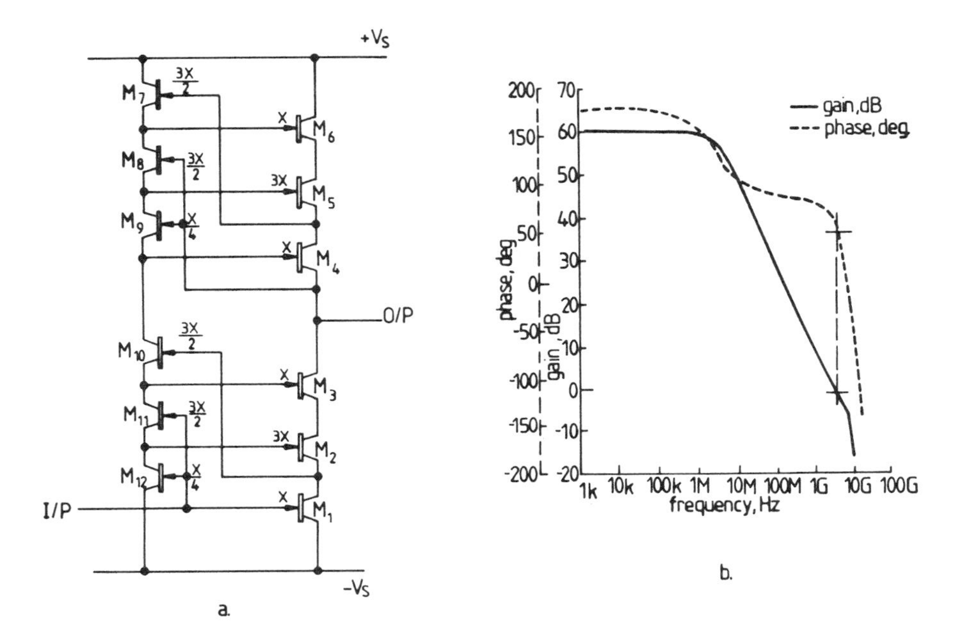

Figure 10.13: High gain double cascode amplifier and performance (a) Amplifier circuit; Vs=5V, (b) Frequency response

figure of 60 dB, which is required for use in SC circuits which are not of the finite gain insensitive type, has been obtained at the cost of some increase in settling time.

10.6.5 Fast settling architectures

The amplifier in Figure 10.13a has a typical minimum settling time for a 1 μm GaAs process of 630 ps, which permits its use in sampled data circuits with switching frequencies up to about 500 MHz, as will be described in Section 10.9.1. We now present an amplifier architecture for higher speed systems. An amplifier formed from the current source in Figure 10.11c has been found to give faster settling than the circuit of Figure 10.13a [29,30]. A possible reason for this is that, in the current mirror used in this circuit (see Figure 10.12b), device M9 is not fully saturated. The fast settling amplifier is shown in Figure 10.14. It uses a double cascode driver, current source of Figure 10.11c with modified M7 gate connection and double cascode current sink to provide I in Figure 10.11c. Simulation indicates that the minimum settling time is close to 200 ps and key performance parameters are listed in Table 10.3.

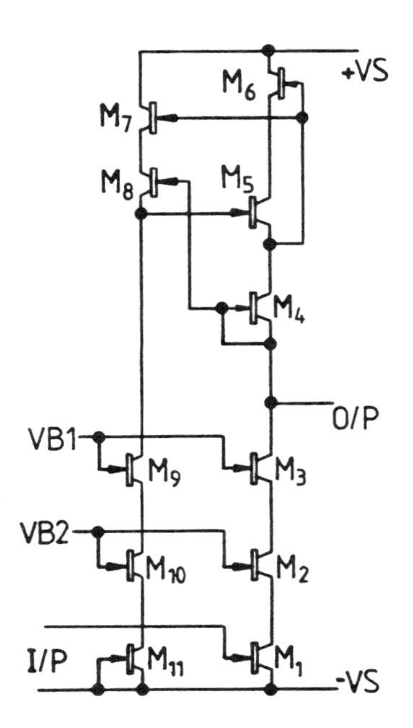

Figure 10.14: Fast settling high gain op–amp circuit; all devices 100 μm except M5, M6 and M8 500 μm; Vs=5V, VB1=–2.2V, VB2=–3.6V

Such amplifiers, which realise the high gain required without sacrificing settling time, will form the basis of future sampled data circuits designed to operate with 1 GHz switching frequencies.

10.6.6 Differential to single ended converters

All the amplifiers presented have a single ended input which is at a different DC level from the output. Such amplifiers can be used satisfactorily in switched capacitor circuits provided care is taken over grounding and power supply regulation. For some situations, however, a differential input is necessary. Figure 10.15 shows a differential to single ended converter which may be used with the above amplifiers [31]. The previously described current mirror is used (M5 and M6), in this case to provide design flexibility, stability and wide bandwidth [24,31]. Key simulated performance parameters for a composite amplifier comprising the converter of Figure 10.15 in cascade with the amplifier of Figure 10.13a are given in Table 10.3. Due to the exceptionally low phase contribution of the converter, internal compensation is unnecessary for such composite amplifiers.

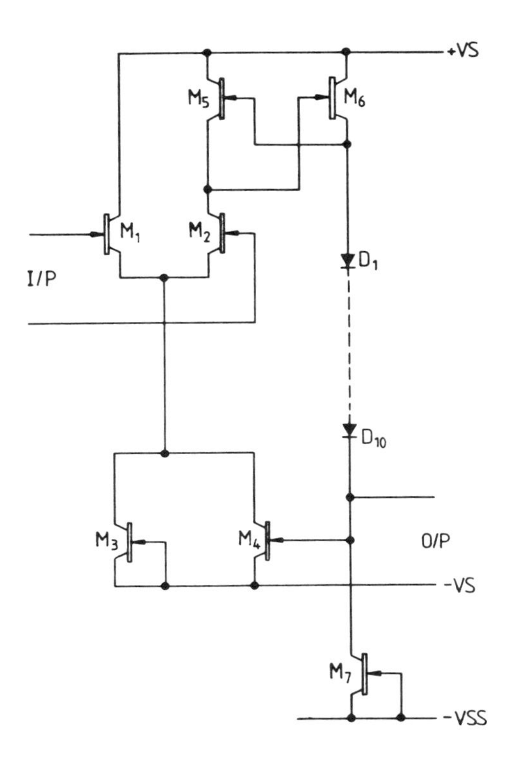

Figure 10.15: Differential to single–ended converter; all device widths 15 μm except M1 and M2 100 μm, M6 and M7 20 μm; Vs=5V, Vss=7.5V

10.7 Amplifier Characterisation

Amplifier settling time is the most important dynamic performance parameter for sampled data circuit design because it determines the maximum frequency limit and affects response accuracy [32]. We now discuss a means of characterising amplifiers in terms of their settling time [33] which reveals the inherent speed capability of an amplifier design and, at the same time, forms the basis of the optimisation of the complete SC circuit in order to maximise operating frequency, as will be described in the next section.

For characterisation, an amplifier is embedded in the test circuit shown in Figure 10.16. The purpose of the feedback capacitor C_f is to establish unity feedback. The ideal unity gain buffer amplifier is included to eliminate the effect of amplifier input capacitance on feedback factor.

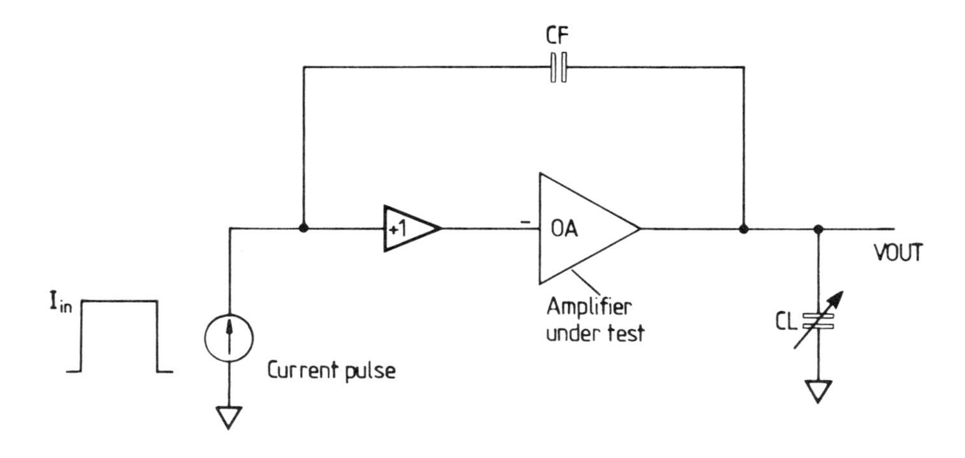

Figure 10.16: Test set-up for amplifier characterisation

The excitation is a finite charge applied to the inverting amplifier input using a pulsed current source to give a required output voltage step $\Delta V = \Delta Q/C_f$, which is normally chosen to be the maximum voltage step the amplifier is to experience in a particular sampled data application. Settling time has been determined using SPICE via a MINNIE graphics interface (see Chapter 4) with the standard JFET model using appropriate parameters (e.g. Table 10.1) for the GaAs MESFET's. Settling time is determined for a range of load capacitance values C_L (C_f does not constitute a load on the amplifier output because one of its terminals is connected to a node of infinite impedance).

The settling time versus load capacitance for the amplifier of Figure 10.13a is shown in Figure 10.17a. Load capacitance has been normalised by dividing it by the gate width factor x (in this case 100 microns). It can be seen that settling time increases for both large and small load capacitances and this is due to under and overdamping respectively. For a given settling time specification, a suitable operating point on the curve of settling time versus load capacitance may be chosen. The fastest settling time of about 630 ps is obtained for a loading factor of 0.008 pF per micron. For the very fast settling amplifier of Figure 10.14, the settling time characterisation curve is shown in Figure 10.17b. The minimum settling time is close to 200 ps and the amplifier has the feature of excellent symmetry between positive and negative settling behaviour.

Such results emphasise the important fact that the settling time cannot be reduced indefinitely by increase of the widths of the gates in an amplifier or by reduction of load capacitance. The minimum achievable settling time is specific to the architecture of the particular op–amp under investigation.

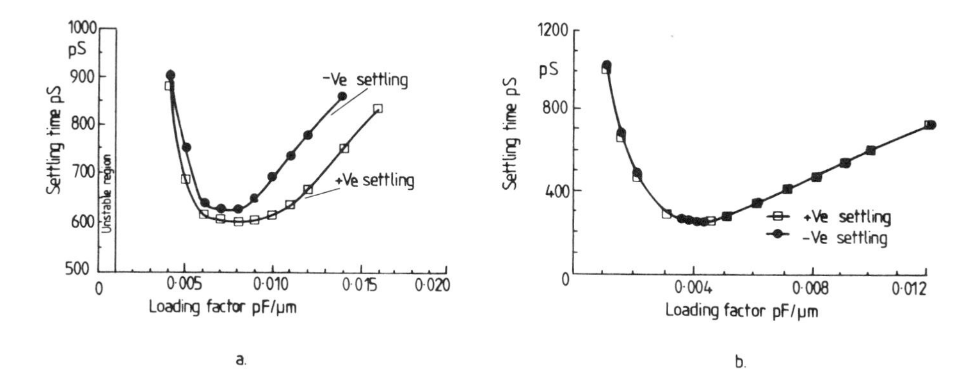

Figure 10.17: Settling time characterisation curves
(a) Characterisation of amplifier in Figure 10.13a, (b) Characterisation of amplifier in Figure 10.14

10.8 SC Circuit Optimisation

When amplifiers are used in SC circuits, the key parameter which determines the maximum switching frequency is settling time. Since suitable amplifiers are generally of the single–stage cascode type, settling time is critically dependent on load capacitance [33]. In general, the various amplifiers in a given SC circuit experience different capacitative loading conditions and, in order to achieve operation at a given point on the amplifier settling time characterisation curve, we have to scale the gate widths for each amplifier independently, according to its load. This scaling of amplifier gate widths is one aspect of the optimisation of an SC circuit for high frequency performance.

Most SC circuits, like that in Figure 10.2, operate with a 2–phase switching scheme, and thus, as shown in Figure 10.1(b), there are three distinct states ('E', 'O' and 'INT') for which amplifier settling characteristics have to be considered. In order to investigate the effect of switching state on settling time, we initially make two simplifications [26]. Firstly, we assume that the switches are ideal, i.e. open or short circuit depending on switching state. Secondly, whereas in certain switching states some of the amplifiers are coupled together, as for example in Figure 10.2 in the 'E' phase, we investigate the settling characteristics of each amplifier in turn with the assumption that the remaining amplifiers are ideal.

Under these simplifications, each amplifier may be considered to have an embedding of the general form shown in Figure 10.18 for each

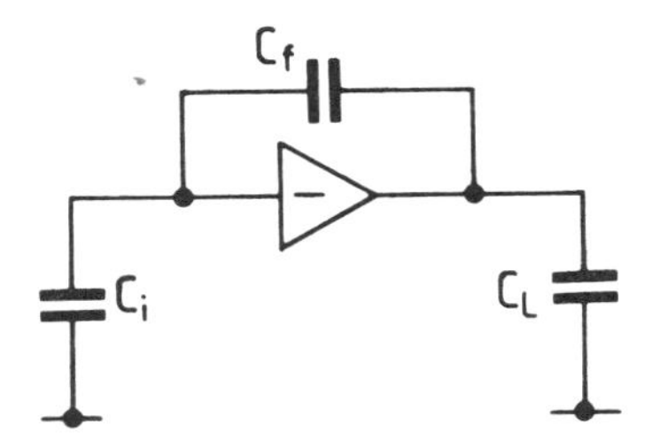

Figure 10.18: General single amplifier embedding

switching state. Matsui et al [34,35] have shown that for the circuit of Figure 10.18, where the amplifier is modelled as an ideal transconductance gm, the time constant which determines settling behaviour is given by

$$T = CLeff/gm \qquad (10.8)$$

where CLeff, the effective load capacitance, is given by,

$$CLeff = Ci + CL + Ci.CL/Cf \qquad (10.9)$$

Switching states may be divided into critical and non–critical switching states. Critical switching states are those in which the amplifier is receiving charge at its input or conveying charge to a subsequent integrator. It is desirable that, in a critical switching state, the amplifier gate widths are scaled in relation to the CLeff value for that state, in order that the operating point on the characterisation curve is close to that for minimum settling time. Other switching states are denoted non–critical and usually include the guard interval (INT state). CLeff is frequently zero in such states since Ci and CL are often both zero. It can be seen from the characterisation curves in Figure 10.17 that this leads to possible unstable oscillation. Although for these cases the amplifiers may be isolated from the rest of the filter by open–state switches, it is often desirable to avoid such instability since it could cause secondary problems and can in fact be avoided without affecting the maximum switching frequency.

A solution to this problem is to introduce grounded capacitance at the amplifier output such that the variation over the switching period of the effective load capacitance, and hence in settling characteristics, is reduced and becomes acceptable [26]. In order to remain at a defined operating point on the settling time characterisation curve, it is necessary to increase the amplifier gate widths. Although this procedure will not increase settling time, it will increase chip area and power consumption and therefore the values of added load capacitance must be carefully optimised.

We now derive general expressions for added load capacitance required to achieve given operating points on an amplifier settling time characterisation curve. For each switching state, denoted i, let the values of CLeff and Ci/Cf before addition of load capacitance be CLeffi and ai, respectively. Let the value of added load capacitance be ΔCLi, the required loading factor fi and the value of amplifier gate width factor to be determined x. Using eqn (10.9) we write,

$$x\,fi = CLeffi + \Delta CLi\,(1 + ai) \qquad (10.10)$$

Since the number of variables exceeds the number of equations by one, we may constrain the added load capacitances to be equal in two switching states, which we denote j and k. Equation (10.10) for these two switching states allows us to solve for ΔCLj = ΔCLk and x. ΔCL for the remaining switching states may then be determined from eqn (10.10) for these remaining states. The expressions obtained for x and the ΔCLi are given in Table 10.4. It can be seen that if fj = fk and either CLeffj = CLeffk or aj = ak, then the values of x or ΔCLj = ΔCLk will become indeterminate. Thus it is important to choose carefully the two switching states for which ΔCL is the same. The application of the expressions in Table 10.4 to optimise an SC circuit such as that in Figure 10.2 given the characterisation curve for the amplifier and chosen operating points in the various phases, will be illustrated in Section 10.9.1.

Table 10.4 : Expressions for Added Load Capacitance

$$x = \frac{CLeffj(1 + ak) - CLeffk(1 + aj)}{fj(1 + ak) - fk(1 + aj)}$$

$$\Delta CLj = \Delta CLk = \frac{CLeffj\;fk - CLeffk\;fj}{fj(1 + ak) - fk(1 + aj)}$$

$$\Delta CLi = \frac{x\,fi - CLeffi}{1 + ai}$$

The above design procedure has not taken fully into account the input capacitance of the amplifier which can influence the amplifier characterisation and hence the SC circuit optimisation [36]. To overcome this, a series of transient analyses can be run on an amplifier embedded in a capacitative network as in Figure 10.18 in order to determine settling time for various combinations of the capacitor values. From these results,

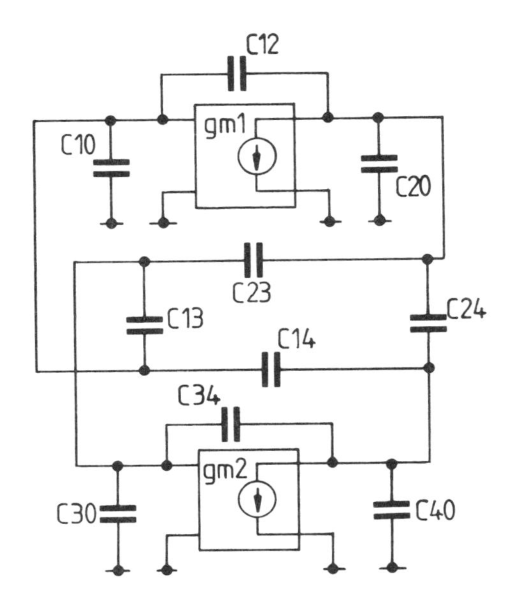

Figure 10.19: 2–amplifier network with general capacitative embedding

graphs can be plotted which allow the determination of the amplifier's input and output capacitances.

Another assumption which has been made is that the amplifiers are not coupled in any switching state. This can be overcome by representing the SC network, for the purpose of determining settling time, as a general capacitatively coupled multiamplifier network, an example of which for a 2nd order filter (e.g. Figure 10.2) is shown in Figure 10.19, where the amplifiers are treated as transconductors [18,19]. Such networks can be analysed using artificial intelligence techniques and analytical formulae obtained for the time–constants, from which the settling–time of the network may be assessed for any set of circuit capacitor values and amplifier gate widths. Available freedom in the choice of filter capacitor values and amplifier scaling factors is then used to minimise the worst–case settling time in the filter and thus maximise operating frequency.

10.9 Biquad Filter Examples

10.9.1 SC filter using high gain amplifiers

In this Section, we describe the design of a GaAs 2nd order bandpass filter [26]. The circuit is based on the structure of Figure 10.2 which is shown

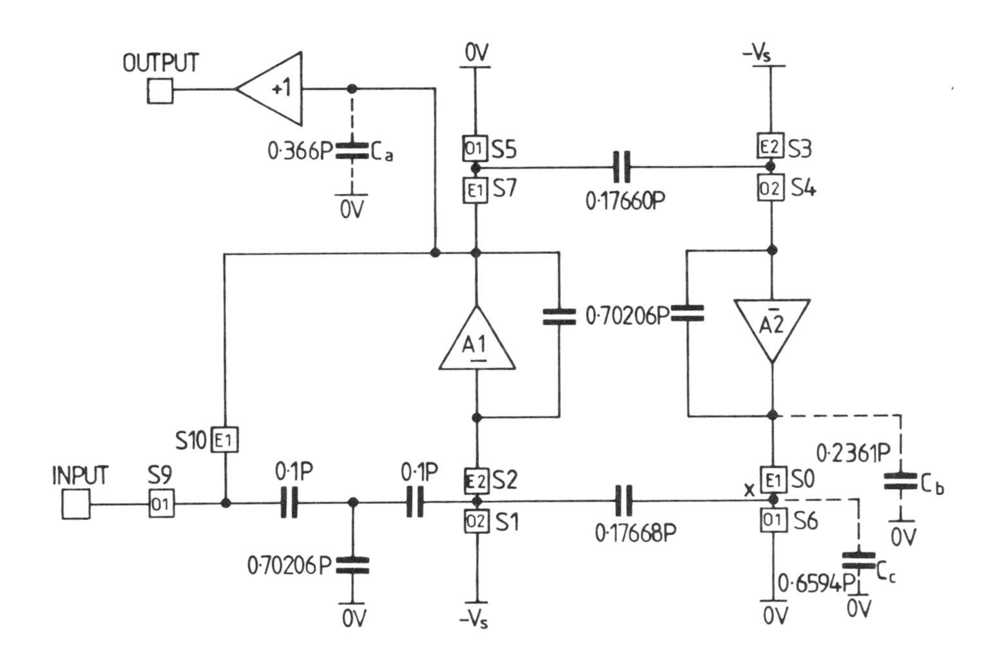

Figure 10.20: 2nd order SC filter for implementation

again in Figure 10.20 with some modifications; capacitor C11 has been replaced by an equivalent capacitor 'T' network in order to reduce capacitor spread and we have added a buffer amplifier for driving off-chip loads and capacitors CA, CB and CC which will be discussed later. The circuit uses the high-gain push-pull amplifier of Figure 10.13a, which has a low frequency gain of over 60dB. The switch control circuit to be used is that shown in Figure 10.6e.

Since the amplifier in Figure 10.13a has the property that the quiescant input voltage is equal to the negative power supply voltage, the ground connections for switches S1 and S3 in the filter of Figure 10.20 have to be connected to the negative power supply. The settling time characterisation curve for the amplifier has already been shown in Figure 10.17a and it is shown again in Figure 10.21a with the addition of chosen operating points. It can be seen that the minimum settling time of 630 ps is obtained for a loading factor of 0.008 pF per micron. Since points to the left correspond to underdamped behaviour it was considered safer, in view of process tolerances, to select the optimum operating point for the critical switch phases to the right of the fastest point and 0.012 pF per micron was chosen. As well as this operating point, denoted 'Opt' in Figure 10.21a, we will also use other operating points corresponding to loading factors of

Table 10.5 : Amplifier Settling Times

Operating Point	Settling time (ps)	
	Positive	Negative
Opt	670	778
P1	881	910
P2	615	643

Table 10.6 : SC Filter Loading for Amplifier 1 in Figure 10.20

Phase	Ci (pF)	Cf (pF)	Cl (pF)	CLeff (pF)	a (Ci/Cf)
E	0.2545	0.7131	0.2545	0.5998	0.3569
O	0	0.7021	0	0	0
Int	0	0.7021	0	0	0

Table 10.7 : Effect of 'O' phase Loading Factor for Amplifier 1 (fe = 0.012 pF/μm

fo (pF/μm)	ΔCL (pF)	x (μm)	Ts (ps)
0	0	50	oo
0.002	0.129	65	5000
0.004	0.366	90	900
0.006	0.930	155	700
0.008	4.190	524	600

Table 10.8 : SC Filter Loading for Amplifier 2 in Figure 10.20

Phase	Ci (pF)	Cf (pF)	Cl (pF)	CLeff (pF)	a (Ci/Cf)
E	0	0.7021	0.1767	0.1767	0
O	0.1767	0.7021	0	0.1767	0.2517
Int	0	0.7021	0	0	0

0.004 and 0.006 pF per micron, indicated 'P1' and 'P2' in Figure 10.21a. The settling curves for the three operating points to be used are shown in Figure 10.21b, illustrating the different degrees of damping. The settling

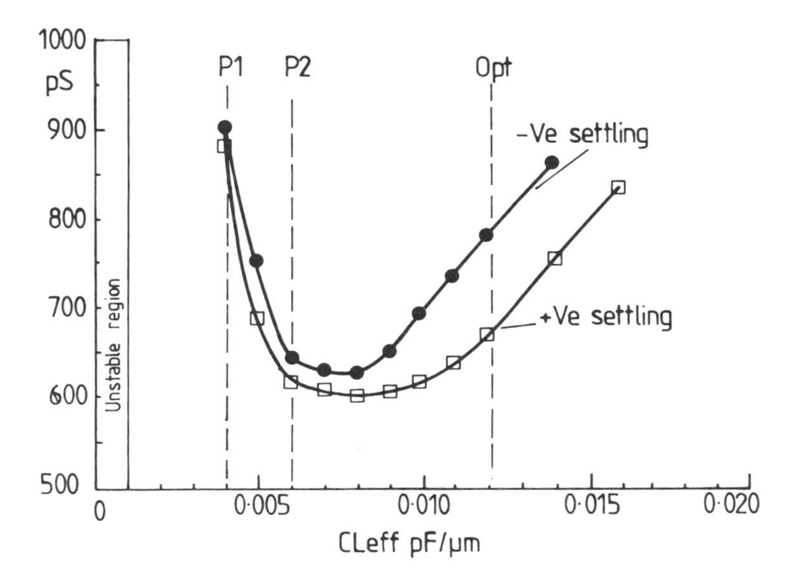

(a) characterisation curve with operating points

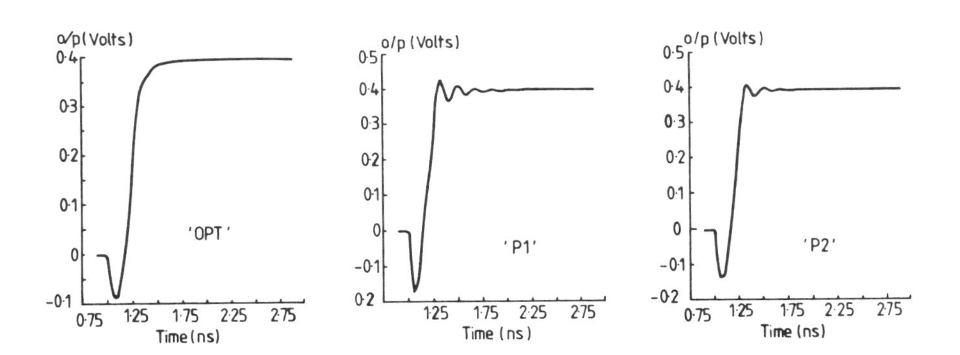

(b) transient settling curves for 'OPT', 'P1' and 'P2' operating points

Figure 10.21: Amplifier settling characteristics

times for the three operating points are given in Table 10.5. We now consider the optimisation of the SC filter circuit, including the choice of amplifier gate width factors x.

For amplifier 1, Table 10.6 gives, for the three switching states, the values of Ci, CL and Cf (as defined in Figure 10.18), CLeff (as defined in eqn (10.9)) and a = Ci/Cf. In the case of amplifier 1, the 'E' phase is the critical phase, and its loading factor fe is to correspond to the optimum operating point on the characterisation curve. We assume that the operating points in the non-critical 'O' and 'INT' switching states are identical and we denote the corresponding loading factor fo. This makes all parameters in the 'O' and 'INT' switching states identical (see Table 10.6) and therefore they may be treated as a single phase. Application of the equations in Table 10.4 to the data in Table 10.6 yields the added load capacitor value and amplifier gate width factor:

$$\Delta CLe = \Delta CLo = \frac{fo\ CLeffe}{fe - fo(1 + ae)} \tag{10.11}$$

$$x = \frac{CLeffe}{fe - fo(1 + ae)} \tag{10.11}$$

Table 10.7 shows values of ΔCL and x for a range of 'O' phase loading factors fo for the optimum fe of 0.012 pf per micron. From Table 10.7, a value of fo of 0.004 pF per micron was chosen, corresponding to operating point 'P1' in Figure 10.21a; this provides a reasonable amplifier device width factor of 90 microns and a ΔCL value of 0.366 pF, which is indicated as Ca in Figure 10.20.

For amplifier 2, the relevant parameters for the three switching states are given in Table 10.8. In this case, phases 'E' and 'O' are both critical phases and the required loading factor, denoted fe, should correspond to the optimum operating point. Table 10.8 indicates that for these two phases, the values of CLeff are also the same and therefore, as discussed in Section 10.8, the added load capacitances in these switching states cannot be made equal. Selecting the 'O' and 'INT' phases as the phases for which the load capacitances are equal and denoting the loading factor in the non-critical 'INT' phase by fint, the equations in Table 10.4 yield:

$$\Delta CLo = \Delta CLint = \frac{CLeffo\ fint}{fe - fint(1 + ao)} \tag{10.13}$$

$$x = \frac{CLeffo}{fe - fint(1 + ao)} \tag{10.14}$$

$$\Delta CLe = \frac{CLeffo\ fe}{fe - fint(1 + ao)} - CLeffe \tag{10.15}$$

A value for fint of 0.006 pF per micron (operating point 'P2' in Figure 10.21a) gave an amplifier device width factor x of 40 microns and values for ΔCLo and ΔCLe of 0.2361 pF and 0.2955 pF, respectively. These load capacitor requirements are met by adding the 0.2361 pF capacitor Cb directly to the output of amplifier 2, as shown in

Table 10.9 : Simulated and Design Filter Settling Times (ps)

Amp	Sign		Int	O	E	
					Iso	Cpl
A1	+	Sim Des	778 881	778 881	689 670	935 –
A1	–	Sim Des	797 910	797 910	739 778	894 –
S2	+	Sim Des	605 615	658 670	652 670	811 –
S2	–	Sim Des	507 643	747 778	772 778	865 –

Figure 10.20, and a 0.0594 pF capacitor Cc to the node ‘x’. If the load capacitances had been made equal in the ‘E’ and ‘INT’ switching states, then the equations would still have been soluble but then ΔCLint would have been greater than ΔCLo; this would have made implementation difficult as no switches are closed in the ‘INT’ switching state.

The design procedure which has been used, as described in Section 10.8, made the assumption that the amplifiers settle independently. However, it can be seen from Figure 10.20 that, in the ‘E’ phase, the amplifiers are coupled together via switches S7 and S4. We now investigate the effect of this assumption. SPICE can be used to simulate the settling times of the two amplifiers for positive and negative output steps, and the results are shown as ‘Sim’ in Table 10.9. The settling times denoted ‘Des’ are the design settling times of Table 10.5 for the chosen operating points. For the results denoted ‘Iso’ in Table 10.9, the interaction was broken by replacing the other amplifier by a voltage source or a short circuit. The results denoted ‘Cpl’ include the coupling. It can be seen that the coupling between amplifiers increases settling time by about 30%. For the ‘O’, ‘Int’ and non–coupled ‘E’ cases, there is reasonable agreement between design settling times and those obtained by simulating the filter.

Since off–chip loads are highly capacitative compared to on–chip loads a buffer, shown in Figure 10.20, is required to provide suitable interfacing. The buffer design used is shown in Figure 10.22 and consists of two unity gain sections seperated by a switching circuit and hold capacitor. The unity gain sections are very high quality modified source followers derived from the operational amplifier of Figure 10.13a. The

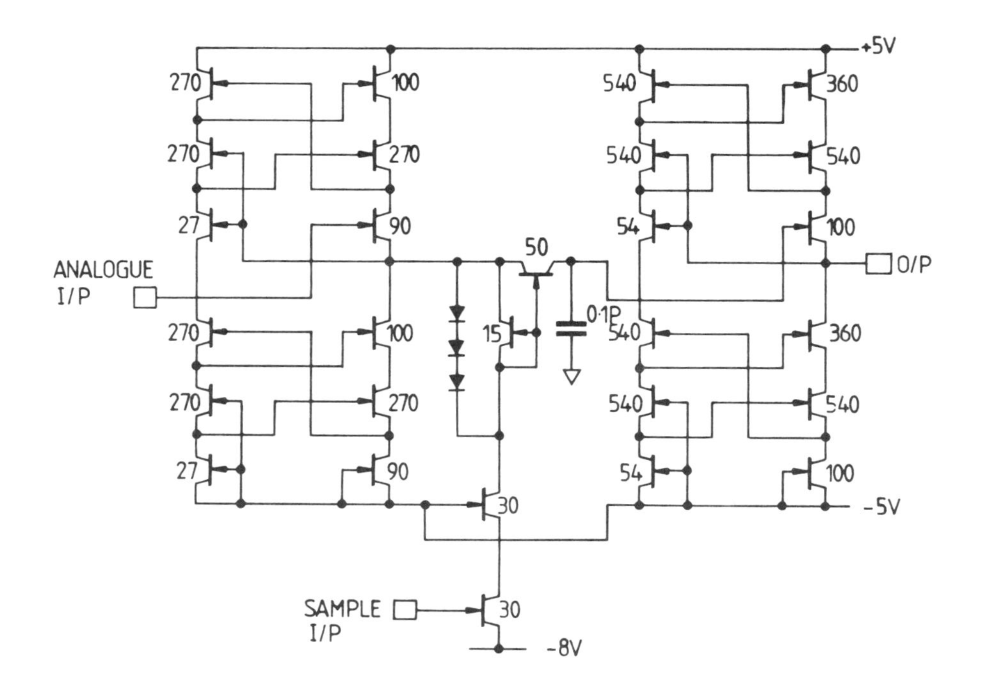

Figure 10.22: Output buffer circuit

switching circuit is similar to that used in the SC filter. By feeding an appropriate clock signal to the sample input, the buffer acts as a sample and hold circuit where the sampling instant may be freely chosen. Alternatively, with a DC signal applied to the sample input, the circuit acts as a non–sampling buffer.

The frequency response of the final circuit was simulated using the SC network analysis program SCNAP [15,16]. The analysis was based on small signal macromodels for the amplifiers and the switches. The macromodels were obtained by curve fitting on the basis of SPICE simulation of the subcircuits. The simulated responses are shown with the ideal curves for switching frequencies of 250 MHz and 500 MHz in Figures 10.23a and b, respectively.

The circuit was layed out using MAGIC with a specially written technolgy file for the Anadigics 0.5 um GaAs process. Seperate lines and pads were used for the analogue and digital grounds and power supplies. Reasonable precautions were taken to avoid backgating effects. A plot of the layout is shown in Figure 10.24. The size of the chip is 3.3 x 2.6 mm and the power consumption is 400 mW. From the simulation results in the

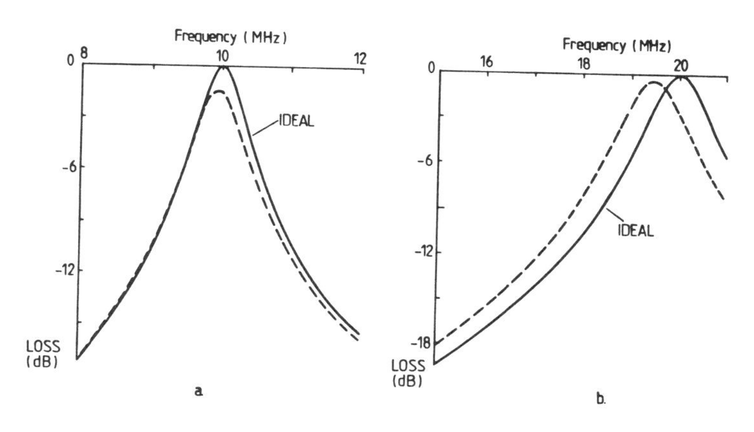

Figure 10.23: Simulated filter gain responses
(a) Performance for 250 MHz switching frequency
(b) Performance for 500 MHz switching frequency

previous Section, the circuit is expected to operate at a switching frequency of 500 MHz, realising a bandpass response centered at 20 MHz.

10.9.2 SC filter design using finite gain insensitive SC circuits

We now describe the design of a high frequency SC filter which uses scaled down geometries with minimum sized devices in order to minimise chip area [18]. The approach will be to adopt the 2nd order SC filter structure of Figure 10.25 which uses the finite gain insensitive (FGI) integrator of Figure 10.3 in conjunction with moderate gain (40 dB) amplifiers. Fo this filter, the design Q–factor is 8, the midband gain unity and the switching to midband frequency ratio is 12.5. The layout plot for the filter is shown in Figure 10.26. The relatively large block INT2 corresponds to the two–input (lower) integrator in Figure 10.25, and the block INT1 corresponds to the the four–input (upper) integrator. The blocks BUF1 and BUF2 in Figure 10.26 are unity gain buffer amplifiers for driving off–chip loads for the bandpass and lowpass output nodes in Figure 10.25 and SCC1 is a switching circuit which will be discussed later. This design is based on use of the Plessey III–V GaAs foundry process which provides a gate length for the D–type GaAs MESFET's to be used of 0.7 microns. The modelling procedure is as described in Section 10.3 with seperate SPICE subcircuits based on a JFET model to describe devices used in digital, analogue and switched circuit modes.

The Silicon Nitride capacitors required are realised by the interconnection of unit–sized elements in order to alleviate the effect of undercut. Consideration of the capacitor values required in Figure 10.25

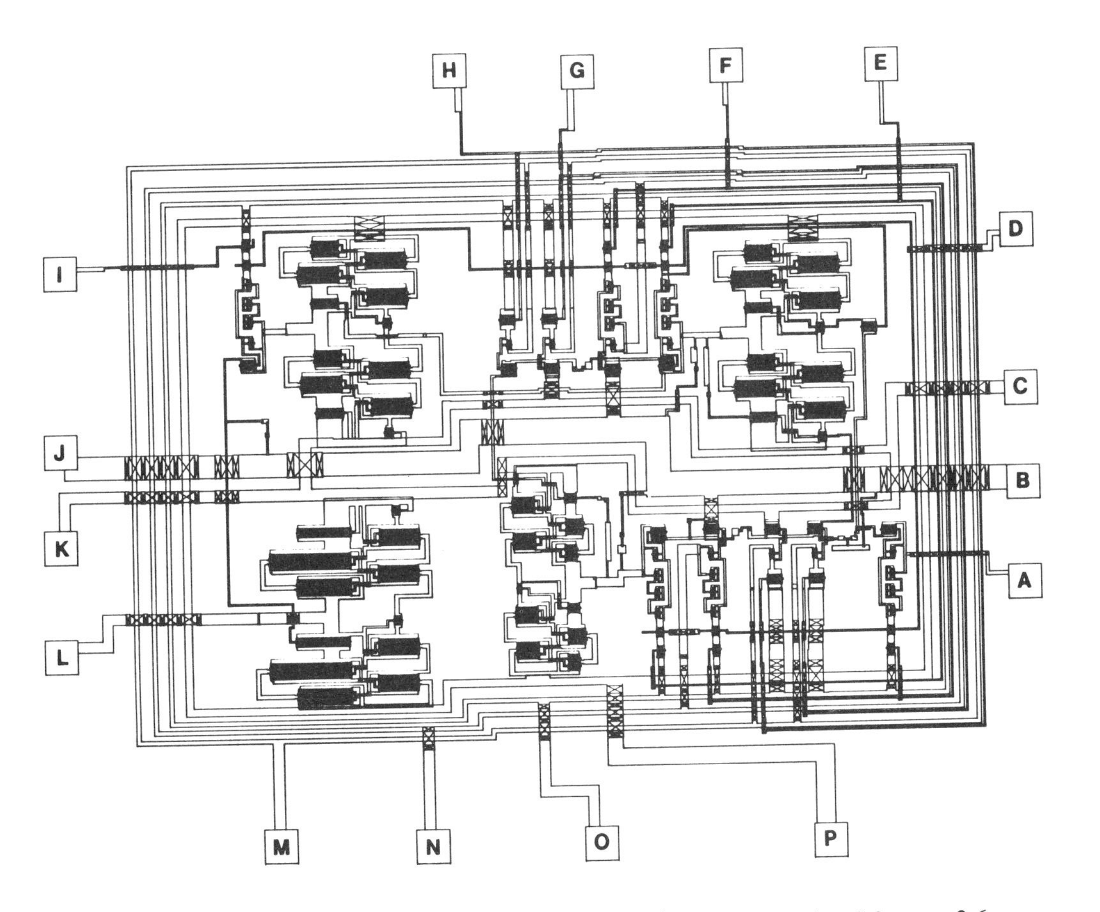

Figure 10.24: Layout plot for filter using high gain amplifiers; total size 3.3mm x 2.6 mm

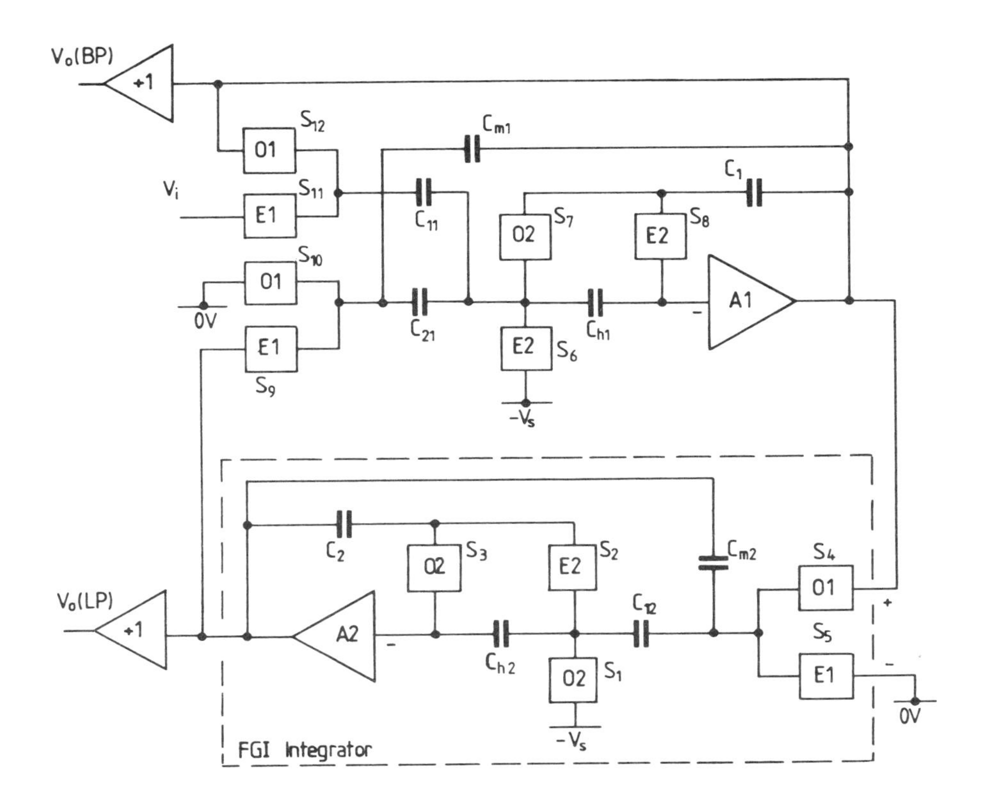

Figure 10.25: SC bandpass filter using FGI integrators; all capcitors = 1 unit, except C1 = C2 = 2, Ch1 = 4, C11 = 1/8

and the achievable ratio precision results in a choice of 0.055 pF as the unit which consists of a series combination of two minimum size capacitors of the type discussed in Section 10.5. The capacitor Cm2 in Figure 10.25 is indicated in Figure 10.26 within the INT2 cell and consists of a series combination of two minimum size capacitors. The capacitor C11 in Figure 10.25 is realised as the series connection of 16 minimum size capacitors which can be identified within the INT1 cell in Figure 10.26.

The amplifier used is the single-stage single-cascode architecture using diode level shifting of Figure 10.10, having a single-ended input. The use of diode level shifting is adopted in preference to techniques relying on device width ratioing on account of the particular nature of the MESFET characteristics, as discussed in Section 10.3. Simulation of a unit scaled amplifier (k=1) using appropriate analogue mode models for the

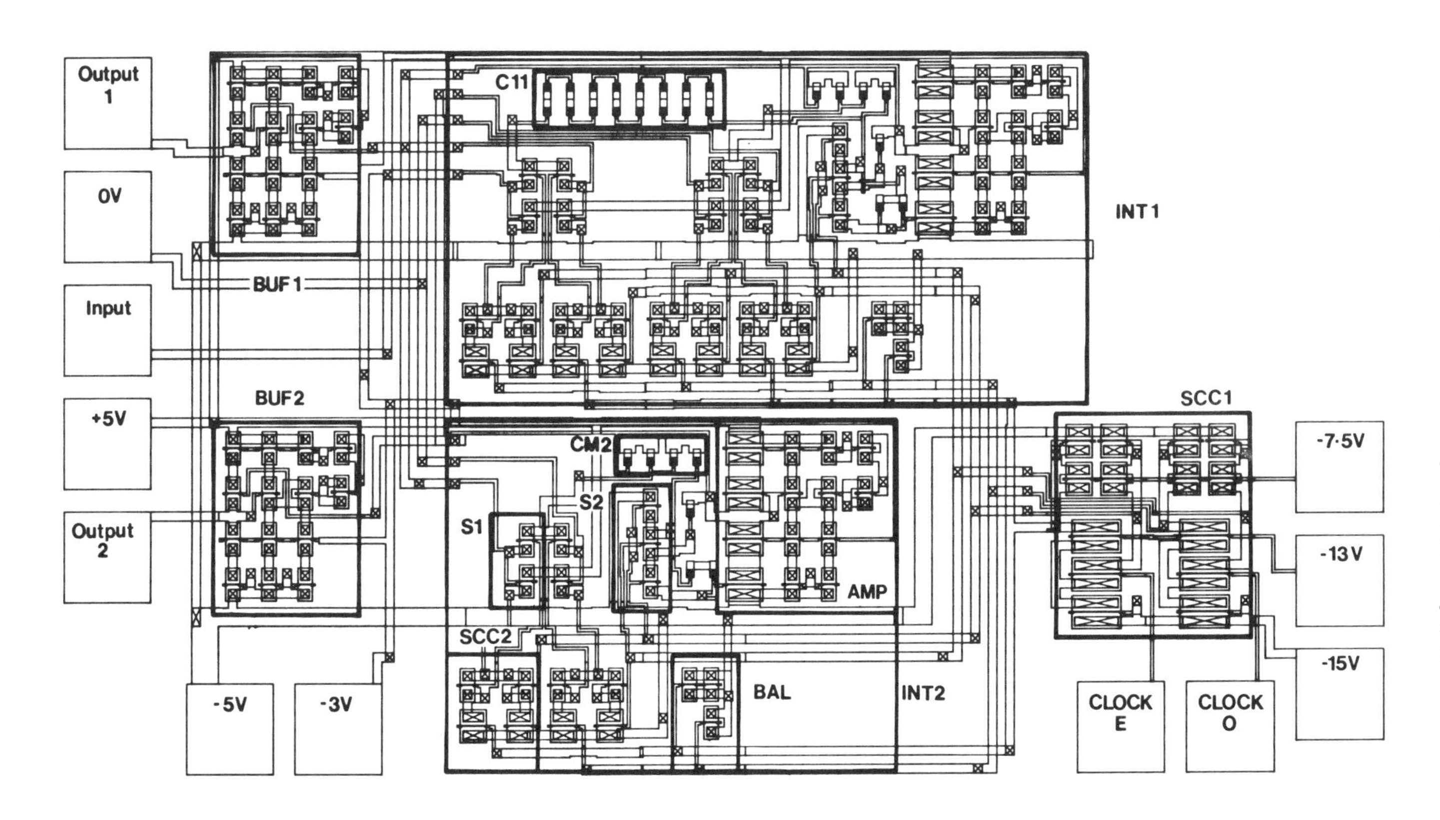

Figure 10.26: Layout plot for filter using FGI integrators (total size 2mm x 1mm)

MESFET's indicates a DC gain of 36 dB that is maintained for quiescent output voltages from –1.5V to +2.0V, and settling time characterisation indicates a minimum settling time of approximately 200pS obtained for an 80 fF load. A consequence of using a single–ended input, with a quiescent voltage of –Vs, is that the switches S1 and S6 in Figure 10.25 must be grounded to –Vs. In the layout plot of Figure 10.26, the amplifier is denoted AMP and can be seen in the upper right corner of the INT2 cell and in a similar position within the INT1 cell. Both of these amplifiers have a scaling factor k of 3, the choice of which will be discussed later (Plates II ans III show the amplifier layout in more detail as seen by the PRINCESS and MAGIC layout editors, respectively). The buffer amplifiers used for driving off–chip loads, BUF1 and BUF2 in Figure 10.26, are derived directly from the amplifier of Figure 10.10 and have a scaling factor of unity.

The switches in Figure 10.25 may be divided into a number of categories. The switches adjacent to the amplifier inputs (S1 – S3 and S6 – S8) are switching signals close to the –Vs voltage and they may each be realised as a minimum size MESFET with gate voltage switched between –Vs (on–state) and –Vs – 3V (off–state); one such group of three switches is indicated S2 within the INT2 cell of Figure 10.26. The remaining six switches in Figure 10.25 are associated with the integrator input terminals and it was decided to introduce SCC's for all of these switches for the reason discussed in Section 10.4. The SCC used is that already shown in Figure 10.6d. As far as simulation is concerned, device M1 in Figure 10.6d is represented by a switch–mode model and the remaining devices by digital–mode models. For layout, the switch and source follower devices M1 and M4 in Figure 10.6d are placed in the upper half of the integrator as indicated by S1 in the cell INT2 of Figure 10.26. The remaining devices are indicated by SCC2 and are placed in a lower digital part of the integrator where higher packing densities are permitted. The complementary clock signals Vinp and Vinm required by the SCC's are generated at one point on the chip by a clock generator denoted SCC1 in Figure 10.26. The circuit consists of two identical channels, one for each clock phase, each of which has the structure of Figure 10.7. The output voltages from the clock generator in Figure 10.7 are suitable for driving, not only the SCC's, but also the switches at the amplifier inputs in Figure 10.25. Since these amplifier input switches are connected only to the Voutp output of the clock generator, a set of dummy switches are connected to the Voutm output to keep the outputs symmetrical; these dummy devices are indicated as BAL within INT2 in Figure 10.26. All devices in Figure 10.7 were simulated as digital–mode models and the device width ratios needed to drive the SCC's, directly driven switches and balancing switches are as given in Figure 10.7. This circuit has the largest devices anywhere on the chip which have gate widths of 64 microns, i.e. four times minimum size.

In order to maximise the frequency at which the filter can operate it is crucial that the amplifier scaling factors and the values of the embedding capacitors in the SC circuit are optimised. The SC filter in Figure 10.25 operates with a 2–phase clock and therefore amplifier settling time has to be considered in three switching states, 'O', 'E' and 'INT'. For each of these switching states, the circuit is represented, for the purpose of determining settling time, as a special case of the structure in Figure 10.19. The settling–time of the network is assessed for any set of circuit capacitor values and amplifier scaling factors. For this filter, additional load capacitance at the amplifier outputs was not introduced as the circuit relies on the intrinsic parasitic capacitances of the components for its stability. The capacitor Ch1 in Figure 10.25, whose value does not affect the filter transfer function, was found to have quite a strong effect on the worst–case settling–time and it was consequently made four times larger than Ch2. The fact that this makes Ch1 twice the size of C21 slightly increases the phase error of the upper integrator, but this was found to have no appreciable effect on the filter transfer function. The minimum worst case settling time of 500 ps was obtained for amplifier scaling factors (k in Figure 10.10) of 7 and a choice of 0.025 pF for the unit capacitance in Figure 10.25. However, due to considerations of power consumption and clock feedthrough, it was decided to limit the amplifier scaling factors to 3 and to adopt a unit capacitance of 0.055 pF consisting of a series combination of two 0.11 pF minimum size capacitors. The predicted and simulated worst–case settling–time for the design is 745 ps assuming ideal switches and 995 ps including switch on–resistance effects. The estimated power consumption of the circuit was 800 mW.

The layout of the filter, already shown in Figure 10.26, is arranged hierarchically, with seperate cells for such entities as amplifiers, buffers and switching circuits. Verification is done by manual extraction of the circuits from these cells, and combining the extracted circuits in a SPICE file using the cell interconnections defined by the top–level layout file. The resulting SPICE file is used in a transient analysis of several signal cycles to verify correct operation. An example of the output from this analysis is shown in Figure 10.27, showing the bandpass output voltage of the upper amplifier in Figure 10.25. The spikes in the guard intervals are due to the relatively small value of the feedback capacitance on the amplifiers in those intervals. In spite of this, the output is seen to settle to its final value well within the time allowed by the 200 MHz clock.

Layout used the MAGIC symbolic layout editor, which has the advantage of on–line design–rule checking. The CIF file produced by MAGIC has been converted to FDR format by a specially written program in C. After transmission of the FDR file to SERC Rutherford Appleton

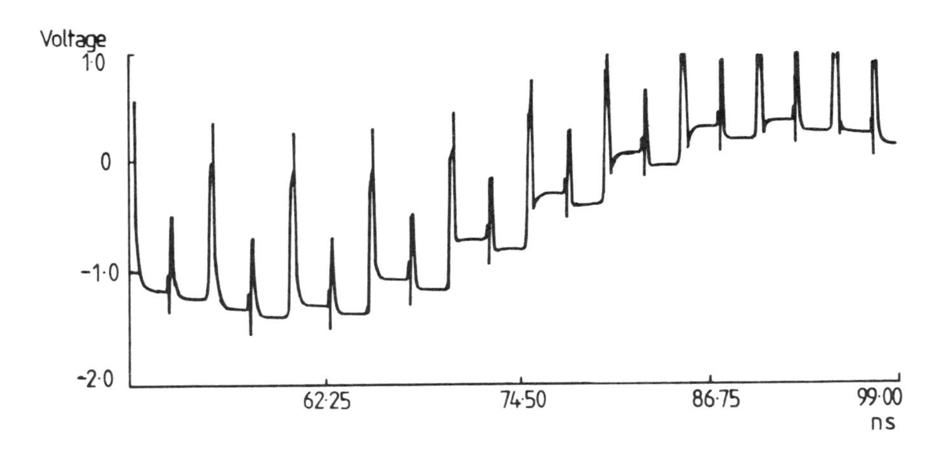

Figure 10.27 SPICE time-domain simulation of filter response

Laboratory, the layout editor PRINCESS has been used to introduce 45 degree features (not available in MAGIC) in the cells defining the MESFET's, and additional design rule checking and error correction carried out. As an example of this procedure, the amplifier cell layout as seen via MAGIC and PRINCESS are shown in Plate II and III, respectively; in Plate II, the 45 degree angles in the drain and source contacts and in the gate structures can be seen. From the verification results in Figure 10.27, it is expected that the filter described in this Section will realise a bandpass filter centered at 40 MHz for a 500 MHz switching frequency with minimum chip area requirement.

10.10 Conclusions

GaAs technologies are maturing rapidly and so, coupled with the advancement in circuit techniques, much larger levels of integration are now becoming possible. In the development of suitable circuit techniques, emphasis has been placed upon the necessary minimisation of power consumption and chip area. Past work has already demonstrated the feasibility of operating GaAs SC filters at switching frequencies of 250 MHz. The filter design procedures presented in this Chapter will lead to higher response precision, switching frequencies as high as 500 MHz and much reduced chip area approaching 0.25 mm^2 per pole pair. These new systems will permit filtering frequencies approaching 50 MHz. The present limitations on larger integrated systems are due to the effect of minimum realisable gate width and minimum unit capacitance on total chip area and power consumption. In the context of the present technological restrictions, we have been able to achieve significant reductions in chip area. Comparable reduction of power consumption is now a necessary step towards the much larger levels of integration warranted for the successful application of GaAs for sampled data analogue signal processing applications.

10.11 Acknowledgements

The authors gratefully acknowledge the invaluable help of numerous colleagues. In particular, Kevin Steptoe, Heloisa Facanha and John Taylor of University College London, Steve Harrold of UMIST, Jack Sewell and Brian Wishart of Glasgow University, and Owen Richards and Graham Johnson of Imperial College. Also, Liza Wong and Steve Newett of the SERC Rutherford Appleton Laboratory, Shiva Carver, David Goldsmith and Steven Hart of Interactive Solutions Ltd, Charlie Suckling and Jim Arnold of Plessey, John Phillips of STC Technology and Robert Bayruns of Anadigics Inc gave valuable support. The preparation of the diagrams by John Milne of Imperial College is gratefully acknowledged. The financial support of the SERC is also gratefully acknowledged.

10.12 References

1 S. Masuda et al, CMOS sampled differential push pull cascode operational amplifier, Procs 1984 IEEE ISCAS (Montreal), May 1988, pp 1211–1214

2 P. E. Allen and C. M. Breevoort, An analogue circuit perspective of GaAs technol-ogy, Procs 1987 IEEE ISCAS (Philadelphia), May 1987

3 E. L. Larson, K. W. Martin and G. C. Temes, GaAs switched capacitor circuits for high speed signal processing, IEEE J Solid State Circs, SC–22,pp 971–981,Dec 1987

4 S. J. Harrold, I A. W. Vance and D. G. Haigh, Second–order switched–capacitor bandpass filter implemented in GaAs, Electronics Letters, Vol 21, no 11, pp 494–496, 23rd May 1985.

5 GaAs Integrated Circuits, ed J Mun, BSP Books (Oxford), 1988

6 R. C. Eden and E. K. Shen, GaAs MESFET modelling for analogue interface and digital IC applications, Procs 1987 IEEE ISCAS (Philadelphia), May 1987

7 Analogue MOS Integrated Circuits, R Gregorian and G C Temes, John Wiley (New York), 1986

8 D. G. Haigh and J. T. Taylor, On switch–induced distortion in switched capacitor circuits, Procs 1988 IEEE International Symposium on Circuits and Systems (Espoo, Finland), June 1988, pp 1987 - 1990

9 A. K. Betts, J. T. Taylor and D. G. Haigh, Spike–free biphase switched–capacitor integrator pair with low sensitivity to non–ideal op–amp effects, Electronics Letters, vol 24, no 4, 18th February 1988, pp 202 - 204

10 C. Toumazou and D G Haigh, Design of a high–gain, single–stage operational amplifier for GaAs switched–capacitor filters, Electronics Letters, vol 23, no 14, pp 752–754, 2nd July 1987

11 L. E. Larson et al, Comparison of amplifier gain enhancement techniques for GaAs MESFET analogue integrated circuits,Electron Lett, 1986, 22, pp 1138–1139

12 C. Camocho–Penalosa and C S Aitchison, Modelling frequency dependence of output impedance of a microwave MESFET at low frequencies, Electronics Letters, 1985, vol 21, pp 528–529

13 W. A. White and Namordi M R, GaAs MESFET model adds life to SPICE, Microwaves & RF, pp 197–200, Sept 1984,

14 D. G. Haigh and C Toumazou, On computer simulation of integrated switched capacitor circuits, Procs 1989 European Conference on Circuit Theory and Design, Brighton (UK), 5th - 8th September 1989

15 J. I. Sewell, The SCNAP series of software for switched capacitor circuit design, IEE Digest 1987/94 on 'Analogue IC Design', November 1987

16 L. B. Wolowitz and J I Sewell, General analysis of large linear switched capacitor networks, IEE Proceedings Part G (Electronic Circuits and Systems), vol 135, pp 119–124, June 1988

17 S J Harrold, A switch–driver circuit suitable for high order switched capacitor filters implemented in GaAs, Electronics Letters, vol 24, no 15, pp982–984, July 1988

18 D. G. Haigh, A. K. Betts, K. Steptoe and J. T. Taylor, The design of switched capacitor filter circuits for GaAs MSI technology, Procs 1989 European Conference on Circuit Theory and Design, Brighton (UK), 5th - 8th September 1989

19 A. K. Betts, PhD Thesis, University of London, To Be Published, October 1989

20 N. Scheinberg, Design of high speed operational amplifiers with GaAs MESFET's, Procs 1987 IEEE ISCAS (Philadelphia), May 1987, pp 193–198

21 L. E. Larson et al, A 10 GHz operational amplifier in GaAs MESFET technology, Digest 1989 IEEE Int Solid State Circs Conf, pp 72–73, February 1989

22 A. A. Abidi, Gain bandwidth enhancement in GaAs MESFET wide band amplifiers, Procs 1988 IEEE ISCAS (Helsinki), pp 1465–1468, June 1988

23 S. I. Katsu, M. Kazumura and G. Kano, Design and fabrication of a GaAs monolith-ic opertaional amplifier, IEEE Trans on Electron Devs, vol 35, no 7, pp 831–838 July 1988

24 C. Toumazou and D. G. Haigh, Analogue design techniques for high speed GaAs operational amplifiers, Procs 1988 IEEE ISCAS (Helsinki), June 1988

25 C. Toumazou and D. G. Haigh, Design and application of GaAs MESFET current mirror circuits, IEE Digest 1989/25 for Colloquium on Current Mode Analogue Circuits, London, 17th February 1989

26 D. G. Haigh, C. Toumazou, S. J. Harrold, J. I. Sewell and K .Steptoe, Design and optimisation of a GaAs switched capacitor filter, Procs IEEE 1989 International Symposium on Circuits and Systems, Portland, Oregon, USA, June 1989

27 B. Gilbert, Wideband negative current mirror, Electronics Letters, vol 11, no 6, March 1975, pp 126–127

28 C. Toumazou, D. G. Haigh and L. Larson, High speed GaAs operational amplifiers, Special Issue on 'Microwaves' of Electronics and Wireless World, August 1989

29 C. Toumazou, D. G. Haigh and O. Richards, Fast settling GaAs operational amplifiers for switched capacitor applications, Procs 1989 IEEE ISCAS (Portland, Oregon), May 9–11 1989

30 D. G. Haigh and C. Toumazou, Fast settling high gain GaAs operational amplifiers for switched capacitor applications, Electronic Letters, vol 25, no 1, 25th May 1989, pp 734 – 736

31 C. Toumazou and D. G. Haigh, Level–shifting differential to single–ended convertor circuits for GaAs MESFET implementation, Electronics Letters, vol 23, no 20, 24th September 1987, pp 1053–1055.

32 H. C. Yang, J. R. Ireland and D. J. Allstot, Improved operational amplifier compensation techniques for high frequency switched capacitor circuits, Procs 30th Midwest Symposium on Circuits and Systems, August 1988, pp 952 – 955

33 C. Toumazou and D. G. Haigh, Some designs and a characterisation method for GaAs operational amplifiers for switched capacitor applications, Electronics Letters, vol 24, no 18, 1st September 1988, pp 1170 – 1172

34 K. Matsui et al, CMOS video filters using switched capacitor 14 MHz circuits, Digest 1985 IEEE Int Solid State Circs Conf, pp 282–283 & 364, February 1985

35 K. Matsui, T. Matsuura and K. Iwasaki, 2 micron CMOS switched capacitor circuits for analogue video LSI, Procs IEEE 1982 International Symposium on Circuits and Systems, Rome, 10th – 12th May 1982

36 A. K. Betts, J. T. Taylor and D. G. Haigh, An improved characterisation technique for amplifiers used in high–speed switched capacitor circuits, Procs 1989 European Conference on Circuit Theory and Design, Brighton (UK), 5th – 8th September 1989

Chapter 11

Integrated Circuits for Optical Fibre Systems

R. P. Merrett

11.1 Introduction

The penetration of optical fibre systems into telecoms networks has grown rapidly since the first operational system was installed in 1980. Most of British Telecom's trunk traffic is now carried digitally and almost two thirds of this is in the form of 140 Mbit/s transmission over single mode optical fibre. This network is being expanded by the progressive introduction of 565 Mbit/s operation and higher bit–rate systems are being envisaged. There is also considerable interest in the development of high bit–rate optical fibre systems for local distribution of TV services and for computer links. Electronics requirements for these systems can be met by GaAs ICs and a wide range of standard parts are becoming commercially available.

Discrete GaAs MESFETs mounted in hybrids were for a long time the backbone of the microwave industry. The early development of monolithic microwave ICs (MMICs) was driven by the needs of the military market. For civil applications, direct broadcast satellite receivers were perceived as the first high volume market but this is still growing sluggishly. Digital and baseband analogue ICs were slower to find applications, partly because of their initially low complexity but also because there was no immediate systems demand for their high performance.

The main potential application areas for high speed digital and baseband ICs were military systems (where the good radiation hardness of GaAs was an additional advantage), computing, instrumentation and optical fibre communications. The last, which is the subject of this chapter, encompasses both telecommunications and data links for computing and control. These can differ in both the type of fibre and the wavelength employed.

For the early silica fibre the minimum loss occurred near 0.85μm. Eventually better control of the moisture content of fibres resulted in the loss minima being moved to 1.3 and 1.5μm. The former also corresponds to a minimum in the wavelength dispersion characteristics but there are fibre structures which enable this dispersion minima to be shifted to 1.55μm.

For all these wavelengths either multimode or single mode fibre can be used. Multimode fibre is of larger diameter and is thus much easier to splice and couple, but mode conversion limits the usable product of bandwidth and link length to less than about 200 MHz.km. By contrast, the corresponding figure for low dispersion single mode fibre is greater than 100 GHz.km

For those computing and control applications which either require short links or low bit-rates it is convenient to use the cheaper and easier to handle multimode fibre. In such cases it is also advantageous to use a wavelength of 0.85μm as this can be provided by either GaAs LEDs or lasers. Any associated high speed electronics can also be provided by GaAs and there is currently considerable interest in monolithic integration of these technologies, i.e. optoelectronic integrated circuits (OEICs).

The first telecommunications application of optical fibres was for long distance transmission and thus both low loss and minimum wavelength dispersion were essential. Single mode fibre operating at either 1.3 or 1.55 μm was rapidly adopted as the standard and, because of its bandwidth flexibility, it is also being introduced for local distribution. At these wavelengths GaAs cannot be used for the optical components. Instead it is necessary to use the smaller bandgap ternary and quaternary semiconductors (InGaAs and InGaAsP) lattice matched to InP substrates. Optoelectronic integration using such materials presents a greater challenge and is still very much at the research stage.

Other communications uses for GaAs include microwave and millimetre-wave radio for the distribution of video signals and mobile services. Although these would appear to be outside the scope of this discussion, they form part of a complete communications system having optical fibre interfaces. There is also interest in the microwave modulation of optical signals (subcarrier modulation) and it is likely that future communications networks will use a blend of radio and optical transmission.

The technology for the present range of optical communications GaAs ICs has been discussed in Chapter 6. Thus it is only necessary to summarize this aspect here before describing the IC requirements of the various application areas. Finally, future technology developments, and their likely impact, will be considered.

11.2 Summary of current status of GaAs IC Technology

Over the last four years there have been a number of demonstrations of the ability of GaAs ICs to provide the high performance electronic functions needed for the next generation of optical communications. These ICs were initially based on application-specific designs processed in research laboratories of vertically integrated companies.

Production of standard parts was hindered by the diversity of functions and bit–rates required by various telecoms organisations, and by the lack of agreement on input/output voltages. During the last two years, however, there has been a dramatic increase in the range of commercially available standard components. Various foundries have also demonstrated their capability to provide application–specific designs using either custom based standard cells or gate arrays. These components are based on the technologies discussed below.

Most of the early circuits used a 1μm depletion FET technology. Threshold voltages are typically about −1.5V although less negative values will give reduced power albeit at the expense of low speed. For applications above 2 Gbit/s this technology is limited to about 300 gates by virtue of its power dissipation, even when low power interstage coupling options are used. Smaller circuits have been shown to be capable of operation up to about 5 Gbit/s.

Enhancement–depletion (E–D) technology offers a much lower power dissipation but the speed is restricted by poor drive capability. Another limitation arises from the use of switching FETs having a very low threshold voltage, i.e. typically 50 mV. This makes the noise margin very sensitive to device–to–device variations in threshold voltage and ultimately limits complexity. Various circuit innovations have been introduced to improve the noise margins and, although these increase the power dissipation, they have made it possible to produce 1000 gate circuits capable of operating up to about 1.2 Gbit/s.

Source coupled FET logic provides an intermediate solution between depletion and E–D technologies. Gates can be stacked as in emitter coupled logic (ECL), and their differential drive capability reduces sensitivity to threshold voltages variations and to changes in temperature. This technology is presently favoured by Japanese manufacturers who have made a late but strong entry into electronics for optical fibre systems and have produced a range of MSI circuits for systems operating at up to 2.4 Gbit/s

Packaging and interconnection of high performance ICs has required careful attention to impedance matching, minimising crosstalk between lines and reducing propagation delays. The pulse rise and fall time is typically about 100 ps and this corresponds to a free space propagation distance of 3cm. Although transmission delays do not present a problem to a chip, they do for a hybrid assembly or a printed circuit board, and there is thus an incentive to increase the level of monolithic integration. Most system demonstrations have been below 3 Gbit/s and it has proved possible to use existing packaging and interconnection technologies albeit with careful attention to detail. Microstrip or coplanar transmission lines are used on the circuit boards and package lead–throughs are designed to have the same characteristic impedance as these lines. A range of surface mounted packages satisfying this criterion are available with up to 64 pins.

11.3 Applications at 1.3 and 1.5 μm

By virtue of its low loss and low dispersion, single mode optical fibre operating at wavelengths of 1.3 and 1.55 μm has become the dominant medium for long haul digital transmission systems for telecommunication networks. Most of the early digital traffic was at about 140 Mbits/s but systems are being enhanced to 565 Mbit/s to service network growth. In the USA the traffic on some routes has already justified the introduction of 1.2 and 1.7 Gbit/s. Similar upgrades are being considered within Europe and Japan, where a hierarchy based on 2.4 Gbit/s is favoured.

Within the UK single mode fibre has also been chosen for the local loop because of the ease with which high bit-rate services can be added in the future. Link lengths are much shorter than for long haul transmission, but the splitting architectures used to route data to multiple destinations result in similar power budgets for both applications.

The flexibility of digital transmission and the optical network has led to the evolution of new approaches to the provision of telecoms services. Instead of having separate circuits for telephony, data, telex etc. the customer is provided with a single network offering a full range of services, i.e. an integrated services digital network (ISDN). At present this service is restricted to telephony and low bit-rate services (<2 Mbit/s) but there is currently considerable research interest in incorporating broadband services, such as high quality TV transmissions, into the ISDN (B-ISDN).

The various high speed electronic elements needed for these systems will now be considered in more details.

11.3.1 Transmitters and Receivers

For wavelengths of 1.3 and 1.5m, InP based components have to be used for the light emitter and detector. Parasitics associated with the interconnect between these components and high speed electronics can seriously degrade performance unless complex and expensive assembly techniques are used. This problem was acute for the early designs because the optical components had to be connected to discrete GaAs electronic devices, but the use of monolithically integrated receivers pre-amplifiers and transmitter drivers has eased this problem.

All operational optical fibre systems are currently of the intensity modulated/direct detection type [1] and they use the receivers and transmitters described in the next two subsections. These systems make no use of the coherence or spectral purity of the optical signal. Coherent transmission offers the potential for improved system performance and there is now considerable research interest in this technique [2]. The receivers and transmitters needed for such a system will be discussed in subsection 11.3.1.3.

Present systems are also all based on a single wavelength, but there is strong research interest in developing wavelength division multiplexing techniques (WDM) to take advantage of the tremendous bandwidth capabilities of optical fibre [2]. Such systems will require receiver and transmitter arrays and this is only likely to be met by the development of opto-electronic ICs. The prospect for these will be considered in Section 11.5.

11.3.1.1 Receivers

The basic elements of a receiver circuit are illustrated in Figure 11.1. A photo-diode converts the incident light into a current and this is amplified by the subsequent pre-amplifier, gain controlled amplifier and limiting amplifier. The output is then fed into a regenerator stage, comprising clock extraction circuit and D-type latch, which retimes the digital signal.

Both avalanche photo-diodes (APD) and PIN diodes are used as detector elements in such receivers. The APD has internal gain and thus the sensitivity of a receiver based on such an element is less dependent on the noise generated by the pre-amplifier than it is for a receiver with a PIN. The disadvantages of the APD are that it requires a high bias voltage (at least 60V) and that, for III-V devices, the technology is not yet mature. For these reasons the PIN diode is preferred for most applications and this sets stringent requirements upon the noise performance of the pre-amplifier. With such a detector the typical signal levels at various points within a receiver are as shown in Figure 11.1.

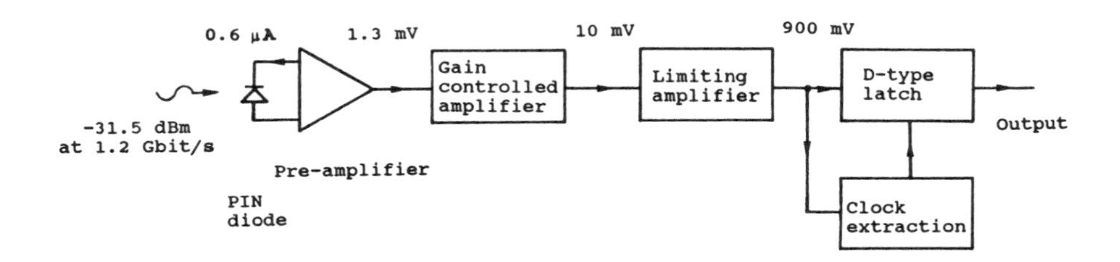

Figure 11.1 Block diagram of an optical receiver showing typical signal levels at each stage

For the transimpedance amplifier a feedback resistor is connected between the input and output of the pre-amplifier. This resistor, in conjunction with the input capacitance, fixes the bandwidth. Thus the latter can be made appropriate for the system bit-rate and it is not necessary to provide subsequent equalization. The disadvantage of this configuration is that the feedback resistor represents an additional noise source which degrades the sensitivity relative to the high impedance amplifier. Nevertheless the transimpedance amplifier is usually preferred because of its better dynamic range and lack of pattern sensitivity.

The relative merits of GaAs MESFET or silicon bipolar inputs for the pre-amplifier have been hotly disputed. A detailed comparison of these two options is outside the scope of this discussion. It is sufficient to point out that 1 and 0.5μm MESFETs are achieving good sensitivities at Gbit/s data rates and that most commercially available optical receivers for such applications are based upon this technology.

No matter which input device is chosen, performance is largely determined by the total capacitance at the pre-amplifier input. For a given PIN diode it is thus necessary to minimise both the interconnect capacitances and the capacitances associated with the input device and the feedback resistor. For this reason monolithically integrated pre-amplifiers are preferred for both predictable performance and low cost. A variety of designs have already appeared and some incorporate features which would not have been feasible with hybrid technology because of the restricted availability of FET sizes. One of these features is the provision of a very small FET shunting the input of the amplifier. This improves the dynamic range by providing the facility of being able to divert a portion of the photo-diode current to ground [3]. Very small FETs have also been used to provide adjustable feedback resistances and thus make it possible to optimise the pre-amplifier for different bit-rates [3]. Monolithic integration is also an attractive option for the optical feedback scheme which has been advocated for some applications [4]. In this case the noise contributed by the feedback resistor is eliminated by using an optical signal and a second PIN to generate the feedback current.

GaAs FETs also provide the high gain and wide bandwidth needed for the gain-controlled main amplifier and limiting amplifier. For example, a four stage, differential amplifier, having a maximum gain of 48dB, a dynamic range of 47dB and a bandwidth of 2.4 Gbit/s has been demonstrated [5].

Regeneration and retiming of the digital signal is performed by the final stage of the receiver comprising a circuit which extracts the clock from the incoming data stream and a D-type latch.

The clock recovery circuit has to produce a signal with a frequency (in Hz) equal to the bit-rate (in bit/s). Data which has been transmitted in the normal, non-return-to-zero format (NRZ), lacks a spectral content at

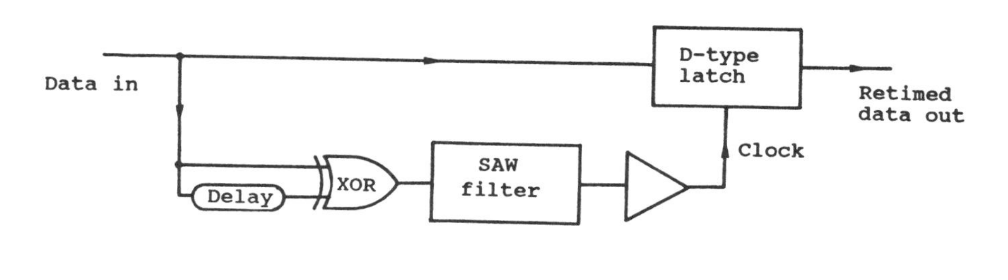

(a) A SAW filter

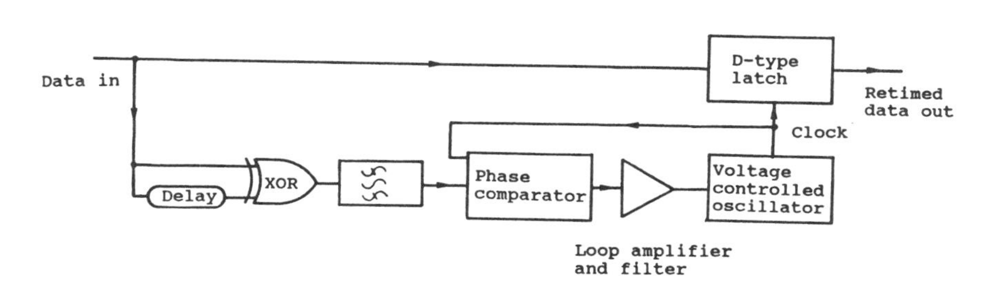

(b) A phase locked loop

Figure 11.2 Configuration of clock extraction and signal retiming units

this frequency. It is thus necessary to modify the signal by performing a non-linear operation on the incoming data stream. The simplest technique is to configure an exclusive-OR (XOR) gate with a 0.5 bit delay on one input as shown in Figure 11.2. This generates a series of short duration pulses, each of less than one bit period, having the desired spectral content; another option is to use a double balanced mixer. Both circuits can be fabricated in GaAs and, for each, clock extraction is accomplished by feeding their output into a filter.

Clock extraction based on surface acoustic wave (SAW) filters have been demonstrated at 2.4 Gbit/s and a typical configuration is shown in Figure 11.2a. these filters have the disadvantage that they cannot be tuned and are thus expensive to produce at a precise frequency. Also, they have high insertion loss and exhibit input–output crosstalk at high frequencies. Moreover, because of their limited Q, the clock cannot be sustained for some data patterns. (The loss of clock can be avoided by coding the data prior to transmission and methods of accomplishing this will be discussed briefly in the next section).

Problems associated with the SAW filter can be overcome by the use of a phase locked loop (PLL) [6] and Figure 11.2b shows a typical configuration. Here the clock component is extracted from the output of the non-linear element by a low Q filter and a phase detector is used to compare it with the output of a voltage controlled oscillator (VCO). (For some applications it is possible to dispense with the low Q filter but doing so increases the sensitivity to noise). A loop amplifier and filter following the phase detector produce a signal which controls the VCO frequency. The time constant of this control loop must be large enough to ensure that lock can be maintained for a wide range of data patterns. Such time constants require the use of either a digital filter [7] or an off-chip capacitor.

Phase detectors may either be analogue, which involves the multiplication of the two signals,, or digital, which is based on a comparison of the times at which both signals pass through a reference voltage [6]. Both can readily be implemented in GaAs but the digital type is preferred because it automatically starts the acquisition cycle if lock has been lost. The simplest digital phase detector is an XOR gate, but this requires a precise 1:1 mark-to-space ratio for both signals. To ensure this condition is met it is usually necessary to divide the frequency of both inputs by two. Another digital option, namely a phase frequency comparator, requires a more complex arrangement of gates but places less stringent requirements on the input waveforms.

The local oscillator used in the PLL must have both spectral purity and a satisfactory tuning range. LC oscillators have such attributes and monolithic versions could be produced using on-chip spiral inductors. Unfortunately, these have insufficient Q to give the desired spectral purity below about 5 GHz and an off-chip resonator has to be used instead. Such a solution has been employed for Gigabit Logic's PLL loop and this is the only commercially available GaAs version at present. Ring oscillators and multivibrators have also been used for GaAs VCOs, but the jitter of these threshold switching circuits is likely to restrict their use to low performance systems.

A D-type latch ultimately controls the timing integrity of the digital signal produced by the receiver. It must have a good phase margin and symmetrical output waveform. GaAs D-types with these attributes have been used in several demonstrations of Gbit/s systems [9]. There are commercially available parts capable of operation up to at least 4.8 Gbit/s, while in research laboratories there have been studies at 8 Gbit/s [10]. It is also worth noting that, for high bit-rate systems, any timing errors between the data entering the D-type latch and the clock can represent a significant proportion of the data bit period and thus lead to impaired systems performance. Automatic timing alignment schemes can be implemented to prevent such timing errors but they represent a considerable increase in complexity.

In summary, GaAs circuits have been used for all the functions within optical receivers. For the early work the level of integration was low and circuits performing single functions were linked together to form the complete receiver. More recently there has been a gradual move to higher levels of complexity and to lower power. A single chip receiver, which includes clock recovery, has recently been announced by Anadigics. One of the advantages of such integration is that it is easier to tailor the response of the amplifier chain in order to optimise the signal–to–noise ratio without degrading the digital signal. To do this it is necessary to minimise the noise bandwidth whilst maintaining both the pulse amplitude, and the degree of intersymbol interference (pulse spreading), at acceptable levels. This requires careful attention to the frequency dependence of both amplitude and phase. Although this can be done when the receiver is assembled from individual modules, the designer has considerably more freedom when tackling a single chip solution.

11.3.1.2 Transmitters

For digital modulation the laser's bias point is set so that the laser current is pulsed from just above threshold. As this threshold current will change during the life of the laser, either due to ageing or shifts in the ambient temperature, the laser driver must incorporate level control features which continuously optimise the bias current. If the bias point rises relative to the threshold current the optical power will not be fully modulated and the system will incur an extinction–ratio extinction penalty. If the bias falls below the threshold current there would be a switch on delay and this would cause a severe penalty at high bit–rates.

Laser bias control arrangements use a monitor photo–diode to detect light emitted by the back facet of the laser. In most cases this photo–current is used in a feedback circuit where its average value is compared with a fixed reference. This scheme is, however, limited to applications where the average value of the data stream does not have a significant pattern sensitivity. To remove this restriction it is necessary to make the reference proportional to the mean of the current modulating the laser. Such a scheme is now implemented in some commercially available laser driver modules such as that from BT&D. Their unit, which is shown in Figure 11.3, provides up to 70 mA bias current and 70 mA pulse current for operation up to 2.4 Gbit/s.

To date the feedback control function has always been implemented with silicon circuits but high speed feedback, in which the bias point is varied during the digital cycle, have been considered for some applications. Such control circuits would have to operate at much higher

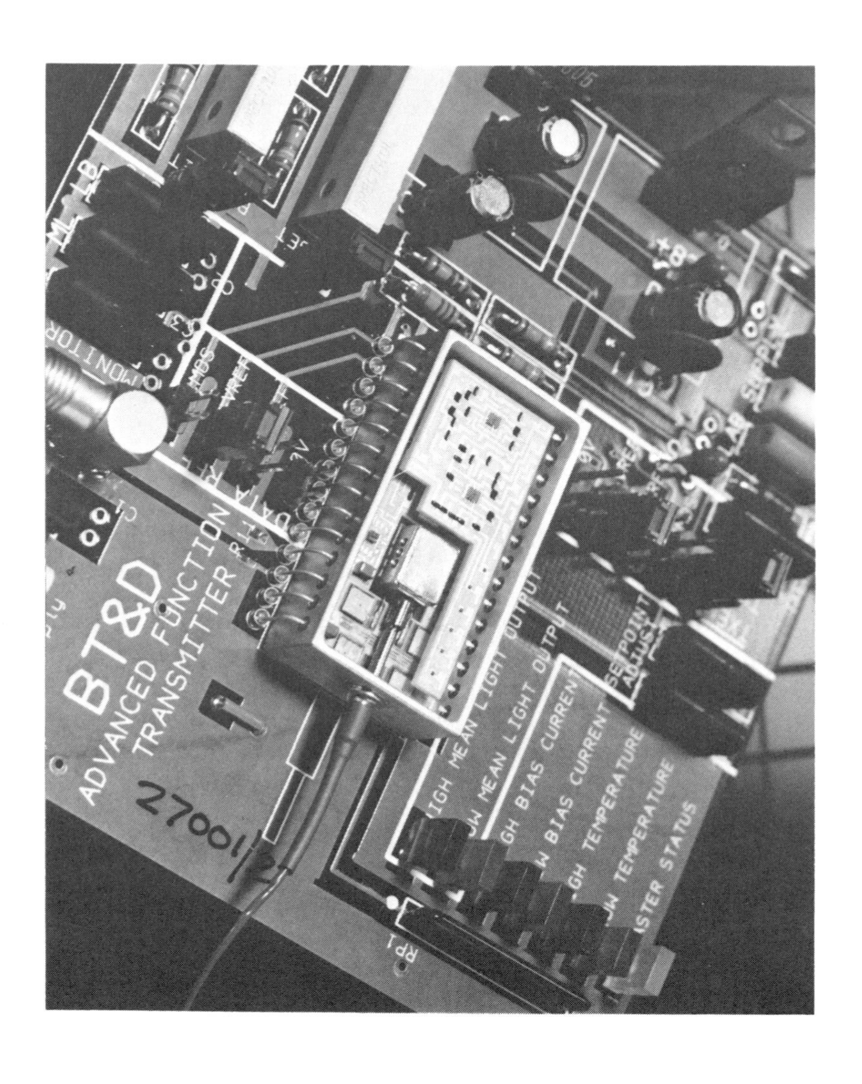

Figure 11.3: The BT & D advanced function transmitter module in the centre of this photograph

speeds than the modulating signal and are thus an ideal application for GaAs.

A small shift in laser wavelength occurs in sympathy with the drive current. This influences the received signal by virtue of the dispersion characteristics of the fibre. The consequences of this laser 'chirp' are dependent upon on the modulation rate and amplitude, as well as on the type of laser and the type and length of fibre. Chirp can be avoided by using an external modulator, such as a $LiNbO_3$ Mach–Zehnder travelling wave structure. These also require high speed drivers and a GaAs IC for one such device has recently been described [8]. This IC was required to provide a 5 V swing into 50 Ohm and operation was demonstrated at 2.4 Gbit/s even though performance was limited by interconnect parasitics. From simulations it was predicted that operation at 3 Gbit/s would be possible if the drive IC was mounted in the same package as the $LiNbO_3$ device.

11.3.1.3 Receivers and Transmitters for Coherent Systems

With the advent of narrow linewidth lasers and external modulators, there has been considerable interest in capitalising on the potential 10 to 20 dB improvement in power budget that coherent systems offer. This could lead to increased repeater spacing, or allow more complex passive optical distribution networks to be developed [2].

For digital coherent transmission, phase shift keying (PSK) and frequency shift keying (FSK) both offer significant systems advantages over amplitude modulation. For PSK an external modulator, similar to the $LiNbO_3$ device described above, is needed. For FSK the wavelength of the laser is modulated by using the mechanism responsible for 'chirp'. For example, with an external waveguide laser the bias current is set near the point at which the output power starts to saturate. Small perturbations of current (typically 1 mA) about this point change the refractive index of the active waveguide of the laser and this changes the frequency of the emitted light. The accompanying amplitude modulation is normally at an acceptably low level, but it can be further reduced by using DFB (distributed feedback) or DBR (distributed Bragg reflector) lasers structured so as to separate the current drive for the gain control and frequency control regions. In all cases the modulation current required is small and specialist laser drivers are thus not required. It is sufficient to increase the resistance of the laser to 50 Ohm by adding an appropriate series resistance and then to feed this combination from a standard GaAs power amplifier.

In a coherent receiver a PIN diode is used to mix both the optical signal and an output of an optical local oscillator. The resultant intermediate frequency is then amplified by a pre–amplifier as for direct detection. The only difference being that, for a given bit–rate, the bandwidth requirement is greater for coherent detection than for direct detection. In principle, the noise performance of the pre–amplifier is not as important as for direct detection, because it can be minimised by increasing the local oscillator power. In practice, local oscillator power is device limited and low noise pre–amplifiers are needed. Subsequent processing of the signal is dependent upon the type of modulation employed but in all cases standard GaAs microwave ICs can be employed.

11.3.2 Long Haul (Trunk) Transmission

Basic elements of a high speed data link are illustrated in Figure 11.4. A multiplexer combines lower speed data tributaries into a single high speed data stream which is fed into the laser driver. At the far end of the link the digital signal produced by the receiver is demultiplexed and the parallel set of low speed data streams is reconstituted. The final feature of the receiver end of the link is the channel assignment circuit. It ensures that data fed into a given channel at the transmitter end appears on the same channel at the other end of the link. To accomplish this, an up–down counter is used to add or remove bits from the clock drive to the demultiplexer until the correct channel assignment is achieved.

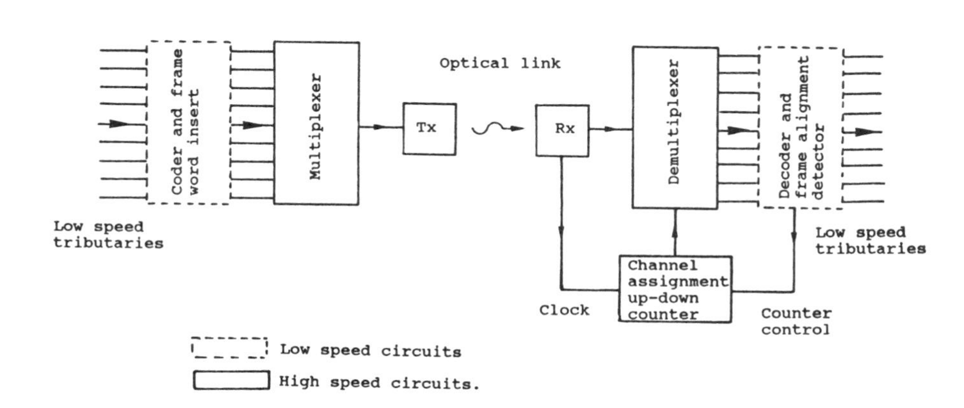

Figure 11.4 Basic structure of a high speed data link

Various multiplexing hierarchies have been implemented for Gbit/s transmission systems, ranging from 4 : 1 to 32 : 1. In some cases two levels of multiplexing have been used, thus a 16 : 1 can be achieved using a three chip solution based on two 8 : 1 multiplexers driving a 2 : 1 circuit. In most cases bit interleaving multiplexing is used, i.e. a bit from each channel is selected in turn. Line rates vary from typically 1.2 to 2.4 Gbit/s depending upon the application, but international standards are now emerging and these will be discussed later.

Suitable chip sets are now offered by several companies. For example, Hitachi have described a 4 : 1 based system for 2.4 Gbit/s [11] and Rockwell a 12 : 1 system for 1.3 GHz [12].

Coding of the data is usually performed at the tributary level where lower speed electronics can be used. Such coding is added, both to ensure that the data is suitable for clock extraction, and to provide error detection and correction facilities. It can either be implemented by using look–up tables or by scrambling the data against a pseudo–random sequence. A prototype 2.4 Gbit/s system based on a 18 : 1 architecture and scrambling of the tributary data has recently been demonstrated in a laboratory trial at BTRL [13] using the in–house 1 μm D–MESFET process [14]. It would also be feasible to perform scrambling at line rate since this would only require a feedback shift register and such a circuit could be implemented to GaAs.

The present diversity in types of high bit–rate systems is unlikely to continue. CCITT have recently recommended an international standard based on a North American development named SONET (Synchronous Optical Network). The new standard, called Synchronous Digital Hierarchy (SDH) defines a framework in which all the present lower bit–rate services can be assembled together into a 155.52 Mbit/s stream. For transmission systems operating at higher rates these tributaries must be multiplexed on a byte–by–byte method ('byte interleaved multiplexing') which preserves the form of the sequence used to indicate the start–of–framing information. At present the CCITT only specify one higher level, namely 622.08 Mbit/s but 1244.16 and 2488.32 Mbit/s have been listed as candidates for further study.

The SDH specification also defines the coding scheme to be used. This is accomplished by scrambling the whole signal, except for the frame alignment bytes. Although such scrambling could be carried out at the line rate, it is more convenient to use lower speed electronics and perform the scrambling/descrambling on a per tributary basis.

At the receive end of the link the channel assignment at the output of the demultiplexer is determined by an alignment circuit which look for a start–of–frame sequence. Because of the tight standards set on the time allowed to establish the correct channel assignment this alignment circuit is likely to have to operate at the line rate. A 1.2 Gbit/s multiplexer/ demultiplex GaAs IC which incorporates these functions is now available commercially from Vitesse.

11.3.3 Local Distribution of High Bit-rate Services

Provision of high bit-rate services within the local loop is currently being considered. Such broadband networks would include facilities such as high definition television. The main characteristics of these multiple access systems is that each user only requires access to a fraction of the total data throughput. Conventional time division multiplexing (TDM) techniques provide an inefficient solution to this problem in that each customer would require a high bandwidth receiver and a demultiplexer. Apart from the complexity of the terminal equipment, the low sensitivity of high bandwidth receivers is a constraint on the size of the network.

Two solutions have been proposed to overcome some of these problems, namely a form of time division multiplexing (TDM) and subcarrier multiplexing (SCM). These will be described in some detail but it must be pointed out that there is still much work being done on designing local distribution architectures. It is possible that a composite solutions involving TDM, SCM and WDM will eventually evolve.

11.3.3.1 Time Division Multiplexing

A form of TDM which dispenses with the need for a demultiplexer in the customer's receiver has recently been demonstrated [15]. Each TV channel is scrambled against its own unique coding sequence prior to multiplexing. At the receiver end, a fast D-type is used to sample the data and this sampling point is progressively shifted until the channel being sampled has the desired coding sequence. Thus most of the digital electronics in the customers premises only needs to operate at the channel rate. Using this approach, thirty-two 69 Mbit/s channels have been distributed to 32 customers over a passive optical network and a laboratory trial at BTRL has run at a line rate of 4.8 Gbit/s by using a GaAs multiplexer fabricated on the in-house D-MESFET process.

Because no additional framing information is needed, the multiplexing function is easy to realise. The D-type needs a fast access time but the toggle rate is low.

11.3.3.2 Subcarrier Multiplexing

Subcarrier multiplexing (SCM) is a scheme in which the optical carrier is amplitude modulated by a microwave signal. The latter carries the required data multiplexed in the frequency domain (FDM) just as in existing microwave radio links. In the first demonstration of such a system, 60 video channels were combined into a 2.7 to 5.2 GHz signal for transmission over 18 km of single mode fibre [16].

Two types of receiver can be used in a SCM system. For the first, a mixer is placed immediately after the detector and is used to select and down–convert a single channel which is fed into a narrow band amplifier. Alternatively, the wideband microwave signal is amplified after detection and then down–converted. The second approach avoids the sensitivity degradation associated with the conversion loss of the input mixer, but there is a penalty associated with the higher noise of the wideband amplifier. Which of these alternatives is preferable depends on the configuration of the system. It should, however, be noted that for the second scheme the bulk of the receiver can be based on a proprietary DBS channel selector.

Wideband receivers optimised for SCM have recently been described [18]. A small inductance is introduced between the PIN and the input FET of the amplifier and is used to optimise the coupling over the microwave band of interest and thereby reduce the sensitivity degradation associated with noise outside this band. This 'resonant PIN–FET' receiver had maximum sensitivity between 2.5 and 5 GHz. It used a discrete PIN and inductor but problems were experienced in bonding to delicate devices and in adjusting the inductance to achieve the required value. It would thus be desirable to produced an integrated version on which the PIN and pre–amplifier could be combined with a spiral inductor.

The depth of modulation required for SCM varies considerably, depending upon application, but it is usually low enough to enable full use to be made of the high bandwidth of present lasers (up to 20 GHz). Specialist laser drivers are not required and, as for coherent transmitters, it is sufficient to increase the resistance of the laser to 50 Ohms by adding an appropriate series resistance and then to feed this combination from a standard GaAs power amplifier. However, some of the future high frequency systems being considered operate at over 8 GHz and for these it would be desirable to use impedance transformation techniques based on transmission lines and these could be incorporated on to a GaAs laser driver IC.

Finally it must be pointed out that although SCM was initially developed in the context of a broadband customer distribution network, it is already finding application as a convenient way of routing microwave signals. Optical fibres provide a light weight, low–cost and flexible alternative to coaxial cable and waveguide and thus SCM is ideally suited for satellite dish feeds and for a number of defence applications on board ships and aircraft.

Both SCM and coherent optical systems make use of the microwave radio techniques and monolithic microwave ICs described in earlier chapters. As operating frequencies rise, the problems of interconnecting these ICs and of connecting them to the optical components will become more severe. There will thus be pressures for higher levels of integration, including the combination of optical and electronic functions on the same chip.

11.3.4 Space Switching

Although the use of optical transmission has become widespread, work on space switching of optical fibres is still at an early stage. The preferred solution would be an all optical device, such as a $LiNbO_3$ matrix switch, which enables the signals to remain in the optical domain throughout [21]. However, such optical switches still need considerable development. $LiNbO_3$ switches are the nearest to becoming commercially available. At present the largest switch matrix in this technology is only 4 x 4 and their physical size is likely to limit them to 8 x 8.

Despite the obvious advantage of the all optical solution, considerable progress has been made on the alternative approach of converting from optical to electrical signals and then performing electronic switching. Two such devices have recently been demonstrated. NEC have developed a GaAs 8 x 8 matrix which can switch 1.2 Gbit/s channels for HDTV distribution [19]. More recently, IBM have described a 16 x 16 crosspoint capable of operation at 1.7 Gbit/s [20]. Neither of these LSI devices is commercially available.

11.3.5 Broadband Integrated Services Digital Networks and Local Area Networks

At present the favoured solution for the broadband integrated services digital network (B–ISDN) is the asynchronous transfer mode (ATM). This variant of fast packet switching [22] is based on transferring information by using fixed length packets. Each packet is prefixed by a small header containing the destination address and this is used to control routing. For comparison, in synchronous transfer mode (STM) routing is controlled by the relative time position of a packet within a frame.

The routing of the packets between their source and destination can be performed by one of four general switching architectures. Only two of these need be considered in order to explain the type of components required, these are the interconnection network and the ring.

Each node of the interconnection network is able to access the address and routing information and then switch the packet to the appropriate output. The technology required for these nodes is thus comparable with that for the space switches described earlier.

The ring structure is similar to those which form the basis of many of the local area networks which are of increasing importance for computing and control applications. Packets are passed sequentially around the nodes of the ring until the destination is reached. An address detector circuit in this node then activates the removal of the packet from the ring and transfers it to the local circuit. A node wishing to transmit data waits for any empty slot to occur and then down loads its packet onto the ring. The high speed operations of these various functions can be implemented using shift–registers as shown in Figure 11.5. Operation of such a configuration at 2 Gbit/s has been demonstrated at BTRL using PISO and SIPO 8 bit shift registers fabricated on the in–house D–MESFET process.

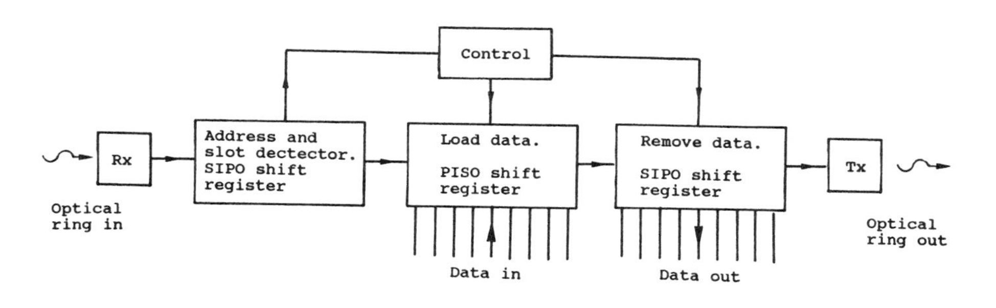

Figure 11.5: Typical configuration of a high speed node on a ATM ring; it is based on parallel in serial out (PISO) and serial in parallel out (SIPO) shift registers

At present ATM demonstrators are all at the laboratory stage and although they use switches capable of handling data at 1 Gbit/s these are realised by having a parallel set of bit streams at much lower data rates, e.g. 140 Mbit/s. There is thus the potential for reduced cost and smaller terminals by having a single fibre between nodes and making use of high speed capability of GaAs.

11.4 Applications at 0.85 μm

At this wavelength GaAs can be used for both the optical and electronic components. Wavelength dispersion of the fibre limits the applications to short haul systems where the bit–rate distance product is less than about 200 MHz.km. Within this limitation there are applications for short hop computer links and LANs as well as board–to–board interconnect and control links.

Present applications can be met by hybrid assemblies of the optical and electronic components. However, for operation at bit–rates as high as 1 Gbit/s, and to achieve lower cost and added functionality, it is desirable to integrate monolithically both the optical and electronic functions. Such OEICs have been produced within several research laboratories and the past year has seen a considerable improvement in both the performance and level of integration.

OEIC receivers present the lesser problem. Excellent results have been achieved by integrating a detector based on interdigitated metal–semiconductor–metal structure (MSM) with a MESFET transistor, and the various semiconductor layers needed for both devices were produced in a single growth run.

OEIC transmitters are a greater challenge because of the incompatibility of laser diode technology with that of a MESFET. Nevertheless such integration has been demonstrated by using separate growth runs for these two types of devices.

The most comprehensive of the OEICs produced to date is IBM's pair of circuits for a 1 Gbit/s data link [7]. The 800 gate transmitter chip accepts a 10 bit input word and produces a DC balanced transmission code to drive an off chip laser. The 1000 gate receiver chip is fully integrated and contains a MSM detector, amplifier, comparator, phase lock loop clock recovery and a deserialiser (demultiplexer). The sensitivity of this receiver is −20 dBm, including a −1.4 dBm penalty due to jitter introduced by the on chip clock extraction.

Although the IBM transmitter IC does not have the laser diode integrated with the drive electronics, such an IC has been demonstrated by Bell Northern [23]. Their combination of a GRINSCH laser with MESFETs was capable of operation in excess of 1 Gbit/s, but further developments are necessary to produce a more efficient laser structure and thus significantly reduce the threshold from its present value of 100 mA.

11.5 Impact of Likely Developments in Technology

Although the GaAs ICs produced to date have already had a significant impact upon the development of high performance optical fibre systems, these components and systems are only a fore–runners of the more advanced types likely to be introduced during the next new years. Future components are going to have to satisfy the needs for speeds above 2.4 Gbit/s, for enhanced functionality and for the composite microwave–optical systems. One micron depletion FET logic has been able to provide 2 to 4 Gbit/s devices but the power–delay product is too great for the LSI requirements of some applications. In contrast, E–D circuits have achieved the desired complexity, but their speed appears to be restricted to below about 2 Gbit/s for large circuits. For this technology the present state of the art is typified by Vitesse's 1.2 Gbit/s 8 channel SONET multiplexer, demultiplexer and frame alignment circuit having a power dissipation of nearly 3 W. Source coupled FET logic combines some of the attributes of both these approaches but there is still likely to be significant complexity limitation at the higher speeds required for the next generation of systems. For example Hitachi's 4 : 1 multiplexer and demultiplexer ICs for use up to 3 Gbi/ts each dissipate 1.5 W [11].

Already it has been necessary to use gates lengths of down to 0.5 μm to achieve acceptable performance for some circuits. These features are near to the limit of optical lithography and any further reduction in size will need to use more expensive e–beam fabrication techniques. The alternative is to move to improved structures based on hetero–junctions (devices comprising combinations of semiconductors having different band gaps).

Probably the simplest hetero–junction device is the high electron mobility transistors (HEMT) which offer both a two fold improvement in speed and significantly lower noise. The latter is an advantage for receiver front ends as well as for microwave oscillators and mixers. The technology for these devices is rapidly maturing and significant levels of complexity have been demonstrated, e.g. a 4.5 Gbit/s 8 : 1 MUX [24].

Hetero–junction bipolar transistors (HBT) have an even greater potential for higher speed although the technology is more complex. There are still major materials and fabrication challenges to be faced; e.g. the low current gain is poor and thus this technology is not yet suited for optical receiver front ends.

All Gbit/s services will benefit from the improved performance, lower cost, added functionality and simplified packaging which are potentially available from optoelectronic integration. Because such systems will be based on single mode fibre, by virtue of the bit–rate, ternary and quaternary materials will have to be used for the optical components. These can be combined with GaAs electronics by growing GaAs on the InP substrates used for the optical components. When this is done it is necessary to minimise the influence of lattice mismatch between these materials. Receiver and transmitters have already been fabricated in this way and they have been used in a 1.2 Gbit/s, 52.5 km transmission demonstration [25].

The alternative OEIC solution is to develop electronic devices within materials lattice matched to InP. The low barrier height of Schottky contacts to these materials precludes the use of a simple MESFET but there are a range of alternatives which are the subject of considerable research interest. These include HEMT and HBT structures which have already demonstrated considerable performance improvements over their GaAs equivalents.

Finally as this review has concentrated upon system aspects it is appropriate to end by mentioning the reliability implications of using GaAs ICs. The potential failure mechanism specific to these circuits is the stability of the Ohmic contacts, the Schottky gates and the GaAs surface. Of these, Schottky gate metal interdiffusion has been reported to be the dominant wearout mechanism. The failure rate is, however, very low and one vendor is claiming a median life of greater than 200 years for a 300 gate circuit of 150° C. Thus there is every expectation that GaAs circuits will be able to meet the most demanding reliability requirements.

11.6 Conclusions

The introduction of 1.3 and 1.5 μm single mode optical fibres into telecommunications has already revolutionised the trunk networks. The first systems were introduced at 140 Mbit/s and are already being upgraded to 565 Mbit/s. GaAs transistors and ICs have been used successfully in the laboratory to develop higher speed systems. From the experience thus gained commercial GaAs ICs have been developed and are now becoming more widely available.

New architectures and techniques for utilising the massive bandwidth of optical fibre systems in local networks are now being developed in the major telecommunications research laboratories. Fibres and GaAs ICs developed for trunk networks are available for this research but the problems are different. The roles of local distribution, switching and routing are becoming merged. At present there are field trials of passive optical networks which, although they avoid the need for electronic switching, still require electronics at the terminals. Microwave/millimetre-wave radio links are also being developed for short distance low cost distribution of broadband data and video, and microwave terminal equipment is being used for optical fibre transmission. In addition, mobile communications shows the potential for considerable growth and this is likely to involve both radio and optical fibre links. Thus the development of local networks is fertile ground for new ideas and there is a need for high performance and specialised components to assess their practicality. Above all, there is a strong emphasis on achieving high performance, in large volume but at very low cost. Much of the feasibility work will require the flexibility of hybrid assemblies or mixed technologies but the need for high performance at low cost will favour the development of OEICs.

11.7 Acknowledgements

The author would like to thank his colleagues in the devices and systems divisions at BTRL who have contributed to this work in many ways. Acknowledgement is also made to the Director of Research and Technology for permission to publish.

11.8 References

1 R. G. Smith and S. D. Personic: Receiver design for optical fibre systems, Topics in Applied Physics, Vol 39, 89

2 M. C. Brain: Coherent Optical Networks, BR Telecom Technol J, 1989, 7, 50

3 G. Williams and H. Leblanc: Active feedback lightwave receivers, J. Lightwave Technology, 1986, 4, 1502

4 B Kasper et al: An optical feedback transimpedance receiver for high sensitivity and wide dynamic range at low bit-rates, J. Lightwave Technology, 1988, 6, 329

5 T. Kinoshita et al: GaAs monolithic gain controllable amplifier, Digest of 1987 Symposium on VLSI Circuits, 63, 1987

6 R. E. Best: Phase locked loops theory, design and applications, McGraw Hill, 1984

7 J. F. Ewen et al: G/bit/s fibre optic link adapter chip set, Proc 10th GaAs IC Symposium, 1988, 11

8 A. R Beaumont et al: $LiNbO_3$ travelling wave electro-optic waveguide devices. SPIE Vol 993, Integrated optical circuit engineering VI, 1988, 94

9 L. Bickers et al: Fully regenerated 2 Gbit/s, 90 km optical transmission system using commercial GaAs master slave D-type flip-flop, Proc 11th ECOC, 1985

10 M. Ohhata: Above 8 Gb/s GaAs monolithic flip-flop ICs for very high speed optical transmission systems, Proc 10th GaAs IC Symposium. 1988, 23

11 Y. Hatta et al: A GaAs IC set for full integration of 2.4 Gbit/s optical transmission systems, Proc 10th GaAs IC Symposium, 1988, 15

12 T. C. McDermott: Product introduction of GaAs Multiplex ICs in Gb/s fibre link transmission systems, Proc 10th GaAs IC Symposium, 1988, 7

13 A. Stevenson et al: A 2.4 Gbit/s long-reach optical transmission systems, Br Telecom Technol J, 1989, 7, 92

14 P. J. T. Mellor: Gallium arsenide integrated circuits for telecommunications systems, Br Telecom Technol J, 1987, 5, 5

15 D. W. Faulkner et al: Novel sampling technique for digital video demultiplexing and channel selection, Electron Let, 1988, 24, 654

16 R. Olshansky and V. Lanzisera: 60-channel FM video subcarrier multiplexed optical communications system, Electron Lett, 1987, 23, 1196

17 T. E. Darcie: Subcarrier Multiplexing for multiple-access lightwave networks, J Lightwave Tech, 1987, LT-5, 1103

18 T. E. Darcie et al: Resonant Pin-FET receivers for lightwave subcarrier systems, J Lightwave Tech, 1987, LT-6, 582

19 S. Hayano et al: A GaAs 8 x 8 matrix switch LSI for high speed digital communication, Proc 9th GaAs IC Symposium, 1987, 245

20 C. J. Anderson: A GaAs MESFET 16 x 16 crosspoint switch at 1700 Mbit/s, Proc 10th GaAs IC Symposium, 1988, 91

21 R. J. Reason: Optical space switch architectures based on lithium niobate crosspoints, Br Telecom Technol J, 7, 83, 1989

22 J. S. Turner: Design of a broadcast packet switching network, IEEE Trans Comms, 1988, 36, 734

23 M. Svilans et al: Monolithic integrated optoelectronic Transmitter on GaAs with fully RIE GRINSCH laser diode, Proc 10th GaAs IC Symposium, 1988, 19

24 B. J. F. Lin et al: A 1 μm MODFET process yielding MUX and DEMUX circuits at 4.5 Gbit/s, Proc 10th GaAs IC Symposium, 1988, 143

25 T. Suzaki et al: 1.2 Gbit/s 52.5 km optical fibre transmission experiment using OEICs on GaAs on InP heterostructure, Electon Lett, 1988, 24, 1283

26 W. J. Roesch and M. F. Peters: Depletion Mode GaAs IC reliability, Proc 9th GaAs IC Symposium, 1987, 27

Chapter 12

Emerging Technologies

J. G. Swanson

12.1 Introduction

A key parameter which limits the intrinsic speed of an active device is the carrier transit time across its active region. In bipolar transistors it is the time to cross the base whilst in FETs it is the source–drain transit time. Many of the recent improvements in well known devices are based on developments in technology which are aimed at reducing this parameter so producing improvements in device speed. Among these are improvements in lithography which make use of short wavelength illumination to minimise fundamental photographic limits set by diffraction to produce lateral feature dimensions considerably below one micron. In the direction which is normal to a substrate surface reductions in dimensions have generally relied on the accurate depth control of diffusions and implantations. Most success has been achieved by new layer growth processes such as MBE and MOCVD which can produce epitaxial layers as thin as 10–100 Å, comparable with electron wavelengths.

This Chapter will outline the properties of GaAs and some of the related compounds which influence device performance relating these to the use of newer processing technologies and the effects which these are having on device performance.

12.2 Carrier Velocity Limitations in Semiconductors

In order to understand the improvements which are being made to devices to ensure higher frequencies and speed it is necessary to describe some of the factors which influence electron velocity with relation to particular materials.

The velocity which charge carriers can acquire when they are accelerated by an electric field E depends on their effective mass m^* in the particular material and the time τ during which they can accelerate frequently without losing energy or momentum. The drift velocity is given by:

$$V_D = \frac{e\tau E}{m^*}$$

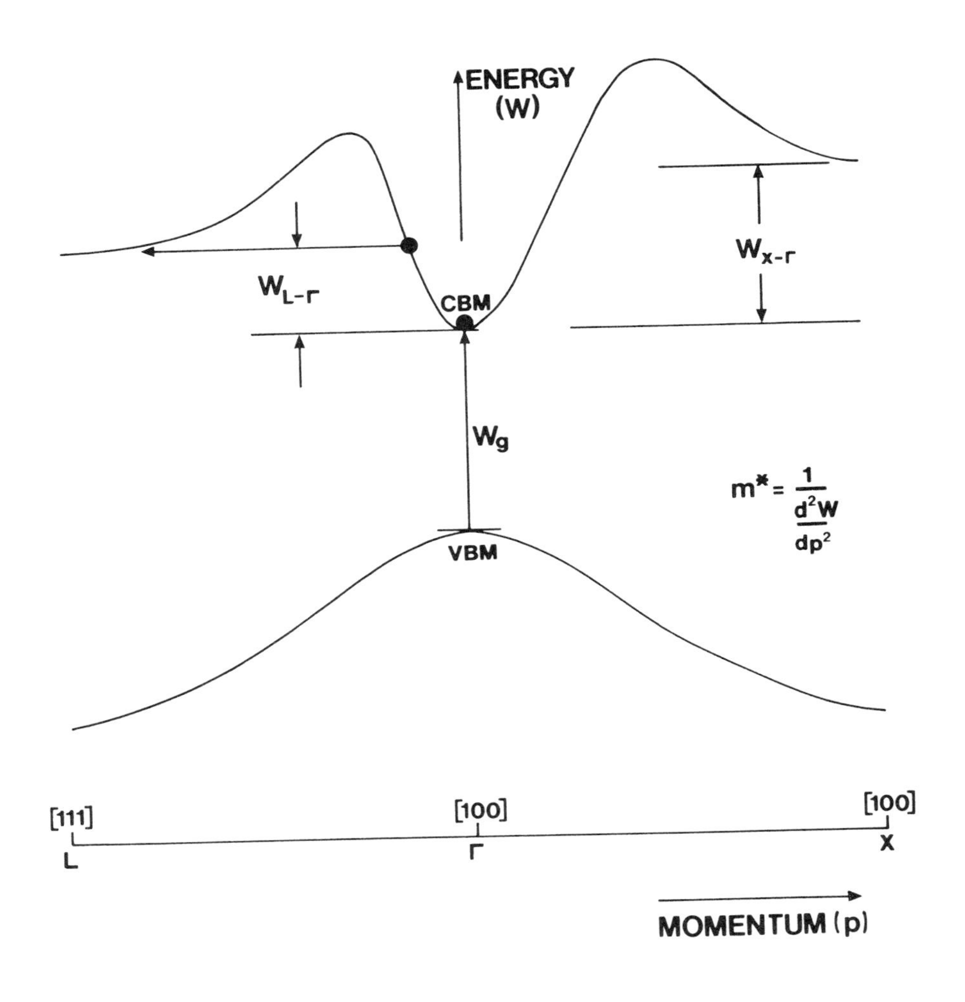

Figure 12.1 Typical energy–momentum relationships for a III–V semiconductor

When low fields are used τ is constant and $e\tau/m^*$ is termed the low field carrier mobility. Measured values of m^* and μ_n lead to a time for electron free motion of about 0.35 ps in GaAs. One might visualise this free motion between energy loss events as ballistic. It is clear that the average drift velocity increases linearly with E. This rate of increase is dependent on m^* and is a property of the band structure of the material. This connection can be depicted graphically by the energy–momentum curves for the conduction bands of III–V semiconductors shown schematically in Figure 12.1. The effective mass depends inversely on the second derivative of these relations. Parabolicity implies constant effective mass.

At thermal equilibrium it is the lowest energy states in the conduction band which are occupied at the point Γ on the momentum scale.

As electrons accelerate they move through states arrayed along the W–p curve. At low electric fields the gain in kinetic energy within the time τ is relatively small compared with the average thermal energy and the electrons remain near to the conduction band minimum (CBM) on the parabolic part of the curve in the lowest valley m^* is constant and the material exhibits a linear variation of V_D with E, it is said to have a constant mobility.

When large electric fields are used to speed the electrons, the time rate of change of momentum is increased and electrons advance further along the W–p curve before energy loss occurs. In GaAs the acquisition of 0.29 eV of kinetic energy would make it possible for an electron to enter other conduction band states having the same energy. These exist at a different momentum labelled L and transfer to them is only possible when a lattice vibration or phonon interacts with the hot electron providing it with the necessary momentum. Electrons in this valley at L have a larger m^* and lower mobility, causing them to slow down when they have transferred. At still larger fields, transfers to the minimum at X can also occur. So in an effort to increase the speed of the electrons, a limit occurs as more and more transfer to the upper valleys, thereafter gaining velocity very slowly. When the time required for a carrier to cross the active region of a device is longer than the time to provide the momentum for transfer, $\simeq$ 0.01 ps, a fundamental limit to carrier velocity occurs when high electric fields are used. This behaviour is summarised by the velocity field curve for this class of materials, Figure 12.2.

As the dimensions of devices are reduced, the time required for transit approaches the momentum supply time. This inhibits transfer to the lower valleys and some of the electrons continue to acquire energy and momentum by moving further along the lower valley of the W–p curve. An increasing number of electrons acquire more velocity than would otherwise be possible. This ballistic effect is called velocity overshoot and is being increasingly referred to in recently developed devices.

The carrier transfer mechanism can be inhibited by using semiconductors with greater energy separations between the lowest valley of the conduction band and its satellite minima. The addition of indium to GaAs to form a ternary alloy $In_xGa_{(1-x)}As$ has the effect of increasing this separation and further improves the situation because electrons in the lowest valley have a lower effective mass and therefore a higher mobility as indium is added. In $In_{0.53}Ga_{0.47}As$, a composition which can be epitaxially grown on InP the separation $W_{L-\Gamma}$ is 0.55 eV and the low field electron mobility is 12000 $cm^2V^{-1}s^{-1}$. The electron velocity peaks at 2 x 10^7 cm s^{-1} at an electric field of 3 x 10^3 V.cm^{-1}. In this case the limit is set not by electron transfer to the satellite valleys but by the generation of

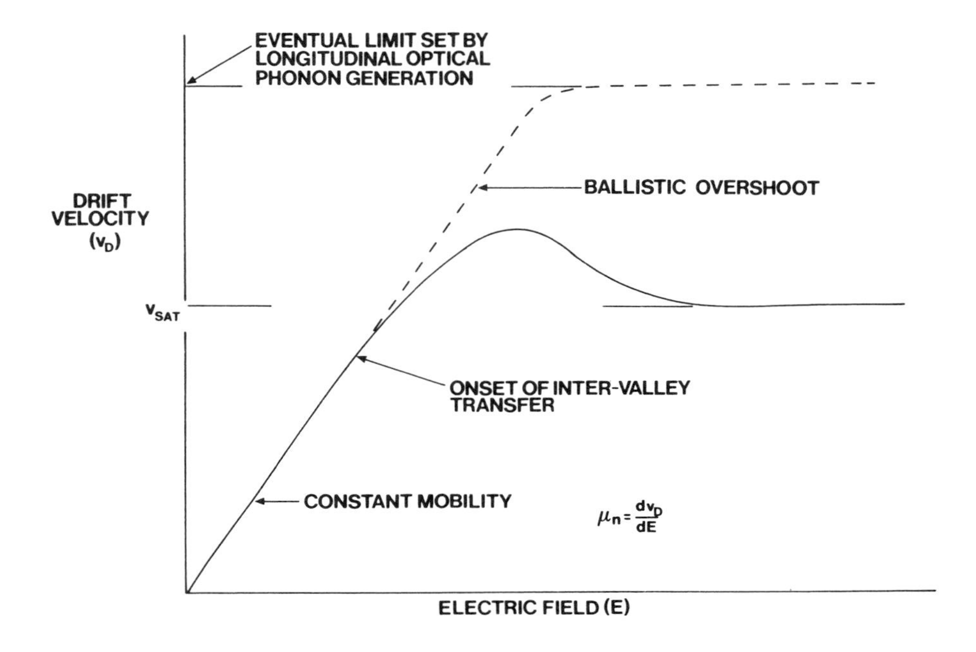

Figure 12.2: Velocity–field relationships in a III–V semiconductor

optical phonons. These require a minimum and well defined energy for their generation. When the electrons in the conduction band have been accelerated sufficiently and have acquired this energy it then becomes possible for them to shed it by launching this type of lattice vibration. It is this mechanism which can limit the velocity before sufficient energy is acquired for transfer to a satellite valley. This is the case in $In_{0.53}Ga_{0.47}As$ which has a sufficiently large valley separation for optical phonon generation to occur first. Further advantage is likely from using even higher indium concentrations to produce ternary alloys having increasing electron mobilities and valley separations. One problem is that as the energy gap W_g increases the impact ionisation of valence band states becomes an increasingly likely energy loss process and may eventually have a counter–productive effect.

12.3 Materials for High Speed Devices

12.3.1 Lattice Matched Materials

The identification of high mobility, high saturation velocity materials is central to the choice of a semiconductor for the active region of a device.

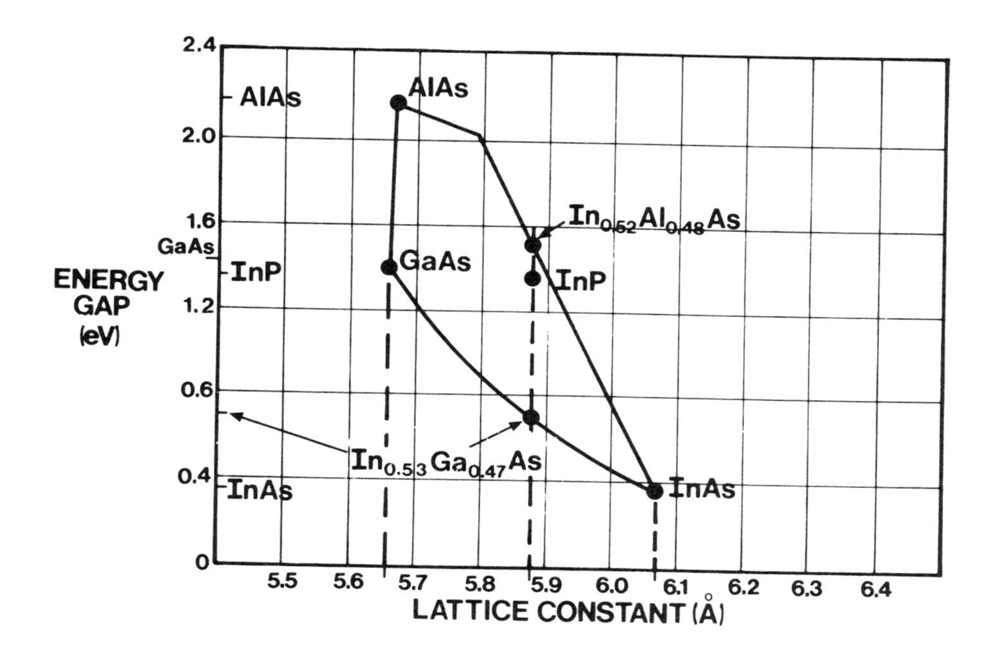

Figure 12.3: Band–gap v lattice spacing for important III–V compounds and their alloys

In practice, however, devices are often formed from multiple epitaxial layers having different doping concentrations. These layers frequently have different compositions in order to vary barrier heights and fundamental energy gaps. Epitaxy implies crystalline order within the multilayer structure. This is achieved when the semiconductor layers have the same basic crystal structure and atomic spacings which are matched to within a percent or so.

Since the epitaxial device structures are grown on a substrate its choice determines the types of material which can be grown upon it. High speed applications require that the substrate itself should be semi–insulating either by the compensation of donors or acceptors with deep level impurities or by growth with a low level of impurity. In practice GaAs and InP are most commonly used and define the class of materials which can be used to form devices upon them. Figure 12.3 provides a convenient summary of relevant III–V compound semiconductor alloys plotted on a domain of lattice spacing and fundamental energy gap.

Important use has been made of the miscibility of AlAs and GaAs which gives a continuous range of compositions providing energy gaps ranging from 2.15 eV to 1.43 eV. The complete composition range is lattice matched to within 1% of GaAs which is used as a substrate. Similar use has been made of InP and the quaternary alloy system $In_xGa_{(1-x)}As_yP_{(1-y)}$. Lattice matching occurs when y = 2.2x. A particular attraction here is that the energy gaps range between 1.3 eV and 0.73 eV and can be used to generate and detect infra-red radiation suitable for transmission along silica optical fibres with low loss and dispersion. The need to combine electrical active devices with optical elements on the same substrate means that this range of materials is of great importance. The advantages of $In_{0.53}Ga_{0.47}As$ have already been mentioned. It forms one extreme of the composition range which can be prepared in InP and satisfies the relation y = 2.2x. Where optoelectronic integration is the main aim materials tend to be InP based.

12.3.2 Lattice Mismatched Materials

There have been important recent developments which aim to overcome the limitations imposed by the need to lattice match materials. The intention is to use alloys with improved electronic behaviour but which unfortunately have the wrong lattice spacing for growth on GaAs and InP substrates. It is found that lattice misfit can be accommodated elastically up to a critical thickness without the formation of edge dislocations. The precise limit depends on the extent of mismatch and on the dislocation density in the substrate, In practice this limiting thickness is a few hundred angstroms. Such layers are said to be pseudomorphic and contain mechanical strain. The strain can sometimes be exploited to modify the band structure by producing relative displacements of band edges and satellite conduction band valleys.

Thicker mismatched layers result in the formation of edge dislocations which propagate through the grown layer. Dislocation densities of 10^{12} cm^{-2} are sufficient to prevent them being used in bipolar devices because of their effect in promoting minority carrier recombination. Carrier scattering is not apparently a serious problem and allows their use in majority carrier devices. Compositional grading has been used to make the transition from one material to another but this requires about 1 μm of growth per percent of composition change; this is not always practical because of the thick layers which are needed.

The superposition of pseudomorphic layers of alternating composition with graded period has been used successfully to make transitions over shorter lengths. An example of this [1] is the use of 64 alternating layers of $Ga_{0.65}In_{0.35}As$ and GaAs ranging in thickness from 10 Å to 200 Å over a distance of 0.4 μm to grow $Ga_{0.65}In_{0.35}As$ as a GaAs substrate. The application of these methods to accommodate the 4.1% lattice mismatch between GaAs and silicon is a key development which it is hoped will eventually allow optical and high speed GaAs based structures to be formed on silicon integrated circuits. Dislocation densities of less than 10^4 cm^{-2} are needed for minority carrier devices. At present the defect densities are around 10^7 cm^{-2} in layers thinner than 4 μm. Thicker layers result in cracking due to differential thermal contraction on cooling after growth. Significant progress has already been made in producing GaAs on Si suitable for high speed majority carrier devices but the reduction of the defect density for bipolar use is still an important problem.

12.4 Devices

The devices to be described in this Section make use of the superior electronic transport properties of compound semiconductors and their alloys. The epitaxial and lithographic methods which are used to form them and to extend their performance are described in Chapter 2. In each case the aim will be to illustrate what it is that limits device performance and to point to the methods which will be used to further improve the performance.

The devices which will be referred to here are of two kinds, the bipolar junction transistor and the field effect transistor. In the first type the active region is a thin layer forming the base. This is bounded by the emitter and collector junctions which have relatively large lateral extents. The problem here is to speed the minority carriers across the thin layer.

In field effect transistors the carrier moves laterally along a thin layer, or channel, between two finely separated electrodes which are the source and drain. In this case the carrier is confined by the plane surfaces of the channel, these can have a controlling effect on the device performance. The intrinsic properties of the materials which are used for the active regions are similar in these two different cases.

12.4.1 The Heterojunction Bipolar Transistor

The maximum frequency of oscillation of a bipolar junction transistor is set by the condition at which the unilateral power gain is unity. This is given by:

$$f_{MAX} = \left[\frac{f_T}{8\pi R_B C_{BC}}\right]^{1/2} \qquad (12.1)$$

Conventionally f_T is the frequency at which the current gain descends to unity; this is often inferred by extrapolating the decrease of β at 6 dB/octave. In terms of equivalent times for charging or transit through regions of the device this can be expressed as:

$$f_T = \frac{1}{2\pi(\tau_E + \tau_B + \tau_C)} \tag{12.2}$$

where

$$\tau_E = \frac{kT}{eI_E} (C_E + C_C + C_P) \tag{12.3}$$

$$\tau_B = \frac{W^2}{2D_n} \quad \text{(for diffusion across base)} \tag{12.4}$$

$$\tau_c = \frac{c}{V_{SAT}} \quad \text{(transit through collector depletion region)} \tag{12.5}$$

A necessary requirement for bipolar transistors is to arrange for efficient injection and collection of minority carriers at the surfaces of the base between which carriers must have a short transit time. This must be accomplished without introducing other predominant charging or transit delays.

(i) Injection and Base Transport

Assuming that the emitter injection efficiency is unity the current gain is given by:

$$\beta = \frac{\tau_R}{\tau_B} \tag{12.6}$$

where τ_R is the recombination time constant in the base.

With this basis it is clear that high current gains require narrow bases and high minority carrier mobilities since the diffusion constant is given by $D_n = \frac{kT}{e} \mu_n$. In order to achieve $\beta = 1000$ using GaAs which has an electron life–time of 1 ns the base width must be less than 0.17 μm. This is not hard to realise using modern layer growth methods and certainly a narrow base helps to ensure a high f_T value. The problem is that a narrow base raises R_B and reduces f_{MAX}. In fact R_B in the f_{MAX} formula is complicated because it consists of a parasitic fixed part $r_{bb}{}'$ and a further resistance which is distributed within the active region of the device. The former lies between the device terminal and the edge of the base active region. Increasing the base doping concentration N_A to lower R_B leads to a loss of injection efficiency γ since

$$\gamma = \left(1 + \frac{N_A\, W\, D_p}{N_D\, L_p\, D_n}\right)^{-1} \quad \text{for a conventional npn device} \qquad (12.7)$$

In such bipolar transistors the base doping concentration is generally made low compared with that in the emitter specifically to ensure a high γ value. Decreasing W aggravates the problem by requiring a high N_A value to keep R_B low. This classical dilemma in the design of bipolar transistors has been overcome by using heterojunction minority carrier emitters, a direct result of the latest layer growth methods but originally proposed by Shockley in 1948.

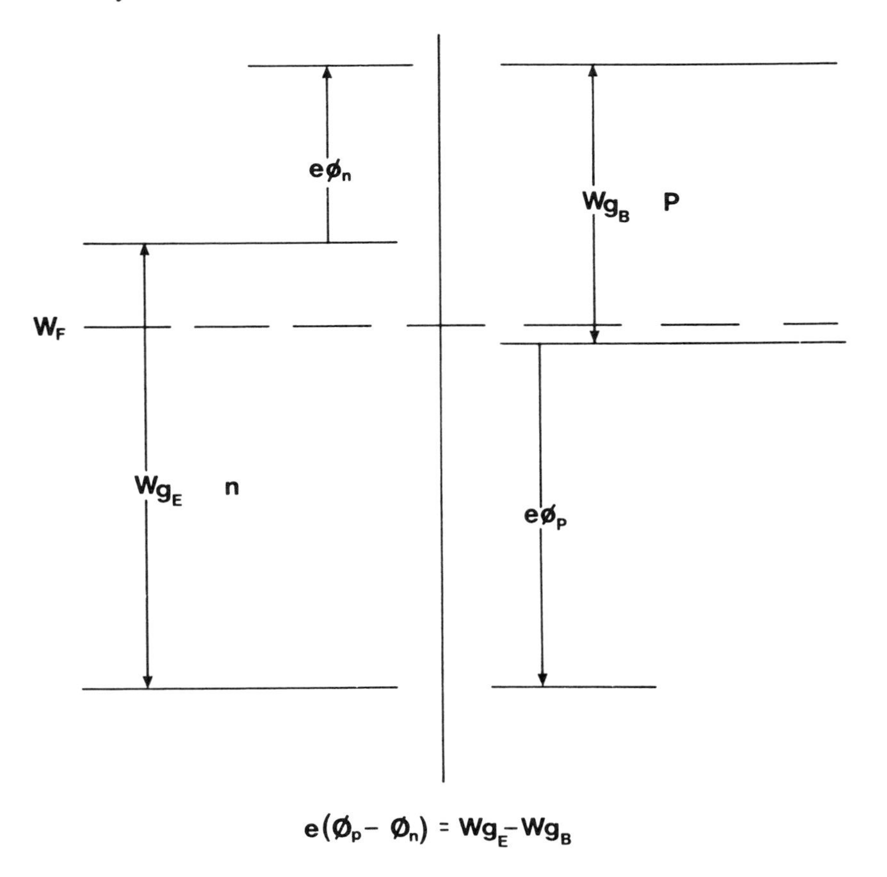

Figure 12.4: A simplified energy description of a heterojunction emitter

In a homojunction the barrier heights for minority carriers are equal and the electron and hole flows are in the ratio $(N_D\, L_p\, D_n)/(N_A\, W\, D_p)$. Figure 12.4 is a simplified energy description of a heterojunction emitter.

Deriving the ratio of carrier flows gives

$$\frac{N_D\, L_p\, D_n \quad \exp(-e\phi_n/kT)}{N_A\, W\, D_p \quad \exp(-e\phi_p/kT)} \tag{12.8}$$

$$= \frac{N_D\, L_p\, D_n}{N_A\, W\, D_p} \exp\left(\frac{\Delta W_g}{kT}\right) \tag{12.9}$$

The heterojunction enables the hole flow to the emitter to be reduced relatively by increasing the barrier seen by these carriers. In practice the electron barrier is also modified but it is the difference in barrier heights which is equal to the energy gap difference that determines the flow ratio and gives a modified injection efficiency given by:

$$\gamma = \left[1 + \frac{N_A\, W\, D_p}{N_D\, L_p\, D_n} \exp\left(\frac{-\Delta W_g}{kT}\right)\right]^{-1} \tag{12.10}$$

A highly doped narrow base can now be used without impairing γ. The real benefit however is in reducing R_B and at the same time narrowing the base. Most work has been performed on HBT's using the GaAlAs epitaxial system in which the emitter is GaAlAs and the base GaAs. The vertical resolution of the layer growth process enables interfaces to be made which are near to being atomatically abrupt giving rise to band edge discontinuities, ΔW_C and ΔW_V. The energy description of the abrupt forward biased heterojunction in Figure 12.5(a) shows that the barrier seen by the electrons is enlarged by the discontinuity and would attenuate electron injection to the base. A less obvious short–coming is that those electrons which pass over the barrier would be launched into the base with a kinetic energy large enough for the electrons to directly enter the low mobility satellite valleys at X and L. These disadvantages of the abrupt interface are overcome by compositionally grading the heterojunction, as in Figure 12.5(b). The energy band edges then change smoothly and the injected electrons enter the high mobility Γ valley immediately passing into the base with a high mobility and diffusion coefficient.

Another important term in the expression for f_{MAX} is the collector–base capacitance C_{BC}. It is obviously desirable to use a small area collector. The extrinsic base resistance $r_{bb'}$ which is the resistance between the base contact and the active region edge has assumed greater importance as the base doping in the active region has been increased. It now constitutes the residue of R_B. As devices have narrowed to minimise C_{BC} the precise placing of the base contacts on both sides of the emitter stripe has become more and more difficult. Self–aligning fabrication methods are being used to achieve this with minimum base–contact to emitter separation [2,3].

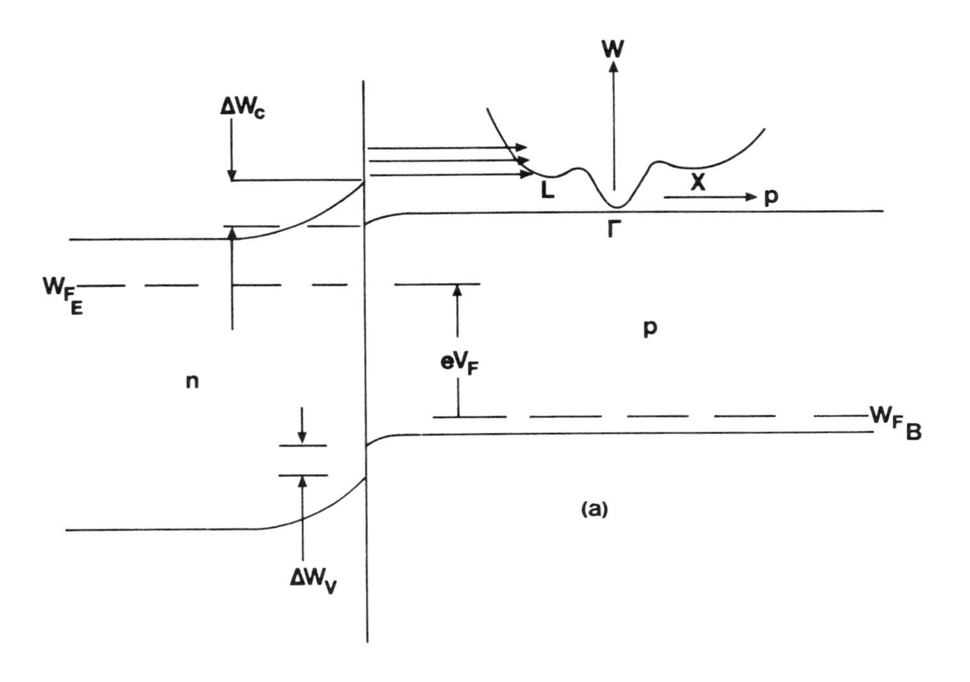

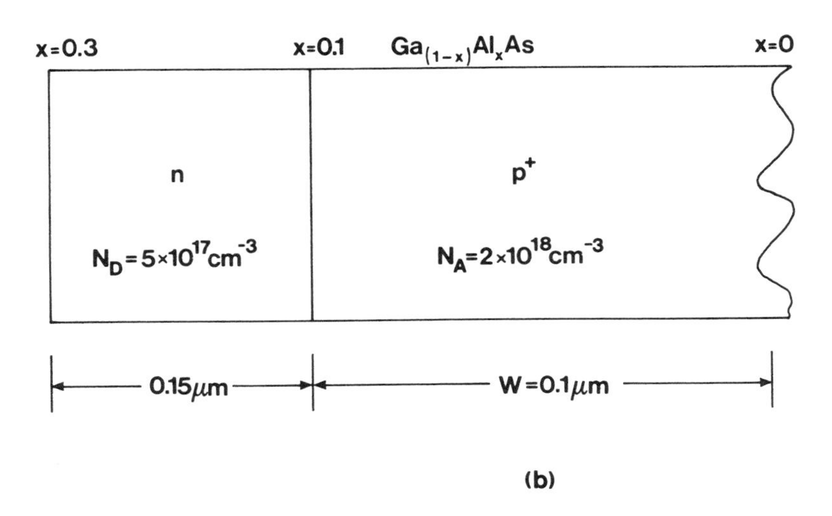

Figure 12.5: A practical heterojunction emitter of $Al_xGa_{(1-x)}As$ (a) without and (b) with compositional grading

Another difficulty caused by dimensional shrinking is the increasing importance of the recombination of minority carriers at the base surface in the extrinsic base region, leading to low current gains. Recombination velocities at GaAs surfaces are about 10^5 higher than for silicon and there is a need for improved passivations for these material systems.

Self–aligned structures exhibit the highest performance so far using the GaAlAs system. Maximum values of 86 GHz and 75 GHz were obtained for f_{MAX} and f_T by extrapolating measurements from 24 GHz. The current density required for this was 10^5 A.cm^{-2}. The emitter base separation was only 0.2 μm. Compositional grading of the base is frequently used now to create a built–in base field to accelerate carriers across the base. It is now felt that the limiting delay which influences f_T is associated with the collector structure [4].

(ii) Minority Carrier Collection

The reversed bias collector base junction is of course the region where the field is greatest and carriers are accelerated most strongly there. It is necessary that C_{BC} should be as small as possible implying that the depletion region within the collector should be wide. The aim however must be to achieve a short collector transit time without raising C_{BC} excessively.

An illuminating view of minority carrier motion in the collector is provided by the modelling work of Rockett [4]. The simulations show that carriers are very rapidly accelerated to velocities which overshoot the saturation value, 8 x 10^7 cm.s^{-1}. Unfortunately the carriers are transferred from the Γ valley to the X valley after travelling only 300 Å, they then slow down continuing at a constant velocity of 1 x 10^7 cm.s^{-1}. the velocity overshoot over such a short length has little advantageous effect. There is a clear requirement for methods which will confine the electrons to the high mobility Γ minimum of the conduction band allowing velocity overshoot to be exploited over a longer length. Intervalley scattering occurs in one hundredth of a picosecond in GaAs once carriers acquire the necessary energy. Efforts have been made to make the electrons move in a lower, but still high, field maintaining the velocity of electrons in the overshoot region for a greater part of their trajectory through the collector depletion region.

The two collector structures shown in Figure 12.6 have been compared [5]. An increase in f_T from 60 to 76 GHz was obtained with the p/n/n$^+$ structure. This was explained by the field weakening caused by the p region which was introduced to slow the acceleration and delay the onset of inter–valley scattering. It is noticeable however that f_{MAX} is higher for the n/n$^+$ collector. This is because the flowing negative charge lowers the

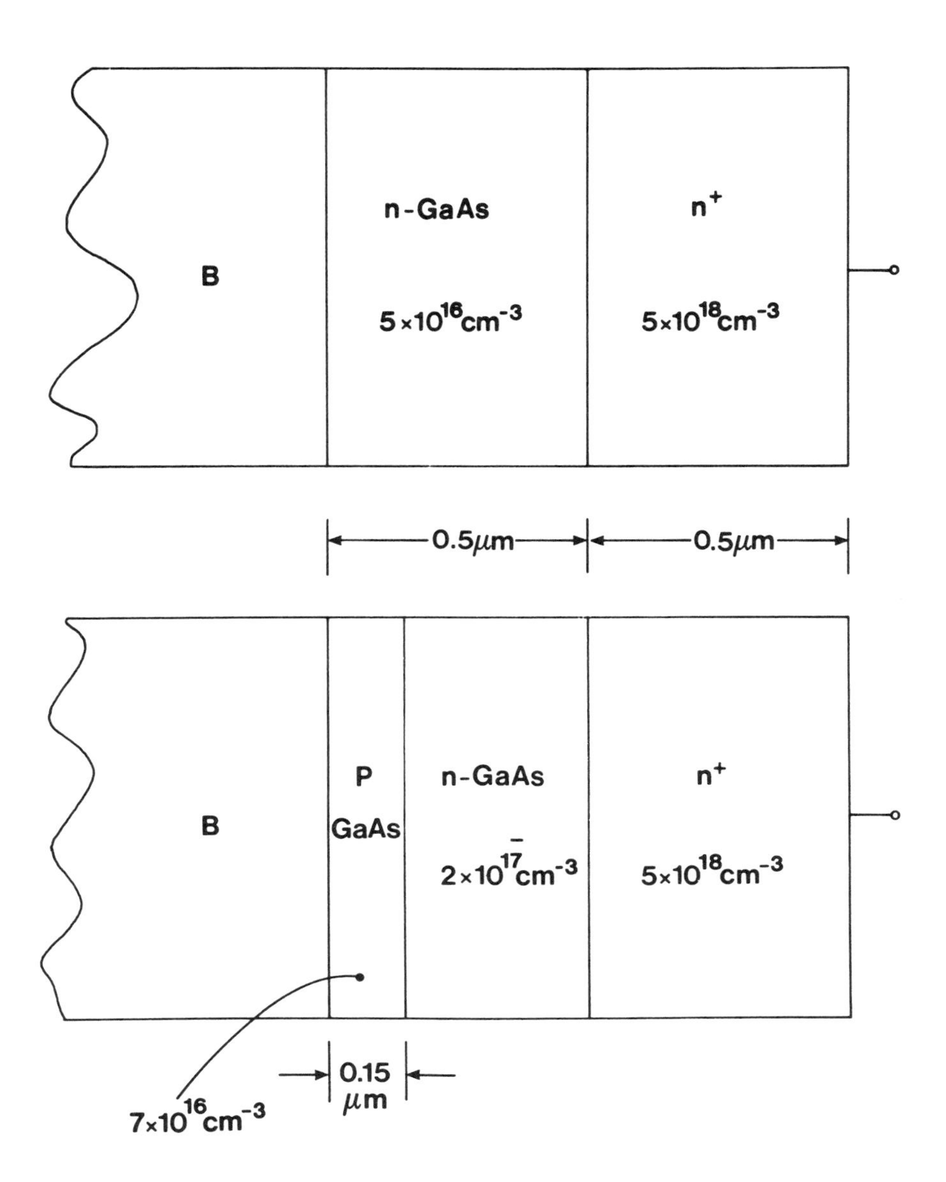

Figure 12.6: Heterojunction collector structures

space–charge density in the collector depletion region causing it to expand, lowering C_{BC} in the n/n^+ structure. The opposite effect occurs in the depletion p region of the $p/n/n^+$ collector and it is the increased capacitance which lowers f_{MAX} in this case [6].

Several factors have combined to shift interest towards HBT's based on the InP epitaxial system. These are opto–electronic compatibility, lower surface recombination and smaller intervalley scattering effects.

It has been shown [7] that HBTs made using InGaAs/InP exhibit current gains which are independent of area. This is not seen when GaAlAs/GaAs is used. As device dimensions are reduced the effect of surface recombination becomes more pronounced. β and h_{fe} values of 7,000 and 11,000 respectively have been observed [8] using InGaAs/InP in which surface recombination is relatively low. This result was based on devices having a base width of 0.2 μm consisting of p^+InGaAs (N_A = 5 x 10^{17} cm^{-3}) grown by MOCVD. Very different behaviour [9] was found in devices prepared by gas–source MBE having a 0.065 μm base with N_A = 5 x 10^{19} cm^{-3}. These exhibited current gains of only 50. This result may not be surface limited and probably has to do with the effect of a high doping concentration on minority carrier lifetime. This suggests that the necessity for very high base doping as base width is reduced might eventually impair the base transport efficiency by increasing recombination within it.

Despite this low current gain these devices justify interest in InGaAs/InP because of the hot electron transport of electrons which is possible in the collector regions of these devices without Γ–X and Γ–L valley transfers. These devices have exhibited extrapolated f_T values of 110 GHz over a V_{CE} excursion from 1.0 to 1.6 V where f_{MAX} was 58 GHz. The constancy of f_T over this voltage range is superior to that seen in similarly optimised GaAlAs–GaAs devices [10] which showed a peak in f_T at 105 GHz. The latter contained an 0.08 μm wide base of GaAlAs of graded composition and doping concentration of 4 x 10^{18} cm^{-3}.

Extracted electrical circuit parameters for the InGaAs device have allowed the emitter and collector charging time constants of 0.43 and 0.53 ps to be subtracted from the overall emitter–collector delay showing that the base–collector transit time was only 0.5 ps, equivalent to an average velocity of 3.7 x 10^7 cm^{-1} in the base and collector regions of the device. This delay represents the ultimate minimum attainable transit time for the structure and is equivalent to f_T = 318 GHz.

12.4.2 Field–effect Transistors

In field effect transistors which are formed from GaAs and related compounds, the source–drain separation is defined by lithography. The processes which are used to do this are described in Chapter 2 and it will suffice here to state that electron beam lithography can now be used to realise channel lengths of around 0.3 μm, and lengths as short as 0.05 μm have been demonstrated.

It is useful here to provide perspective by tracing the development of FET's made of silicon in order to justify the evolution of GaAs FET's. The initial concept of an FET involved an insulating layer superimposed on a semiconductor having a low carrier concentration. Applying a voltage to an overlaying gate establishes an electric field normal to the semiconductor surface within the insulator. This induces extra charges on the semiconductor surface which should enhance conduction between the source and drain electrodes. The first experiments in the 1940's showed that very few of the induced charges were mobile, and those that were caused an extra current to flow which diminished with time. There was little enhancement and a stable quiescent condition could not be maintained. This is now known to be due to carrier trapping in surface states at the insulator–semiconductor interface. These problems were not solved using silicon until the 1960's. This step was the key to the rapid emergence of MOS and now CMOS technologies which are complementing bipolar methods for many applications.

Because of these difficulties, attention turned towards realising silicon bipolar and junction field effect transistors as alternative active devices with considerable success. The subsequent emergence of MOS devices was attractive because there was no need for complicated area consuming isolation methods; there was also the prospect of low power complementary circuits. JFET's were not convenient for logic because they are depletion mode devices and require level shifting circuitry between successive stages. Enhancement mode devices do not require this.

The parallels in GaAs device development are close. Early attempts were made to make enhancement mode IGFET's which were equivalent to silicon n–MOS devices. It was soon discovered that surface states at insulator–GaAs interfaces are three to four orders of magnitude more abundant than on silicon. Unfortunately this means that induced electrons are almost entirely trapped in surface states. It has been demonstrated that trapping is very fast and occurs in times shorter than 200 ps [11]. Attempts to realise inversion mode n–channel FET's were abandoned by most large companies and activity has continued at a fairly low level with little success. There is now a great deal of prejudice against this approach using any compound semiconductor. It is tempting to ask whether success would have been achieved if the same investment had been made as had been made in silicon MOS technology. Nevertheless there are important indications that an inversion mode technology is possible using InP [12] and $In_{0.53}Ga_{0.47}As$ [13]. These materials can have relatively low surface state densities. There is, however, only a relatively low level of research effort.

An inherent feature of any insulated gate inversion mode FET is that the induced carriers move in a narrow channel about 50 Å wide which is bounded on one surface by the insulator–semiconductor interface. Scattering at the bounding surface leads to carrier mobilities which are about three times lower than the bulk value in silicon–SiO_2 devices. It would be surprising if the high bulk mobility values of more exotic materials could be realised fully in the surface channels of comparable devices even if the surface state problem were solved.

As in silicon these difficulties in realising an n–channel inversion mode FET caused attention to turn to depletion mode FETs and more recently to the HBT. The depletion mode GaAs MESFET is now the work–horse of production technologies for digital and analogue applications. The emerging HEMT technology makes use of new layer growth methods to overcome the intrinsic limitations of GaAs MESFET technology.

(i) High Electron Mobility Field Effect Transistors (HEMT)

(a) Intrinsic Problems with MESFETs

The capabilities of MESFET's for use in high speed digital and high frequency analogue circuits will have been described in other Chapters. It will be necessary first to refer to some of the intrinsic problems with the MESFET.

MESFET's, whether they are depletion or enhancement mode, use channels which are modulated by a depletion region on the gate side and are bounded by the substrate or buffer layer on the lower surface. The channel is said to be buried and is generally a few thousand angstroms thick. The boundaries are entirely within well ordered GaAs material and the mobility is not significantly reduced by scattering at the channel boundaries. The carriers move within bulk material and their motion is controlled by the mobility and saturation velocity limiting mechanisms of bulk GaAs.

Modulation of the channel conductance is through variations of the channel thickness, not through carrier density. A high carrier concentration in the channel is assured by employing as high a doping concentration as is consistent with effective depletion thickness modulation at practical gate voltage excursions. This requirement leads to doping concentrations of about 10^{17} cm^{-3}. Figure 12.7 shows the most important mobility components arising from the separate electron scattering mechanisms in GaAs. Their composite effect is obtained by combining them according to Matthiesen's Law:

$$\frac{1}{\mu_n} = \frac{1}{\mu_{II}} + \frac{1}{\mu_{PO}} \tag{12.11}$$

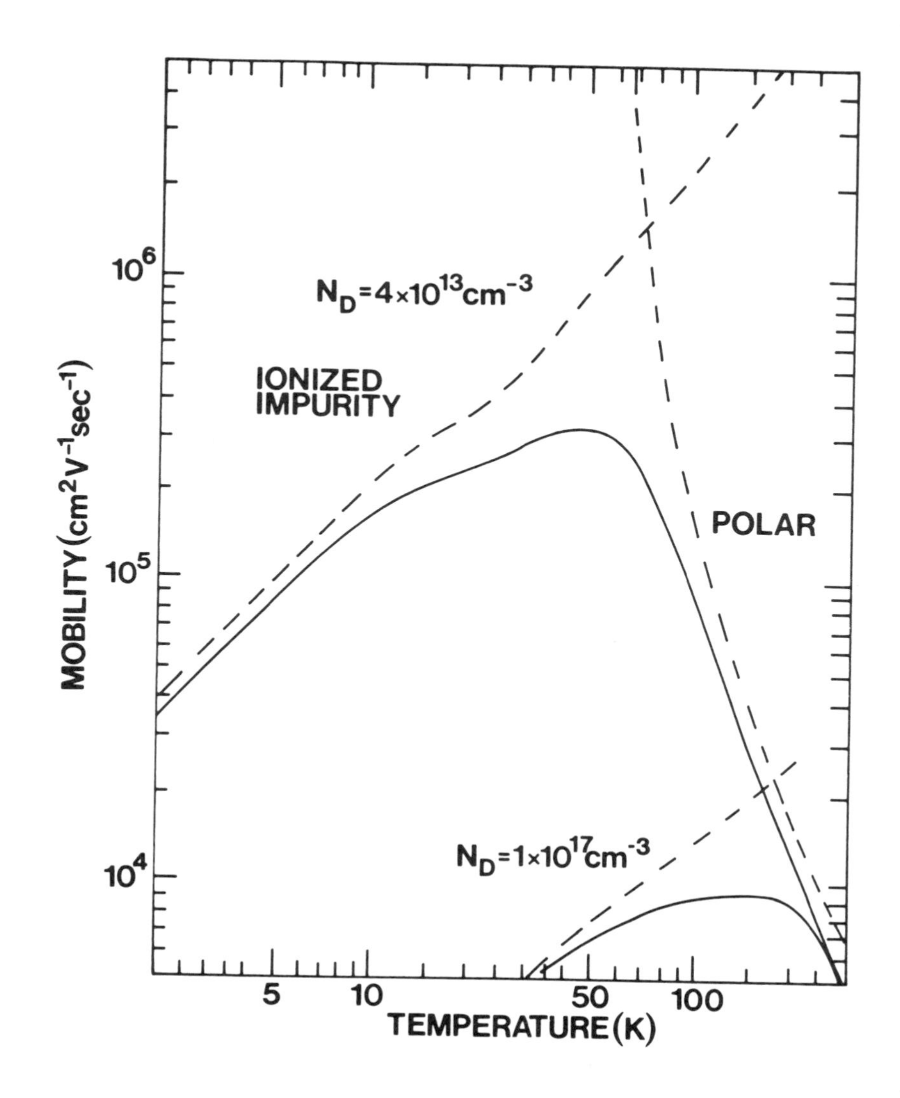

Figure 12.7: Mobility components in GaAs versus temperature

The general form shows that ionised impurity (II) scattering is predominant at lower temperatures becoming less important compared with scattering by polar lattice vibrations as temperature is increased. The cross-over between these two effects gives a maximum mobility at an intermediate temperature. As the donor doping is increased the maximum shifts downwards and occurs at higher temperatures. This is because the ionised impurity ions become more concentrated and give a scattering effect which is comparable with lattice scattering even at high temperatures. The graphs illustrate the situation for GaAs appropriately doped for MESFETs at 10^{17} cm^{-3}. The ionised impurities have an important but not dominant effect at 300K depressing the mobility value from 9,000 to 4,000 cm^{2}V^{-1}s^{-1}. An appreciable benefit would clearly occur if the ionised impurity scattering effect could be removed. The curve plotted for N_D = 4 x 10^{13} cm^{-3} shows that this reduction in doping would produce a dramatic mobility improvement at 77K near to the new cross-over between the two effects. The peak mobility value is in the region of 2 x 10^5 cm^{2}V^{-1}s^{-1}. Little benefit, however, comes from cooling highly doped material. it is necessary to create a structure in which the induced carriers move in a lightly doped region. A two-fold improvement in mobility would then occur when operating at room temperature. It was this that provided the impetus for the HEMT device which is sometimes called a MODFET (modulation doped FET) or TEGFET (transferred electron gas FET).

The GaAs mobility assumes particular importance in the low field regions of a MESFET where velocity saturation does not occur. In particular, the inevitable lengths of channel between the source and gate electrode edges, sometimes called the access regions, contribute to parasitic channel resistance. Since

$$g_m = \frac{g_{mo}}{1 + g_{mo}R_s} \tag{12.12}$$

it is clear that negative feed back due to R_s reduces g_m. This parasitic element also impairs the noise performance since:

$$N_F = 1 + 2.5\omega C_{gc} \left(\frac{R_s + R_D}{g_m}\right)^{1/2} \tag{12.13}$$

In practice these effects are minimised in MESFETs by increasing the doping in the parasitic channel regions or by using thicker access regions with the gated region in a recess. These methods become more and more difficult to implement as the channel length L is reduced to shorten the device transit time. The problem is compounded by the tendency for the ratio of parasitic and active length to increase as L reduces and is strikingly similar to the base resistance problem in HBT's.

In the drain region the field is high enough in device saturation to saturate the drift velocity, this happens because carriers transfer to the upper conduction band voltage.

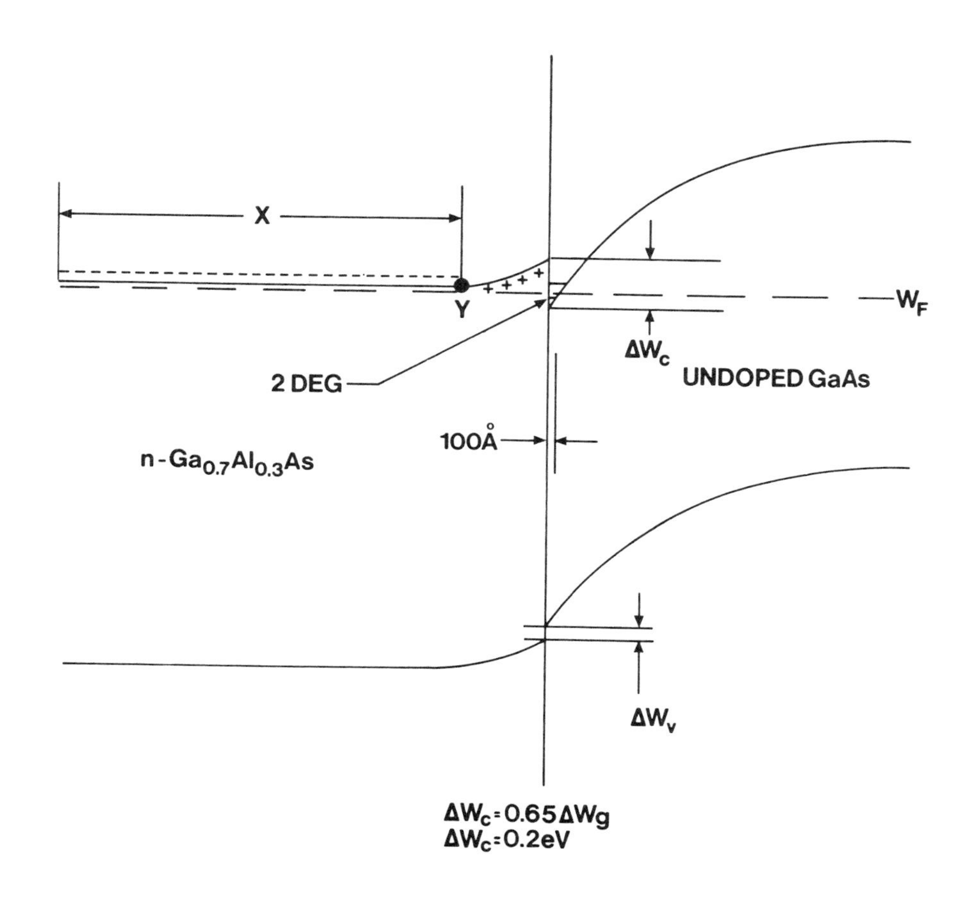

Figure 12.8: Channel formation using a heterostructure in a HEMT

(b) Channel Formation in the HEMT

Freedom from ionised impurity scattering is accomplished by spatially separating the donors which supply the channel electrons from an adjacent potential well formed in undoped material. The well must contain states which are lower in energy than the conduction band edge of the material which hosts the donors. Figure 12.8 shows how this is accomplished at a AlGaAs–GaAs heterojunction incorporating materials which have different energy gaps and a suitable conduction band edge discontinuity. Electrons are able to transfer by tunnelling to the well from the doped material. The well is sufficiently narrow to quantise the electron energies within it and electrons occupy discrete energy levels occupying the lowest first. A two-dimensional electron gas (2-DEG) is created just within the undoped material. In order to completely suppress the coulombic field of the dopant ions at the mobile charge plane, undoped spacing layers of the larger gap material are interposed between the doped and undoped material. The spacer layer is typically only 20-50 Å thick and mobilities of about 2×10^5 $cm^2.V^{-1}s^{-1}$ are normally attainable at 77K in GaAs. At room temperature the limit is set by lattice vibrational scattering and is limited to 9000 $cm^2.V^{-1}s^{-1}$, a two-fold improvement over a doped channel in a GaAs MESFET.

A practical structure based on the AlGaAs materials system is shown in Figure 12.9. The energy diagram shown in Figure 12.8 indicates that the conduction band within the doped n-GaA*l*As is also populated with electrons in the region of thickness X. As Figure 12.9 shows, if this were the case, it would provide a conducting path in parallel with the 2-DEG. The doping concentration of this region, its thickness X and the conduction band edge discontinuity must be chosen to ensure that this region is completely depleted by the Schottky barrier formed by the metal gate on the n-AlGaAs surface. If the depletion extends to just meet the hetero-junction depletion region at point Y parallel conduction in the n-AlGaAs layer is suppressed and only the 2-DEG high mobility channel will exist at zero gate voltage. Further depletion by applying a negative gate voltage extends the depletion region into the 2-DEG and the device can be turned off, the device then operates in the depletion mode. If, on the other hand, the depletion region is designed to accomplish this at zero gate voltage the 2-DEG can be re-established by applying a positive gate voltage to withdraw the depletion region turning the 2-DEG channel on. The device can be designed to function either as a depletion or an enhancement mode FET.

If the positive gate voltage is increased to completely populate the quantised 2-DEG energy levels, the depletion region contracts back into the n-AlGaAs and the parallel conduction path re-appears. In practice this can set a limit on the forward gate voltage which will, in any case, be limited eventually by forward gate conduction as the Schottky gate contact becomes forward biased, a problem which is common to MESFETs.

(c) Channel Conduction

The two-dimensional electron gas created by the heterojunction provides an environment in which electrons can be accelerated by the source-drain field with a much reduced mean time between scattering collisions. This provides reduced resistances in the parasitic channel regions and leads to important improvements in the noise behaviour of these devices. Noise figures of 0.4 dB have been attained at 12 GHz in HEMT devices made of AlGaAs-GaAs operated at 100K with L = 0.5 μm [14].

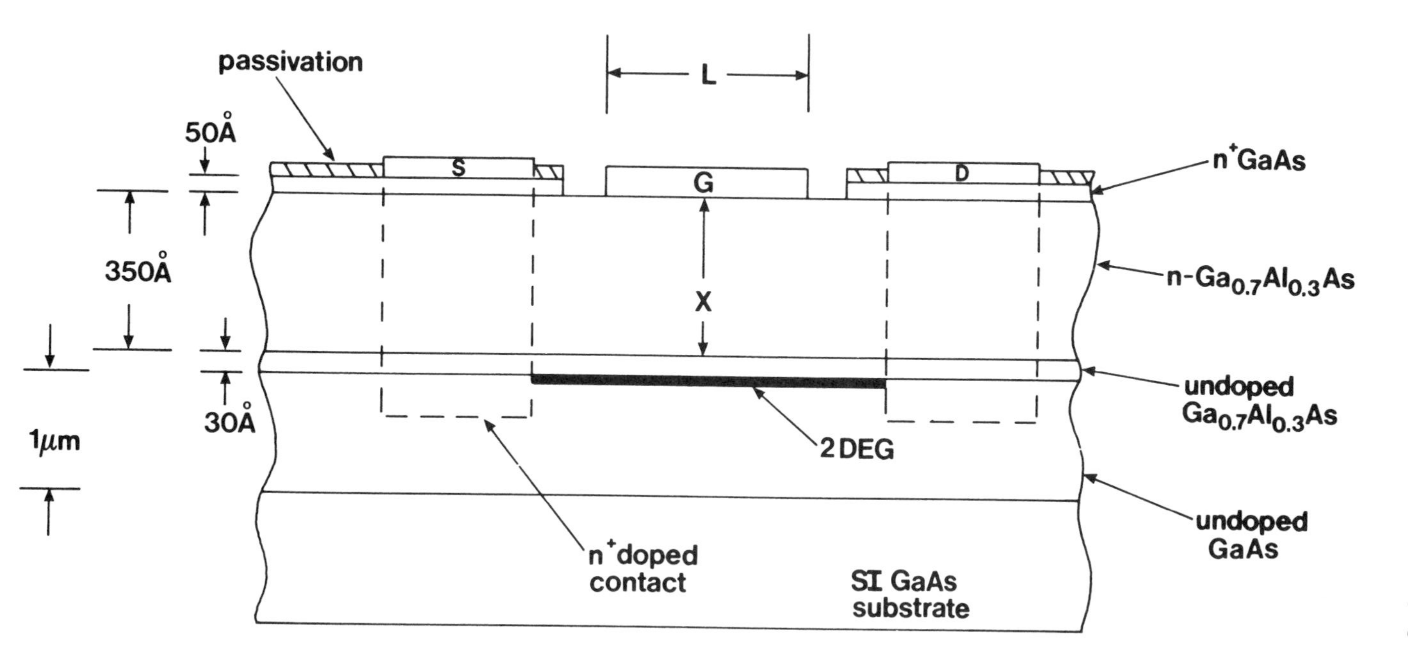

Figure 12.9 The structure of a practical HEMT based on the AlGaAs system

As in other FETs increasing the source–drain voltage reduces the gate–drain electric field; this increases the channel resistance near to the drain causing an increase in channel field at this end of the channel.

The accelerating electrons gain momentum and energy at an even faster rate because of their high mobility, giving them a higher average velocity over the length of the device. As carriers move near to the drain the high field causes inter–valley transfer as energies reach 0.29 eV above the conduction band minimum and the velocity saturates.

In these devices there is another effect which causes carriers to slow down which is inherently related to the device structure and does not occur in heterojunction bipolar transistors. It will be recalled that the electrons in the 2–DEG are contained in a potential energy well at the heterojunction and that this is undoped and free from ionised impurities. If the energy acquired by acceleration in the channel field is sufficient, the carriers may spill over the edges of the well and move through the device within the bounding material. The carriers spilling on the gate side now encounter ionised impurities and slow down because of increased scattering. On the substrate side, carriers still move in undoped GaAs but are subject to scattering in three dimensions. Within the well the quantisation of the states in the normal direction reduces the number of states into which electrons can be scattered and it is this which contributes, in part, to the high mobility. These effects, due to spatial transfer, can appear in the saturated part of the device output characteristic as a negative differential resistance [15]. In typical $Al_xGa_{(1-x)}As$–GaAs HEMT structures the conduction band energy discontinuity at the well boundary is only 0.2 eV so real space electron transfer into AlGaAs is comparable in importance to inter–valley transfer in degrading carrier velocity. Limiting electron velocities in 1 μm HEMT devices are typically 2×10^7 cm s^{-1} at 300K and 3×10^7 cm s^{-1} at 77K compared with values as high as 1.5×10^7 cm s^{-1} in MESFETs. Since the saturated HEMT channel current is given by:

$$i_D = en_s vW \qquad (12.14)$$

the average velocity v has a controlling effect. In practice the number of charges per unit area of device is limited by the number of states in a quantised energy level within the well and is typically 1×10^{12} cm^{-2}. The charge in the 2–DEG can be treated simply as that charge induced by the gate voltage applied across the depletion AlGaAs layer which is typically 350 Å thick (d) leading to:

$$n_s = \frac{\epsilon(V_{GS} - V_T)}{ed} \qquad (12.15)$$

Inducing a charge greater than the 2–DEG area concentration causes extra carriers to move in the parasitic MESFET in parallel with the 2–DEG. It is easily shown that this occurs when $V_{GS} - V_T = 0.5$ V, placing an important limit on the analogue dynamic range of the device. It can similarly be shown that:

$$g_{mo} = \frac{evW}{d} \tag{12.16}$$

For digital applications it is necessary that the logic swing should be at least 20 standard deviations in the spread of manufactured threshold voltage. In practice $V_T = 0.1$ V and $\sigma = 12$ mV leading to a minimum swing of voltage equal to 0.24 V. The maximum gate voltage would then be 0.34 V, safely below the voltage which would cause parallel conduction.

The conventional definition of f_T:

$$f_T = \frac{g_{mo}}{2\pi C_{GS}} \tag{12.17}$$

leads to

$$f_T = \frac{v}{2\pi L} \tag{12.18}$$

If v is 2×10^7 cm s^{-1} at 300K and L = 1 μm a fundamental limit to f_T is set at 300 GHz. Improvements on this value require materials which can yield a higher average channel velocity or reduction in channel length. One of the best f_T values for AlGaAs – GaAs is 70 GHz for a 0.25 μm device [16]. This result falls short of an expected scaled value of 120 GHz and suggests that the benefits expected from ballistic velocity overshoot are not evident at L = 0.25 μm. This is consistent with the scepticism expressed by Hess and Iafrate [17] about ballistic velocity enhancement and suggests that the velocity degrading effects of inter–valley and real space transfers are predominant at least in GaAs–AlGaAs.

(ii) Iso– and Pseudo–morphic HEMT Devices

(a) Introduction

Until quite recently most work on HEMT devices has centred on the $Al_xGa_{(1-x)}As$ system. In seeking other materials it is clear that a high mobility is desirable, not only to minimise the resistances of parasitic channel regions which contribute to noise, but also to help maintain a high carrier velocity over a greater length of the channel. A higher saturation velocity is desirable and this can be ensured by using large inter–valley energy separations and using heterojunctions with large conduction band energy discontinuities to confine the electrons to the 2–DEG. This latter parameter also increases the amount of charge which a 2–DEG can accommodate per unit area, helping to increase the current carrying capability of the channel and the device g_m.

(b) GaAs Based Epitaxial Materials

There is a continuous trend as x increases within the system $In_xGa_{(1-x)}As$ towards higher electron mobilities and saturating velocities. These materials however do not lattice match to GaAs. MESFETs have been made in this system with x = 0.08. this material was grown epitaxially on a GaAs substrate by using a compositionally graded layer to accommodate the 0.6% lattice mismatch. The buffer thickness needed was 1.8 μm. Useful improvements in performance were achieved providing a cut–off frequency f_T of 36 GHz for a 0.4 μm device [18].

A similar pseudo–morphic approach has been pursued [19] to incorporate a channel layer with x = 0.15 into a HEMT structure based on AlGaAs. In this case the layer was only 125 Å thick allowing it to be formed pseudo–morphically and the mismatch accommodated elastically. A channel length of 0.35 μm produced an f_T value of 60 GHz. These devices were integrated into GaAs based low noise amplifier circuits. A 5.3 dB noise figure was achieved over two stages of 59.5 GHz with a gain of 10.2 dB. There is no doubt that further benefits could result by increasing the indium fraction but the pseudo–morphic approach requires that the layer thickness must reduce as the lattice mis–match is increased in order to remain below the critical thickness for elastic matching. It remains to be seen how far this trend can be continued.

Simulations have been performed [20] which predict the electrical performance of these devices quite closely. The calculated velocity variation along the channel predicts a peak at 4 x 10^7 cm s^{-1} indicating velocity overshoot at the drain edge of the gate. At that point only 15% of the charges are in the 2–DEG and 70% have escaped into the adjacent InGaAs material. The remainder are in the GaAs buffer or the doped $Al_{0.2}Ga_{0.8}As$ layer above the 2–DEG. It seems very likely that the real benefit here is not from the 2–DEG but from the higher mobility $In_{0.15}Ga_{0.65}As$ layer itself. It is not clear how free from dopant the InGaAs layer is nor whether the carriers are any better contained than in the $Al_xGa_{(1-x)}As$ HEMT structure.

(c) InP Based Epitaxial Materials

Putting on one side the relevance of the InP system for opto–electronic integration it appears to offer inherent performance advantages. This is partly because of the higher mobilities and saturation velocities which larger valley separations provide and also because of the better spatial confinement provided by larger conduction band edge energy discontinuities.

Lattice matched HEMTs have been realised on InP substrates [25] where the 2–DEG was formed in $In_{0.53}Ga_{0.47}As$ and a superimposed InP layer was used to make the channel forming heterojunction. This structure provided an energy discontinuity of 0.3 eV larger than is used in the AlGaAs system (0.2 eV). In this case a channel length of 1.2 μm gave f_T = 24 GHz.

Much better confinement has been achieved by replacing InP with $In_{0.52}Al_{0.48}As$ a composition which lattice matches InP. In this case a conduction band edge discontinuity of 0.51 eV was obtained [22]. Devices [23] with L = 0.2 μm and 0.1 μm gave f_T values of 120 GHz and 170 GHz respectively. The former device was used in a single stage low noise amplifier and gave a noise figure of 0..8 dB at 63 GHz with an associated gain of 8.7 dB. This data is claimed as a world record for any three–terminal device.

The pseudo–morphic technique has also been used [24] in this materials system in order to add more indium with the aim of raising the mobility and saturation velocity even further. The 2–DEG layer was formed in an $In_{0.65}Ga_{0.35}As$ layer only 150 Å thick. The deviation from the x = 0.53 lattice matching composition inevitably introduced elastic strain. The layer was bounded by a spacer layer of undoped $In_{0.52}Ga_{0.48}As$ 100 Å thick and then a doped layer of the same material. This arrangement was repeated above and below the pseudomorphic layer to form two parallel 2–DEG channels. An increase in confining barrier was anticipated at this composition to ensure a higher channel charge concentration though no value was cited. The output characteristics showed an unusually low saturated output conductance (13 mS/mm) suggesting that the carriers were well contained. These devices were 1 μm long and yielded an extrapolated f_T value of 37 GHz. The trend towards higher indium concentrations in pseudomorphic layers is subject to similar critical thickness–composition limitations.

A useful summary of HEMT results for the lattice matched InGaAs/InAlAs system, the pseudomorphic InGaAs/AlGaAs system and the $Al_xGa_{(1-x)}As$ system is provided in a paper by Hikosaka et al [25]. the data is reproduced and supplemented in Figure 12.10. The authors suggest the following empirical figures of merit for the three materials systems: f_T = 25.8, 20.9 and 17.6 GHz μm respectively.

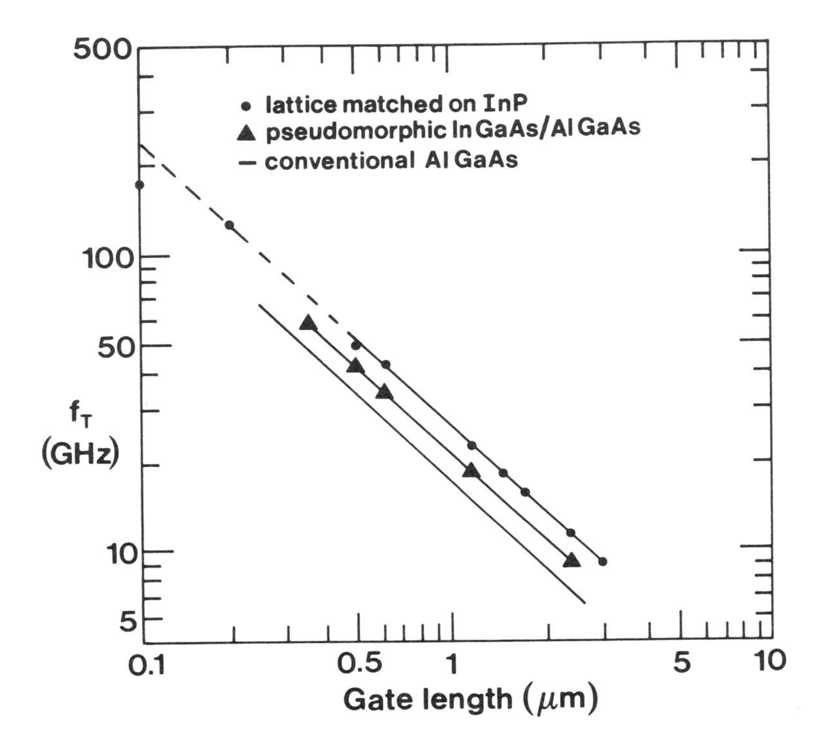

Figure 12.10: HEMT cut-off frequencies v channel length for the lattice matched InGaAs/InAlAs, pseudomorphic InGaAs/AlGaAs and $Al_xGa_{(1-x)}As$ systems

12.5 Conclusions

The HEMT and HBT are still research devices and little use has been made of them so far in electronic systems. Their principles of operation are well established but there are undoubtedly problems in ensuring adequate yield and consistency in manufacture. The methods used are very specialised and are available in relatively few companies. Their availability will certainly be crucial to the competitiveness of companies offering advanced electronic systems.

The path to achieving even higher performance is indicated by the trends in advanced semiconductor materials. There is a clear need to be able to make use of materials in which electrons can be speeded or heated without scattering into low mobility energy bands. The trends towards InAs in the $In_xGa_{(1-x)}As$ system is part of this but it is inhibited by the need for epitaxy on existing highly developed substrate materials such as GaAs and InP. Developments in layer growth methods will probably allow the lattice mismatches associated with larger indium concentrations to be accommodated. In the HEMT, containment needs to be improved by creating heterojunctions with larger conduction band discontinuities.

The choice of substrate material places an important constraint on the range of epitaxial material options. In the longer term may be necessary to use completely new substrates to make use of alloys based on InSb which has a room temperature mobility of 78,000 $cm^2\ V^{-1}\ s^{-1}$ compared with 33,000 $cm^2\ V^{-1}\ s^{-1}$ for InAs. These materials have low energy gaps, 0.16 and 0.33 eV respectively. This suggests that while they may be suitable for majority carrier devices their use in bipolar structures at room temperature is problematical.

The growth of GaAs in silicon is an exciting development which will permit new circuit configurations providing improved overall performance. There are still important growth problems to be solved before this method can be extended to include bipolar devices.

Parallel developments in lithography are taking place which will shrink active region dimensions in FET's below 0.1 μm. This is of the same order as the mean free path of electrons moving with thermal velocities at room temperature in GaAs. It corresponds to the mean time between collisions of 0.35 ps and suggests that electrons will increasingly be able to cross devices without making collisions, moving ballistically. Under these conditions inter–valley scattering or optical phonon generation will limit the energy which the electron can gain and sets their maximum velocity. Only when transit times fall below the momentum transfer time, 0.01 ps, can the motion be said to be truly ballistic. This limit will occur when intrinsic f_T values approach 1000 GHz. Further progress will then require a means for inhibiting the transfer of phonon momentum to the hot electrons. In the meantime, if maximum carrier velocities remain at about 3×10^7 cm s^{-1}, the ballistic situation will not result until channel lengths approach 0.003 μm. It seems very likely that effects which are presently regarded as second order will play an important role in modifying device behaviour well before this situation is reached.

12.6 References

1 P. D. Hodson, R. H. Wallis and J. I. Davies: Elec. Letts. 23,pp273–275, (1987)

2 K. Morijuka et al: IEEE, EDL–9, (11), p598 (1988)

3 N. Hayama et al: IEEE ED–35, (11) p1771 (1988)

4 P. I. Rockett: IEEE ED–35, (10) p1573 (1988)

5 K. Morijuka et al: IEEE, EDL–9, (11), p585 (1988)

6 K. Morijuka et al: IEEE, EDL–9, (11), p570 (1988)

7 R. N. Nottenberg et al: IEEE, EDL–7, (11), p643 (1986)

8 O. Suguira et al: IEEE, EDL–9 (5), p253 (1988)

9 R. N. Nottenberg et al: IEEE, EDL–10, (1), p30 (1988)

10 T. Ishibashi and Y. Yamauchi: IEEE ED–35, (4) p401 (1988)

11 W. S. Lee and J. G. Swanson: Thin Solid Films, 103, (103), p177 (1983)

12 L. J. Messick: IEEE ED–31, (6) p763 (1984)

13 P. D. Gardner et al: Thin Solid Films, 117, p137 (1984)

14 K. Joshin et al: IEEE Microwave and Millimeter Wave Monolithic Circuits Digest pp563–565 (1983)

15 Y–K Chen et al: IEEE, EDL–9, (1), (1988)

16 P. C. Chao et al: Electronics Letts. 19, pp894–895, (1983)

17 K. Hess and C. J. Iafrate: Proc. IEEE, 76, (5) (1988)

18 H–D Shih et al: IEEE, EDL–9, (11) p604 (1988)

19 G. M. Metze et al: IEEE, EDL–10 (4) p165 (1989)

20 D. H. Park et al: IEEE, EDL–10 (3) p107 (1989)

21 Y–K Chen et al: IEEE, EDL–10 (4) p162 (1989)

22 U. K. Mishra et al: IEEE, EDL–9 (1) p41 (1988)

23 U. K. Mishra et al: IEEE, EDL–9 (12) p647 (1988)

24 G. I. Ng et al: IEEE, EDL–10 (3) p114 (1989)

25 K. Hikosaka et al: IEEE, EDL–9 (5) p241 (1988)

Chapter 13

Digital Optoelectronic Devices and Systems

G. Parry, M. Whitehead and J. E. Midwinter

13.1 INTRODUCTION

13.1.1 The Role of GaAs

Gallium arsenide technology has played a crucial role in the development of optoelectronics from the early days of semiconductor lasers to the current developments on novel quantum well and superlattice structures. The three characteristics which have been particularly significant in promoting the use of gallium arsenide have been its direct band–gap energy, its near lattice match to gallium aluminium arsenide and the quality of epitaxially grown and processed devices. Material with good optical and electrical characteristics has been produced at a number of laboratories and good control over processing and fabrication has been obtained with various techniques. Consequently gallium arsenide has frequently been used to demonstrate device concepts even if the ultimate aim has been to produce a device in a different III – V material. For example, lasers and detectors for low loss and high data rate optical communication systems are now fabricated from ternary or quaternary materials based on indium phosphide. However, many of the concepts used in these lasers were first demonstrated in gallium arsenide. Of course, there are many other applications where gallium arsenide is still the first choice technology. In optical storage systems gallium arsenide lasers are preferred to the longer wavelength indium phosphide based devices and for local area communications gallium arsenide sources may be selected.

In this chapter we will focus on some of the more recent developments of novel gallium arsenide devices which are being designed for use in digital optoelectronic systems. Here we use the term digital optoelectronics to refer to the use of optical or optoelectronic techniques to enhance the performance of a digital signal processing or switching system. Whilst in principle it would be feasible to use a variety of III–V materials, in practice it seems likely that gallium arsenide will be the preferred technology for the optoelectronic devices. The devices which we will discuss exploit the optical and optoelectronic properties of quantum well structures formed in the gallium arsenide – gallium aluminium arsenide system. These structures are leading to new classes of optoelectronic devices some of which do not have equivalents in bulk material.

13.1.2 The Role Of Optics In Digital Electronic Systems

There have been many proposals and simple demonstrations of ways in which optics can play a role in signal processing or computing systems varying from switching with all–optical logic elements (Gibbs 1985)[6] to the use of optics for interconnection in high speed computing systems (Haugen 1986)[9] and neural networks (Psaltis and Farhat 1985)[20]. The relative merits of an all–optical approach and a hybrid approach have been considered previously (Midwinter 1988)[15] so they will not be repeated here in detail. We will summarise our view on this by stating that it is difficult to envisage that even simple all–optical logic elements will be produced which can operate with both low optical powers and high speed. Devices produced to date generally have slower switching speeds than electronic devices and certainly require higher powers. The real benefits which optics brings are the potential for complex interconnection. Parallel interconnection of processors via crossing, but non–interfering optical paths, is the key benefit of optics. Figures 13.1 and 13.2 illustrates schematically the basic idea of such an optical interconnect. Optical fibre interconnection between subsystems is now quite common. The benefits gained there are primarily data rate and isolation. We are considering here a far more direct and closer interaction between optics and electronics involving the use of optics to interconnect electronic "islands" on a wafer or mother–board.

What is particularly important to note is that the interconnection problems which optics has the potential to solve are among the main problems arising in the design of very high speed electronic circuitry. It is clear from recent developments of both silicon and III–V electronic devices, that producing exceptionally fast individual devices is not a problem (Sai–Halasz 1989)[21]. However connecting them electrically and operating complex circuits at high speed is certainly a problem. There are, of course, a number of issues to be considered before prototype circuits using optical interconnects can be demonstrated. The key issues to be considered are interface technologies (materials and devices), interconnect technology, and circuit design or architecture.

It is important to note that there is no established view on the optimum choice of any of these. We will be presenting an approach that seems to us to be a particularly promising one.

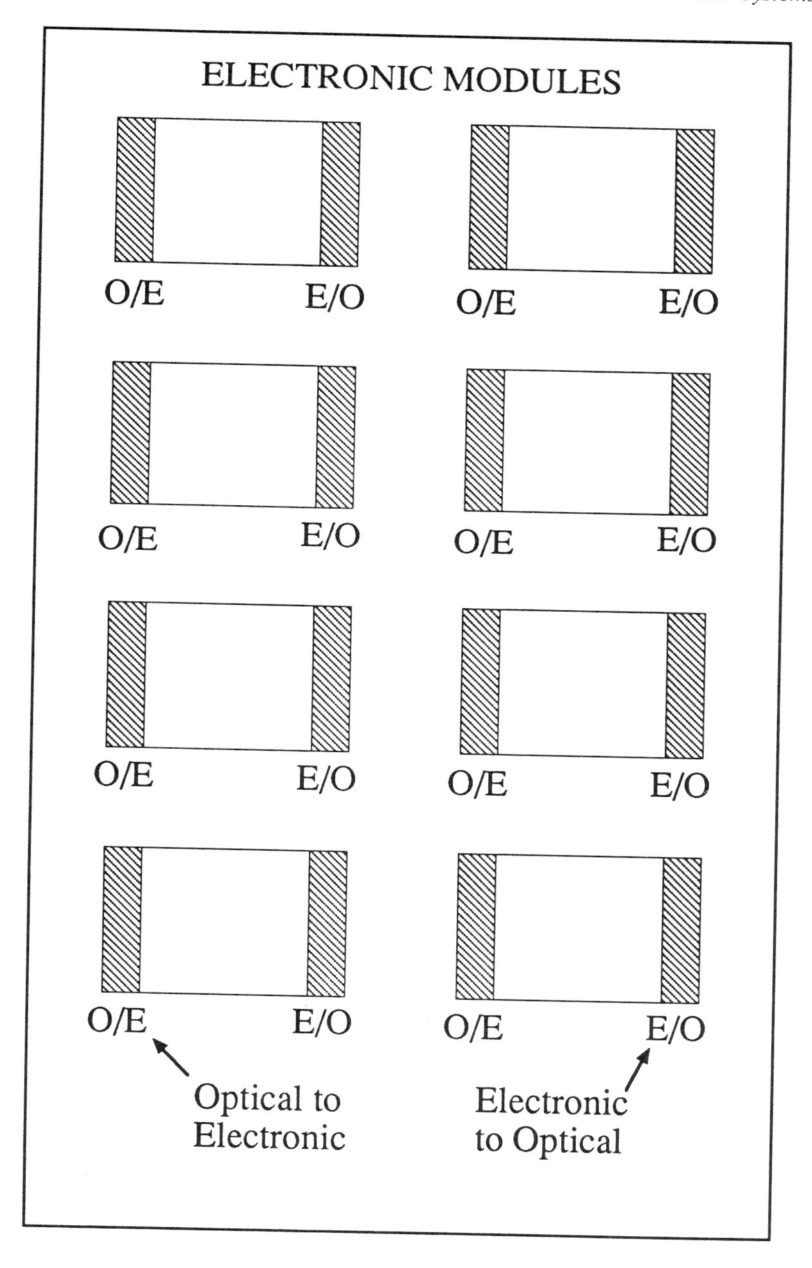

Figure 13.1: Schematic diagram illustrating a possible arrangement of electronic modules or islands which may be connected optically. Each island would be small enough to permit very high speed electronic device operation and local interconnection. Optical to electronic and electronic to optical interfaces would be required on each island.

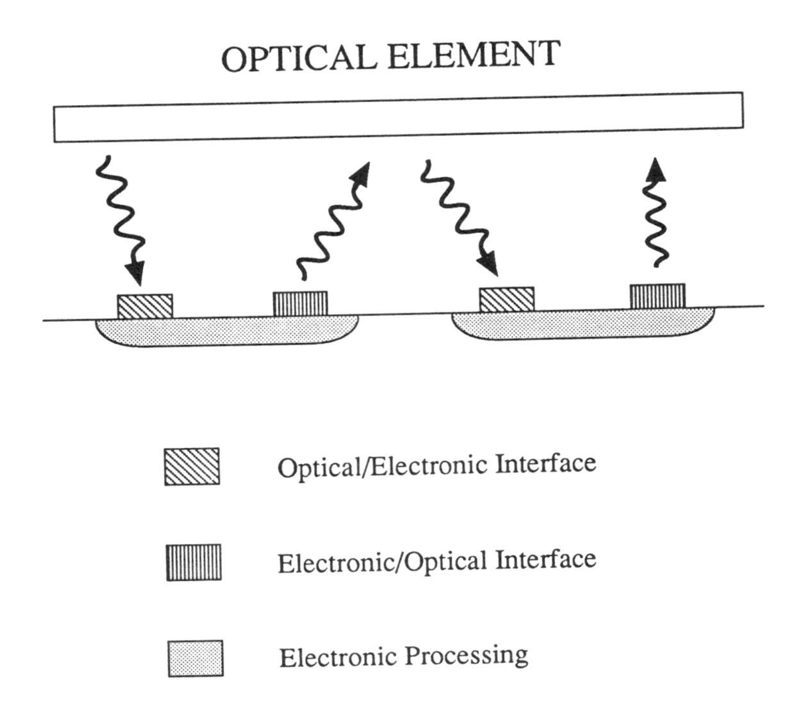

Figure 13.2: Out-of-plane interconnection of electronic islands. The optical element may be bulk optics or a holographic element and would normally need to operate in a reflective mode.

13.2 Interface Devices For Optically Interconnected Electronics

Any optically interconnected circuit has to include interfaces between electronics and optics (E/O) and interfaces from optics to electronics (O/E). There are two factors to be considered at each interface, a materials choice and a device choice. If the electronic circuitry is gallium arsenide based then optoelectronic devices can be fabricated using the same material system. (There are still design and processing considerations, of course, but the materials problem is straightforward.) If the electronic circuitry is in silicon then either a hybrid approach needs to be adopted or gallium arsenide devices must be grown directly on silicon. There is no clear answer as to which approach is likely to be the more promising at present, though the GaAs on silicon technology is of course, less advanced at present (Fan 1989)[4]. The choice of interface devices is also still open but there are some fairly clear benefits which can be achieved from particular device choices.

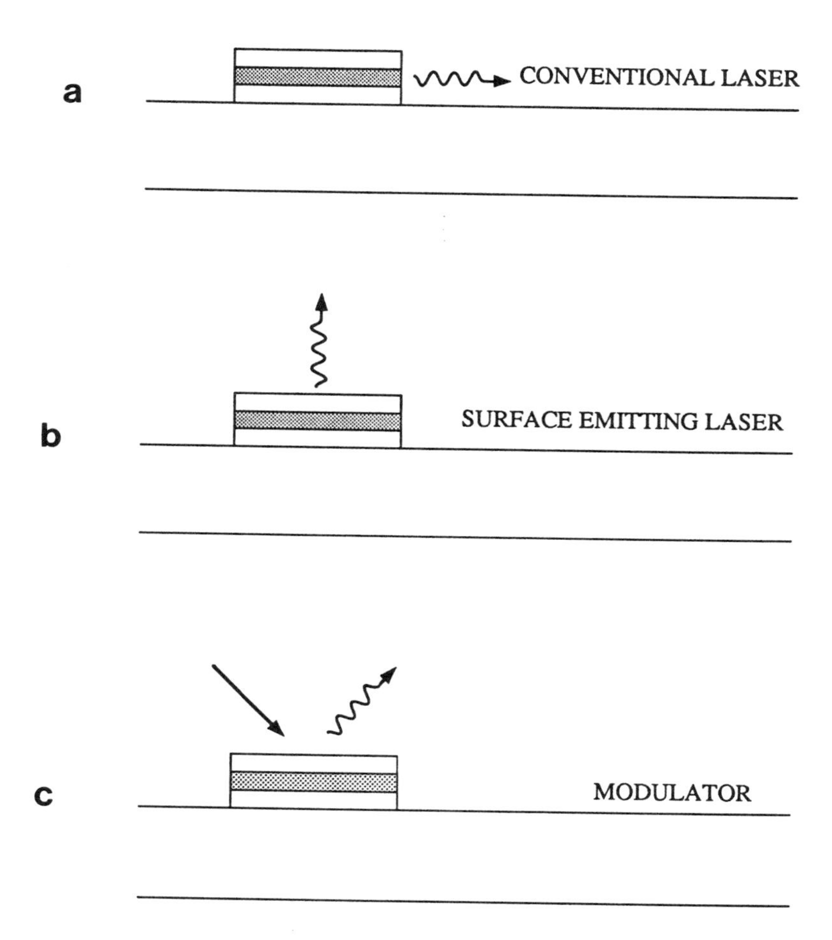

Figure 13.3: Three options exist for electronic to optical interface devices. The conventional laser (a) emits in the plane of the substrate, the surface emitting laser (b) does provide out of plane interconnection, and the modulator (c) is addressed by an external source.

13.2.1 The Electronics To Optics Interface

We have three options here. A laser could be fabricated on chip and modulated electrically to transmit a stream of data in much the same way as one would in a telecommunications network. Most lasers are planar devices so this configuration would only permit planar communications or interconnects (see Figure 13.3(a)). Surface emitting lasers (Uchiyama and Iga 1986)[25] do permit transmission out of the plane of the substrate but they are not so advanced in technology terms and would require further

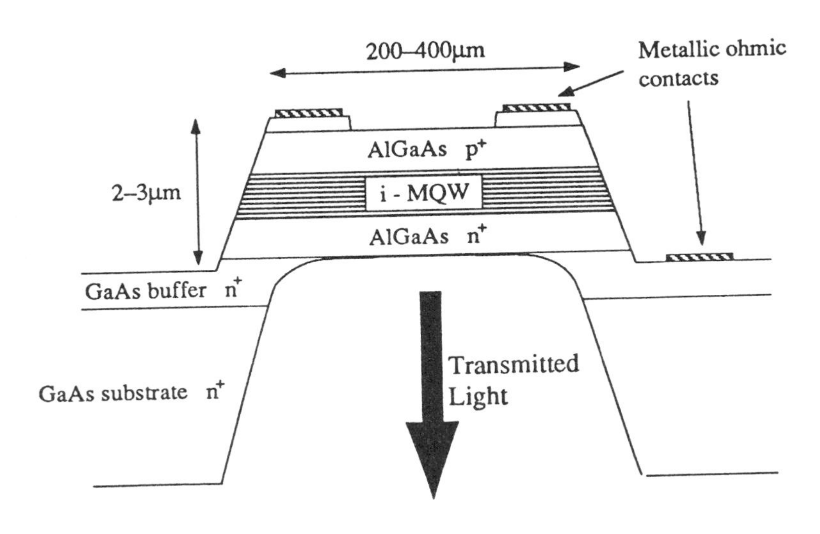

Figure 13.4: Typical structure of a MQW (multiple quantum well) transmission modulator. The MQW consists of a large number (typically about 50) of layers of GaAs and AlGaAs grown epitaxially by MBE or MOCVD. Reverse biasing the p–i–n structure provides a high electric field across the well for modulation. The GaAs substrate has to be removed in the case of the transmission modulator because it is absorbing at the operating wavelength

developments before this option was feasible (see Figure 13.3(b)). A major disadvantage with both lasers is that a significant amount of power may be dissipated on chip causing local temperature variations. This causes particular difficulties in the case of GaAs on silicon wafers because there is a large mismatch in the thermal properties of the two semiconductors and differential expansions can induce strain and consequently limit the lifetime of the device. The third option would involve positioning a modulating device on chip which could be addressed by an externally located laser as in Figure 13.3(c). The low on–chip power dissipation, and the possibility of optical interconnection out of the plane of the substrate tends to favour the modulator approach. (The possibility of using an LED on chip is, of course, feasible but since optical interconnects are only likely to be of value at very high data rates, and since it is not realistic to consider modulating an LED at a sufficiently high frequency, this option can be disregarded).

The possibility of this approach has been apparent for some time but it is only in the last few years that a suitable modulator has been developed. The device which offers most promise at present is the gallium arsenide multiple quantum well electro–absorption modulator (Wood et al 1984)[31]. Figure 13.3 shows the basic structure of this modulator. The well and barrier widths are typically about 50–100 Å. The device operates by an electric field induced shift of the absorption edge of the quantum wells so that the absorption at a selected optical frequency close to the absorption edge is increased or decreased by an applied voltage. As can be seen in Figure 13.4 the quantum wells are located in the intrinsic region of a p–i–n structure so a large electric field can be applied using a low voltage reverse bias. The absorption edge of gallium arsenide quantum wells is dominated by a sharp excitonic absorption which produces a significantly larger absorption change than could be achieved in bulk material. Indeed both the mechanism responsible for the shift of the absorption edge (the quantum confined Stark shift) and the presence of strong exciton absorption at room temperature are associated with the quantum confinement in the well. (Miller et al. 1985)[17]. Figure 13.5(a) and 13.5(b) demonstrate the strong excitonic absorption and the quantum confined Stark shift in measurements of the transmission through two different devices (Whitehead et al. 1988)[28]. The devices differ in the width of the quantum wells used. It is clear that the narrower well sample show a clearer shift of the absorption edge but the wider well sample show a similar transmission change at a lower operating voltage. Because these devices are simple p–i–n structures they may be small in size (area < 700 μm^2) and consequently may operate at high frequencies. The highest reported operating frequency is 5.5 GHz (Boyd et al. 1989)[3]. The main limitation of the devices operated in transmission mode is the low contrast (on to off ratio). A 3 dB contrast ratio or a transmission change of 40% is typical of what can be achieved in an optimised structure at 5 volts (Stevens and Parry 1989)[22]. A further limitation with gallium arsenide devices is that the substrate is absorbing at the operating wavelength so a window has to be opened up to permit transmission. A reflection geometry in which a multilayer reflector stack is grown epitaxially on the substrate before the quantum well device (see Figure 13.6), removes many of these drawbacks and provides a more convenient geometry for the interconnection. We have shown that by exploiting the Fabry–Perot structure which is formed between the high reflector and the natural reflector at the semiconductor air interface, a large enhancement of the contrast ratio is obtained, potentially at 5 volts or less (Whitehead and Parry 1989, Whitehead et al. 1989)[29,30]. Contrast ratios in excess of 20 dB are achieved over a narrow wavelength range but, more important, better than 10 dB is achieved over several nanometers. Figure 13.7 illustrates this result. This structure has virtually all the attributes which we would desire of an optical interface element for the electrical to optical interface.

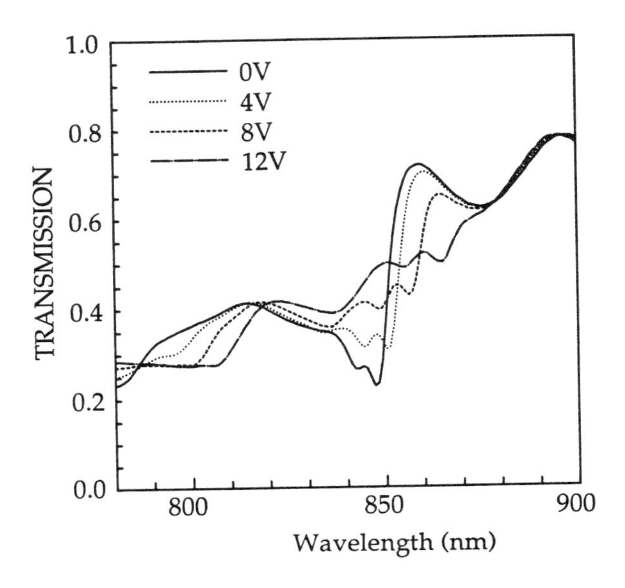

(a) i–region thickness of 0.88 microns with 87 A width wells (GaAs) and 60 A thick barriers (AlGaAs)

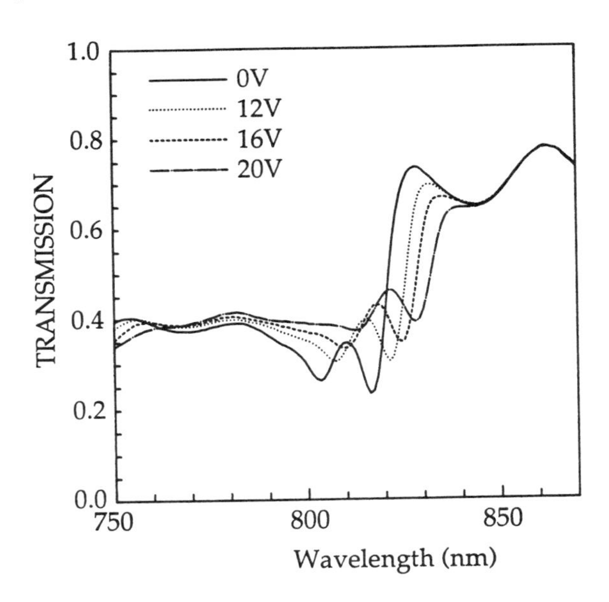

(b) well width of 47 A but all other parameters identical to (a)

Figure 13.5: Transmission spectra of MQW modulators as a function of reverse bias

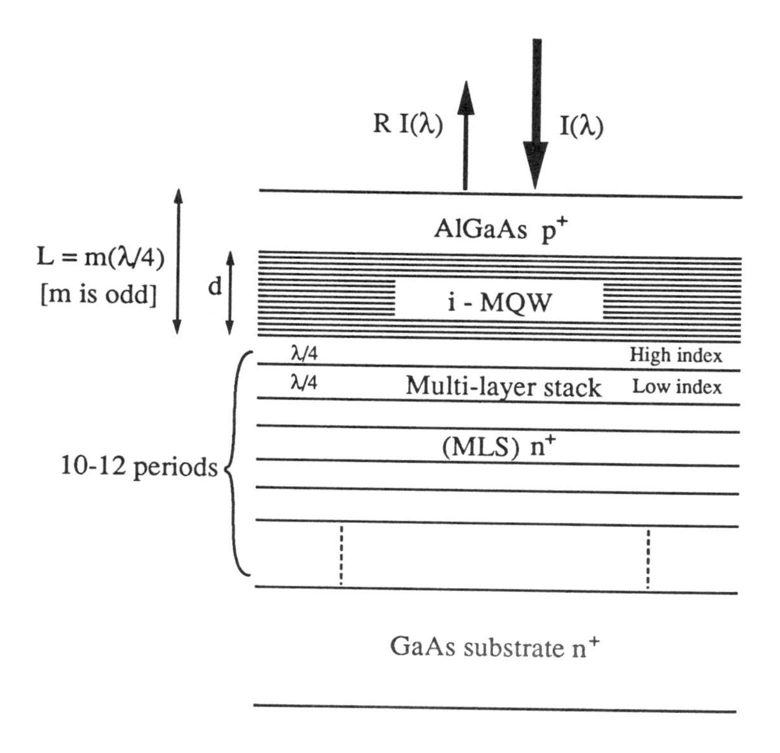

Figure 13.6: Structure of a reflection modulator incorporating a quarter wave reflector stack formed using AlAs and AlGaAs (10% Al). The total thickness of the layers grown on top of the reflector stack is selected so that the device operates as an asymmetric Fabry Perot structure with a resonance dip at the operating wavelength. Reverse bias provides a high electric field across the quantum wells

It is worth spending a little time considering some of the benefits which arise from the use of GaAs as opposed to some other III – V materials. First of all the exciton absorption at room temperature in GaAs/AlGaAs quantum wells is significantly larger than is observed in quantum well structures grown using GaInAs/InP for example. This means that the change in absorption which can be induced electrically is considerably greater in GaAs than in InGaAs. Table 13.1 summarises some of the experimental data which supports this point (Tipping et al 1989)[24]. In practice this means that improved contrast devices and lower operating voltages are achieved with GaAs quantum well devices. A further interesting benefit that is gained from the use of GaAs is associated with the multi–layer reflector stacks which are used in the reflection devices. The stacks are formed from alternate layers of AlAs and $Al_{0.1}Ga_{0.9}As$. In designing a reflector stack $Al_{0.1}Ga_{0.9}As$ it is necessary to choose layer

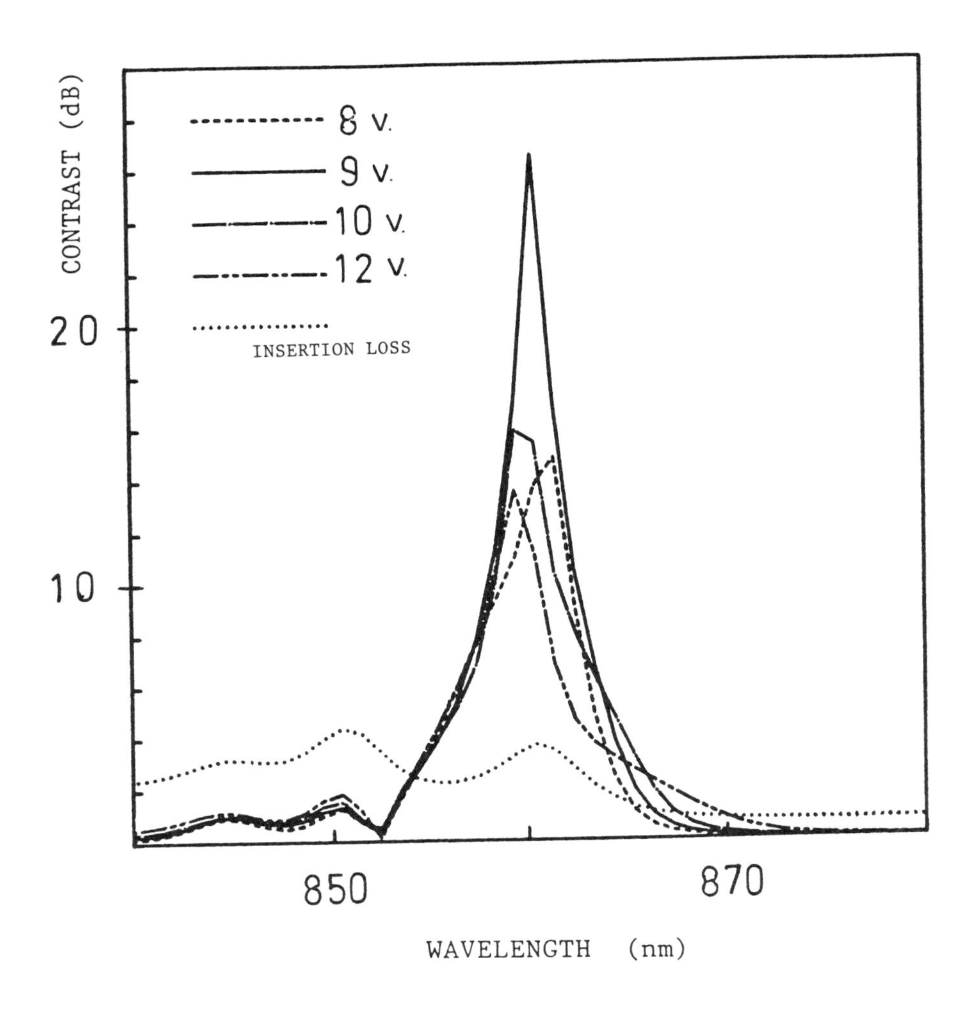

Figure 13.7: Contrast ratio and insertion loss achieved with an asymmetric Fabry – Perot structure as a function of operating voltage. Data from (Whitehead et al 1989)

thicknesses to be one quarter of the desired operating wavelength and to choose alternate layers of high and low refractive index materials. The larger the refractive index difference, the fewer the number of layers needed to give good reflectivity. The refractive indices of the AlAs and the $Al_{0.1}Ga_{0.9}As$ differ by about 0.5, which is sufficiently large to produce good reflectors with about 12 layers. Figure 13.8 shows the reflection spectra from a 12 layer stack grown by MOCVD at Sheffield University by J S Roberts and C Button. A high reflectivity is achieved over a wide

Table 13.1 : A comparison of the Optical Modulation arising from Electro–absorption in Multiple Quantum Well Structures

Material	Structures	Source of data	Operating wavelength (nm)	Optical bandwidth (nm)	Modulation per μm of MQW (dB/μm)	Absorptive Insertion loss (dB/μm)
$In_{0.53}Ga_{0.47}As$ InP	100A well 199A barrier	Bar Joseph et al 1987 [1]	1610 1650	23+ 12+	1.0+ 0.6+	0.3 < 0.1
GaAs $Al_{0.3}Ga_{0.47}As$	87A well 60A barrier	Whitehead et al 1988 [28]	847 857	10+ 5+	3.9 2.3	1.7 < 0.1

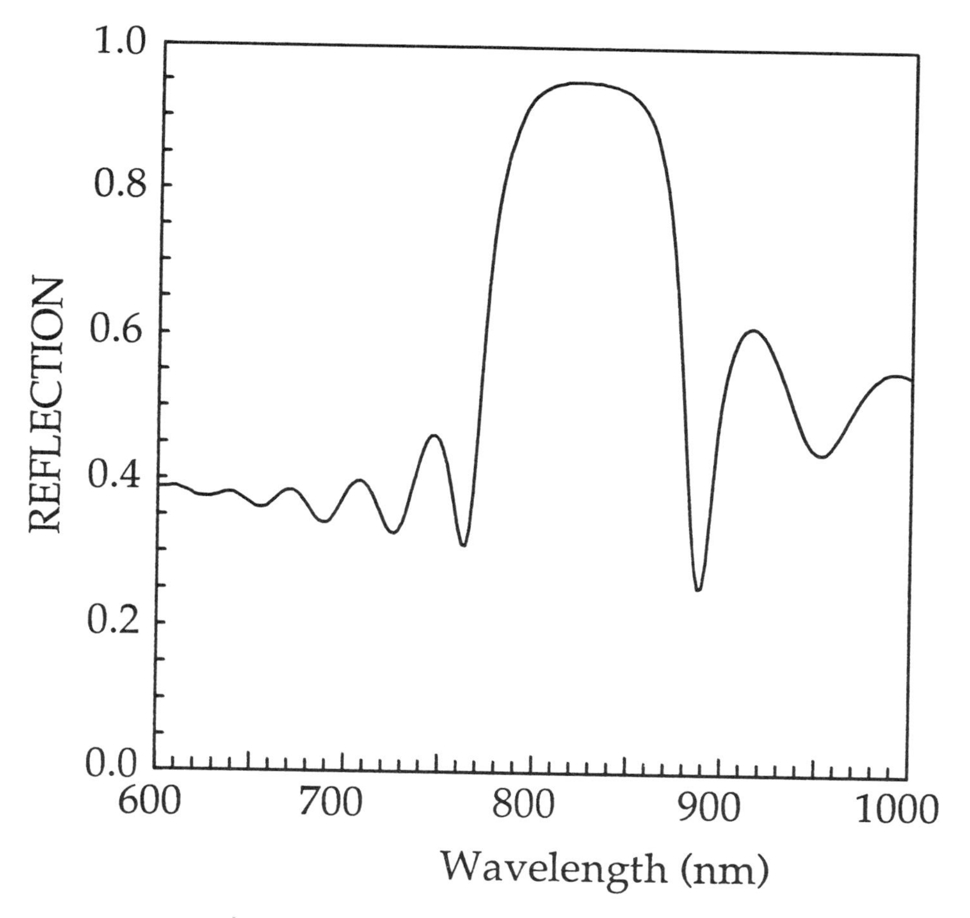

Figure 13.8: Reflection spectrum obtained with a multi–layer reflection stack of quarter–wave layers of alternating high and low index. AlAs and $Al_{0.1}Ga_{0.9}As$ are the two components and 12 periods are used to form the full reflector

wavelength range. In other III–V materials particularly InGaAs/InP it is difficult to achieve a sufficiently large refractive index difference (and convenient absorption bands) to produce this type of reflector stack.

The modulation results shown Figures 13.5(a), 13.5(b) and 13.7 were obtained with Gallium Arsenide substrates. Preliminary results obtained with silicon substrates and reflector stacks have also shown good optical modulation but there is considerable scope for further optimisation of these devices. (Barnes et al 1989, Goosen et al 1989)[2,8].

13.2.2 The Optics To Electronic Interface

An optical to electronic interface element is simply a detector. Detector technology is advanced and well established in the optical communications context. However there are additional considerations which have not yet been fully explored in the interconnect context. For example, the optical power available at the detector may be considerably greater than that typically available in a communications link and this could lead to a simplification of the receiver circuitry necessary. Such a simplification could be valuable if it leads to a reduction in the amount of space occupied on chip by receiver circuitry. Space is likely to be more of an issue in this context than in the case of a communications receiver. For gallium arsenide sources operating at around 850 nm, silicon or gallium arsenide detectors may be used. In fact, the Fabry–Perot quantum well devices described above as modulators will act as detectors because they are used in a p–i–n structure. In many situations we envisage that it would be useful to simply use the gallium arsenide quantum well devices as either detectors or modulators (Midwinter 1988)[15]. The high quantum efficiencies achievable with narrow intrinsic regions could offer significant advantages over silicon detectors which generally require thick intrinsic regions because of the low absorption coefficient of silicon. The schematic diagram shown in Figure 13.2(b) illustrates such an arrangement.

13.3 Digital Optoelectronic Logic Devices (SEEDs)

We have earlier dismissed optically controlled logic elements on the basis of speed and power considerations. There is one interesting class of devices which we wish to discuss in a little more detail because they offer a totally new set of characteristics and their use in systems and subsystems has not yet been fully explored. These devices are generally referred to as SEEDs (self–electro–optic effect devices) and they are hybrid in the sense that they are optically controlled but involve electronic elements in addition to optical ones. Figure 13.9 illustrates the concept of the SEED

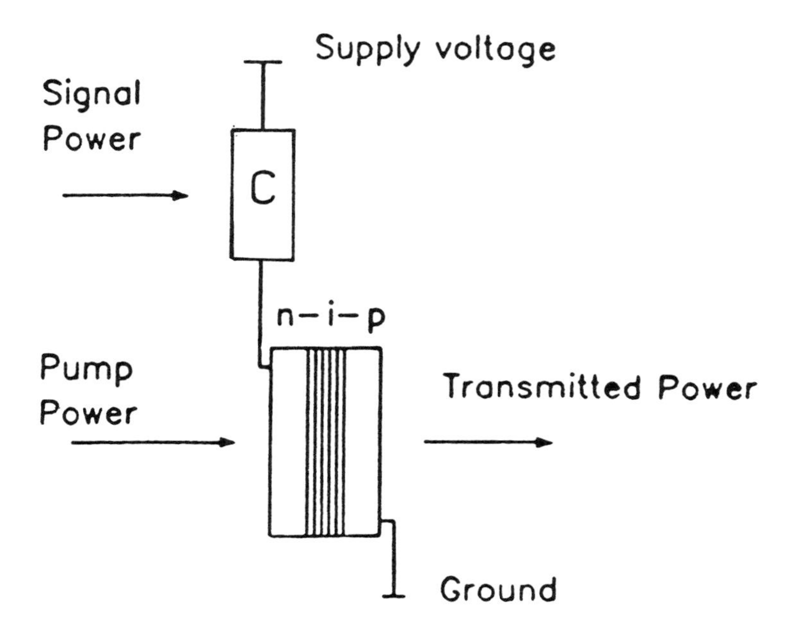

Figure 13.9: Schematic diagram of a SEED. The components C may consist of a resistor, photodiode, phototransistor, MQW modulator or an FET. Light from the signal can be used to control the pump transmission or simply to act as a current source driving the modulator. The n–i–p device is an MQW electro-absorption modulator

(Miller et al 1984, Miller et al 1985)[16,18]. It consists simply of a quantum well electro-absorption element in series with a photodiode (D–SEED), or phototransistor (T–SEED), another quantum well modulator (S–SEED), or a field effect transistor (F–SEED) (Miller et a 1989)[19]. (It is also possible to use just a resistor). The devices are all electrically biased but only receive or transmit information via optical signals incident on one or more of the components making up the complete device.

13.3.1 Diode SEEDs

The basic operation of the SEED can be explained with reference to the D–SEED shown in Figure 13.9 and the corresponding modulator transmission spectrum shown in Figure 13.5(b). An optical signal incident on the photodiode generates a photocurrent and acts as a current source for the MQW modulator. A second optical signal is incident on the modulator at a wavelength selected to be at the heavy–hole exciton absorption peak and a bias voltage applied across the two devices as shown.

The optical beam on the photodiode is kept at a constant intensity but the power of the beam on the modulator is varied. If the power is increased from zero, the transmitted power increases linearly with incident power up to a critical value when the power transmitted drops to a lower value. This behaviour arises from the interaction between the diode and the modulator and can be explained as follows. Because the two devices are connected in series the same current must flow through each device. The power absorbed in the modulator generates a photocurrent (since we are dealing with a p–i–n structure), the magnitude of which will depend on the absorption coefficient and quantum efficiency of the device. If the photocurrent is less than that produced by the series photodiode then the voltage drop across the photodiode will be small to minimise its quantum efficiency and hence equalise the current flows through the devices. The full bias voltage will appear across the modulator and maximum transmission is obtained. If the power of the beam on the modulator is increased then the photocurrent generated by the modulator will increase and the voltage drop across the diode will increase slightly to compensate. At the critical power level a positive feedback mechanism comes into play. The photocurrent generated by the modulator is sufficiently large that the full bias voltage no longer appears across the modulator since the photodiode now requires a voltage bias to generate sufficient current. However reducing the bias across the modulator leads to an increase in the absorption coefficient at the operating wavelength through the quantum confined Stark effect. Consequently the photocurrent increases leading to a further reduction in the bias voltage and hence a further increase in absorption. This positive feedback process results in the sharp drop in transmission observed. A full analysis of the behaviour shows that there is a hysteresis in the curve giving bistable operation over a certain optical power range.

Clearly the D–SEED can be operated as an optical logic element by optically biasing the device to the region of bistability. The AT&T group who have pioneered the development of this class of device, have fabricated fully integrated two dimensional arrays of D–SEEDs to evaluate their potential in optical computing and switching applications.

13.3.2 Transistor SEEDs

The transistor – modulator SEED (Figure 13.10) is similar in many ways to the D–SEED but it is operated at a wavelength in the tail of the heavy–hole exciton absorption (Wheatley et al 1987)[26]. At this wavelength the transmission decreases with applied bias so the positive feedback used in the D–SEED cannot occur. Instead the device is used as a three port device with the optical beam incident on the phototransistor being used to control the transmission through the modulator. Again the

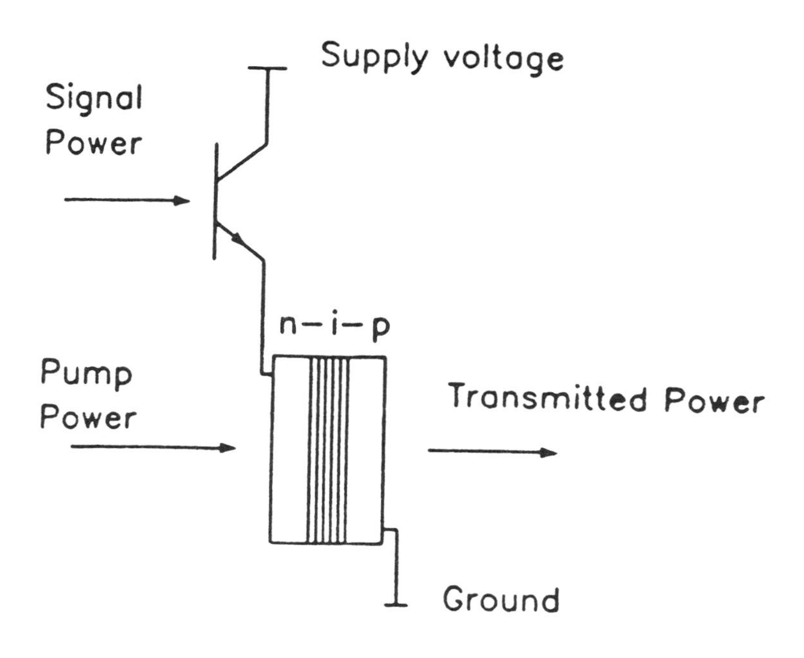

Figure 13.10: The T–SEED showing the transistor connected in series with the n–i–p modulator. (Wheatley et al 1987)

interaction between the current generated in the transistor and the current generated in the modulator is crucial in determining the characteristics of this device. Three regimes can be identified.

(a) The optical power on the phototransistor is low so that the current generated in the modulator results in a biasing ratio such that the full voltage appears across the transistor. Consequently the modulator is in its maximum transmission state.

(b) The optical power on the phototransistor results in a photocurrent which varies linearly with optical power incident and results in a corresponding linear change in the intensity of light transmitted through the modulator. In this case the bias across the modulator varies with the current on the phototransistor resulting in the change in transmission observed.

(c) The optical power on the phototransistor is such that at peak quantum efficiency the transistor would generate too much current for the modulator to match. Consequently the balance of voltage bias is now set so that the full voltage appears across the modulator and the quantum efficiency of the transistor is minimised. In this case the transmission of the modulator is also minimised.

These three ranges are clear in Figure 13.11. Interest in this device as a digital optoelectronic device has focussed on the fact that gain is achieved

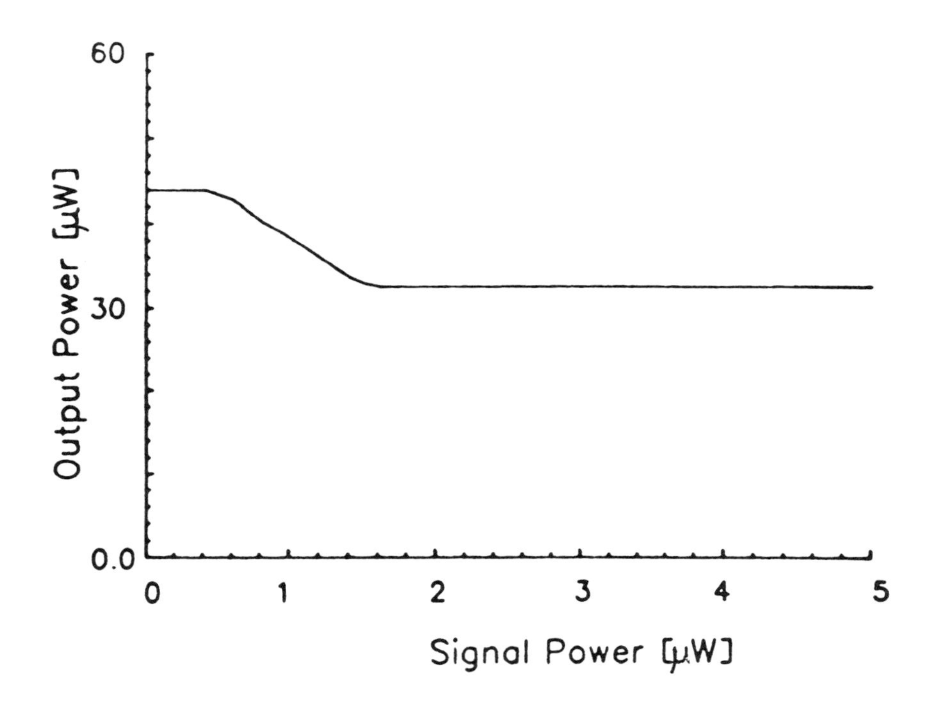

Figure 13.11: Optical power transmitted through the modulator as a function of optical power incident on the transistor for the T–SEED shown in Figure 13.10. (Wheatley et al 1987)

via the transistor and the three terminal configuration is more attractive for systems applications. The device clearly operates as an inverting gate in the mode discussed above. Bistable operation can be achieved in this device by selecting a different operating wavelength relative to the exciton peak (Wheatley et al. 1987)[27].

13.3.3 Symmetric SEEDs

The development of the T–SEED highlighted the attractions of a three–terminal rather than a two–terminal device. The importance of achieving gain in an optical logic device was also appreciated particularly since any practical form of optical computing arrangement would require the interconnection of several elements and the necessary fan-out and unavoidable interconnection losses would have to be compensated. The S-SEED or symmetric SEED, which is formed by connecting two MQW modulators in series (Lentine et al 1988)[11], has some of these features but has the additional attraction that it involves horizontal integration of devices rather than vertical integration as required in the D and T-SEEDs. Figure 13.12 illustrates schematically the structure of the S-SEED.

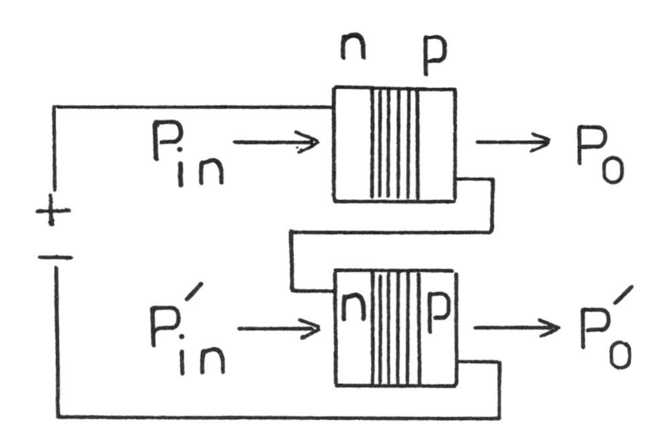

Figure 13.12: Schematic diagram of the S-SEED demonstrating the symmetrical connection of MQW modulators. The quantum wells are in the intrinsic region of the n–i–p structures.

The S-SEEDs have to be operated in a region of decreasing absorption, as discussed for the D-SEED. Because of the symmetric structure we can consider the device as a D-SEED but with either one of the MQWs acting as modulator and the other as the photodetector. It is important to note that because of the constraint that the same current flows through both devices it is the ratio of the optical powers incident on the two MQWs that determines the operation of the device. Following the arguments presented above for the D-SEED it can be seen that the outputs from the two modulators have to be complementary i.e. when one is "on" the other will be "off".

The three examples of the SEED family illustrate the range of characteristics which are achieved through the integration of just two devices. The best reported results have been achieved with the S-SEEDs where switching energies of 3.6 pJ have been achieved and by using a 6 psec pulse switching times below 1 nsec have been obtained (Lentine et lt 1989)[12]. Whilst these are impressive for optically controlled elements, these switching times are not faster than electronic devices. As we have stated earlier, it is not yet clear how such devices will be used in systems but it seems likely that if optically controlled elements do find a key systems role then the SEED approach will prove more attractive than the all optical approach.

13.4 Optical Interconnection

It is clear that the benefits of using optics in the interconnection of circuits will depend most of all on the interconnection scheme which can be implemented using optics. As we have mentioned previously, it is the parallel interconnection of electronic modules using the dimension perpendicular to the electronic wafer, which offers most scope for ideas. The wiring schemes which have been considered to date include:

(a) One to one connections where optics is used primarily to provide a very high data rate communication link.

(b) One to many connections where information must be distributed to many processors or when, for example, clock synchronisation is required across a wafer (Goodman 1984)[7].

(c) More complex configurations such as shuffle interconnection. This involves an interconnection scheme as illustrated in Figure 13.13 and is one which can be used in sorting routines, Fourier transforms and other algorithms. It is a scheme which has attracted attention because it can be implemented optically using bulk optical components quite simply and can probably be implemented in a holographic component. It is not appropriate to discuss the details of the optical designs or technologies here. Further detail can be found in (Lohmann et al 1985, Feldman and Guest 1987)[13,5].

13.5 System Design And Target Specification For A Photonic Switch Using Optical Interconnect And GaAs Technology

We will consider how the hybrid design can be used to produce a wideband switching matrix suitable for routing high speed packet data through a switch. The architecture is similar to that used in the StarLite switch (Huang and Knaner 1984)[10], but in the design discussed here we propose the use of optical interconnection between electronic modules to implement the design. The task to be carried out can be stated as follows. Data enters the switch on a one dimensional array of optical fibres and leaves the switch on a separate one dimensional array of fibres. Each data stream carries an address which indicates the output fibre to which the data must be sent – irrespective of the data port at which the data arrives. The

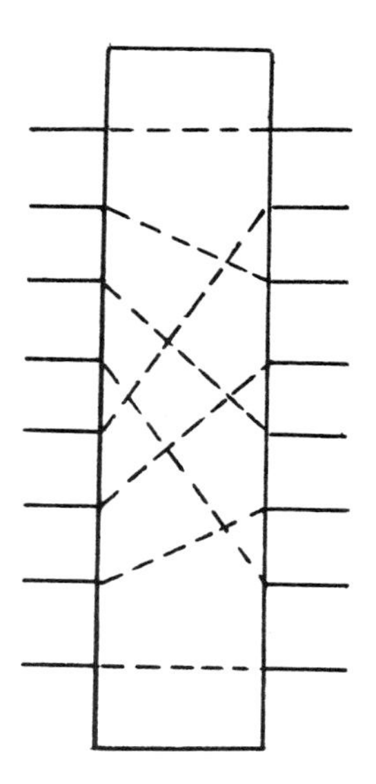

Figure 13.13 Perfect shuffle interconnection scheme

switch has to read the address and sort the data streams so that the output fibres can collect the data for distribution to the correct destination. Full details of the optical system have been given elsewhere (Midwinter 1987)[14] so we will concentrate on principles rather than details here.

The algorithm for the sorting operation is the perfect shuffle sort algorithm (Stone 1971)[23]. The algorithm will sort a one dimensional array of random input addresses into an ordered form – the highest address appearing at one end of the array and the lowest at the other end. The algorithm involves a sequence of logical operations followed by a shuffle interconnect, followed by another logical operation and another shuffle, and so forth. The logical operations are exchange or bypass operations; the two inputs from neighboring data inputs are compared and, depending on the relative magnitudes of the address codes, may be re–ordered (exchanged) or left unchanged (bypass). The shuffle operation mentioned above involves taking the spatial input data, dividing it in two, magnifying the spatial separation by two and overlaying (a direct analogy to shuffling a pack of cards). That is, if we label a pixel element by j in the one dimensional array of N spatially separated inputs, then j will map to $k = 2j - 1$ for j in the range 1 to N/2 and $k = N - 2(N - j)$ for j in the range $N/2 + 1$ to N. The attraction of the perfect shuffle sort algorithm in this context is that the shuffle operation can be implemented optically (Midwinter 1987)[14] so the complete sorting operation can be carried out using a matrix of exchange–bypass modules with each row interconnected by an optical shuffle connection. We could envisage such an arrangement using the two dimensional array of electronic islands in Figures 13.1 and 13.2 with the optical interconnection occurring via the plane perpendicular to the array of electronic logical elements. Each logical element would have an optical receiver, and a quantum well modulator as the electronic to optical interface element. The logical operations involve relatively simple electronics and could be implemented using high speed GaAs electronics to ensure maximum data rate through the switch. The sequence of operations for 4 input ports is shown in Figure 13.14.

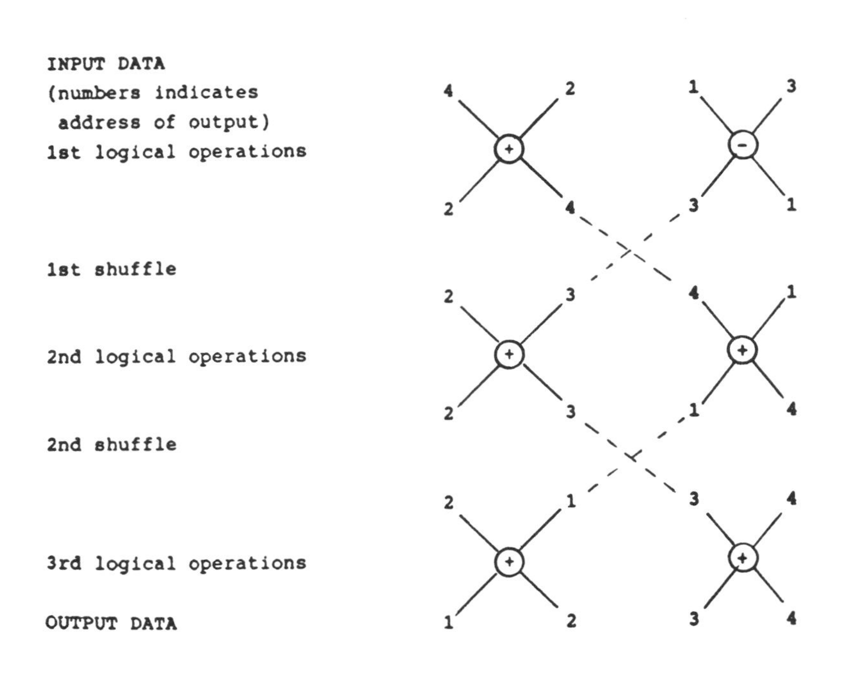

Figure 13.14: Sorting of data at 4 input ports according to address by using the shuffle sort algorithm. the logical exchange bypass modules + and – receive two inputs and re-order (or leave) depending on the relative magnitudes of the input addresses.

In practice a large number of input ports would be needed. A detailed analysis of a 128 input port switch shows that a matrix of exchange–bypass modules 43 x 64 would be required, involving 42 shuffle operations. For a data rate of 1 Gbit/s each modulator would require approximately 300 microwatts of optical power per modulator and the total optical power of 860 milliwatts could be provided by an array of semiconductor lasers (Midwinter 1987)[14].

13.6 Conclusions

In this chapter we have considered some ways in which GaAs optoelectronic devices and systems may be linked with electronics to provide an additional degree of flexibility in the design of high speed circuits and sub–systems. We have proposed a hybrid approach on the basis that optics primarily offers an interconnect technology rather than a logic technology. However we have also highlighted some of the advances achieved using SEED devices. It remains to be seen whether these or any other optical logic elements will prove us wrong in placing our emphasis on a hybrid approach. The ideas arising from merging the optical and electronic technologies are still at early stages. However the prospect of not just using optics to solve problems in electronics, but of achieving a more powerful system or a more flexible architecture provides a strong driving force for research work in this field. It seems likely that GaAs technology will play a major role in encouraging this exciting interaction.

13.7 Acknowledgements

We wish to acknowledge the contribution made by Dr J. S. Roberts and colleagues at the University of Sheffield to the growth and fabrication of some of the devices described here. We also wish to thank Dr A. K. Tipping for preparing Table 13.1.

13.8 References

1 I. Bar–Joseph, C. Klingshirn, D. Miller, D. Chemla, U. Koren and B. Miller: Quantum confined Stark effect in InGaAs/InP quantum wells grown by organo–metallic vapour phase epitaxy, Applied Physics Letter, 1978, Vol. 50, No. 15, pp1010–1012

2 P.Barnes,P. Zouganeli, A. Rivers, M. Whitehead, G. Parry, K. Woodbridge, and C. Roberts (1989): A GaAs/AlGaAs multiple quantum well optical modulator using a multilayer reflector stack grown on a Si substrate, Elect. Lett. to appear July 1989

3 G. D. Boyd, J. E. Bowers, C. E. Soccolich, D. A. B. Miller, D.. S. Chemla, L. M. F. Chirovsky, A. C. Gossard and J. R. English (1989): 55 GHz multiple quantum well reflection modulator, Elect. Lett. 25, 558

4 J. C. C. Fan (1989): New anatomies for semiconductor wafers, IEEE Spectrum, April, p34

5 M. R. Feldman, C. C. Guest (1987): Computer generated holographic optical elements for optical interconnection of VLSI circuit, Appl. Optics, 26, 4377

6 H. M. Gibbs (1986): Optical bistability controlling light with light, Pub. Academic Press, New York

7 J. W. Goodman, F. I. Leonberger, S. Y. Kung and R. A. Athale (1984): Optical interconnections for VLSI systems, Proc. IEEE, 72, 850

8 K. W. Goosen, G. D. Boyd, J. E. Cunningham, W. Y. Jan, D. A. B. Miller, D. S. Chelma and R. M. Lum (1989): GaAs multi-quantum well reflection modulators grown on GaAs and silicon substrates, PD18, CLEO '89, Baltimore April 1989

9 P. R. Haugen, S. Rychnovsky and A. Husein (1986): Optical interconnects for high speed computing, Optical Engineering, 25, 1076

10 A. Huang and S. Knaner (1984): Starlite a wideband digital switch, Proc. IEEE Global Telecom. Conf. Atlanta Georgia, 121–125

11 A. L. Lentine, H. S. Hinton, D. A. B. Miller, J. E. Henry, J. E. Cunningham, L. M. F. Chirovsky (1988): Symmetric self electro-optic effect device: optical self-reset latch, APL. 52, 1419

12 A. L. Lentine, L. M. F. Chirocsky, L. A. D'Asaro, C. W. Tu and D. A. B. Miller (1989): Energy scaling and subnanosecond switching of symmetric self-electro-optic effect devices, IEEE Photonics Tech. Lett., 1, 129

13 A. Lohmann, W. Stork and G. Stucke (1985): An optical implementation of the perfect shuffle, IEEE OSA Topical Meeting on Optical Computing Indie Village Nevada, March 18th

14 J. E. Midwinter (1987): Novel approach to the design of optical activated wideband switching matrices, IEE Proc. 134, Pt. J, 261

15 J. E. Midwinter (1988): Digital optics smart interconnect or optical logic?, Physics in Technology, 19, Part 1, May 1988, Part II, July 1988, pp153–165

16 D. A. B. Miller, D. S. Chemla, T. C. Damen, A. C. Gossard, W. Wiegmann, T. H. Wood and C. A. Burrus (1984): Novel hybrid optically bistable switch: The quantum well self -electro-optic effect device, APL, 45, 13

17 D. A. B. Miller, D. S. Chemla, T. C. Damen, A. C. Gossard, W. Wiegmann, T. H. Wood and C. A. Burrus (1985): Electric field of optical absorption near the band gap of quantum well structures, Phys Rev. B 32, 1043

18 D. A. B. Miller, D. S. Chemla, T. C. Damen, T. H. Wood,C. A. Burrus, A. C. Gossard, W. Wiegmann (1985): The quantum well self-electro-optic effect device: optoelectronic bistability and oscillation and self-linearised modulation, IEEE J. Quant.. Elect. QE-21, 1462

19 D. A. B. Miller, M. D. Feuer, T. Y. Chang, S. C. Shunk, J. E. Henry, D. J. Burrows, D. S. Chemla (1989): Field effect transistor self-electro-optic effect device: integrated quantum well modulator and transistor, IEEE Photonics Tech. Lett. 1, 62

20 D. Psaltis and N. H. Farhat (1985): Optical implementation of the Hopfield model, IEEE/OSA Topical Meeting on Optical Computing, Incline Village, Nevada, USA, March 18–20

21 G. A. Sai-Halasz (1989): Silicon FETs at 0.1 m gate-length, Proc. Picosecond Electronics and Optoelectronics Conf. Salt Lake City, March 1989 (IEEE/OSA)

22 P. Stevens and G. Parry (1989): Limits to normal incidence electroabsorption modulation in GaAs/(GaAl)As multiple quantum well diodes, to appear in J. Lightwave Technology

23 H. S. Stone (1971): Parallel processing with the perfect shuffle, IEEE Trans. C-20, 153–161

24 A. K. Tipping, G. Parry and P. Claxton (1989): A comparison of the limits in performance of multiple quantum well and Franz Keldysh InGaAs/InP electroabsorption modulators, IEE Proc Part J (Optoelectronics), 136(4), 205–208, Aug 1989

25 S. Uchiyama and K. Iga (1986): Consideration of threshold current density of GaInAsP/InP surface emitting lasers, IEEE J. Quant. Elect. QE-22, 302

26 P. Wheatley, P. J. Bradley, M. Whitehead, G. Parry, J. E. Midwinter, P. Mistry, M. Pate and J. S. Roberts (1987): Novel nonresonant optoelectronic logic device, Elect. Lett. 23, 92

27 P. Wheatley, M. Whitehead, J. E. Midwinter, and J. S. Roberts (1987): 3–terminal noninverting optoelectronic logic device, Opt. Lett. 12, 784

28 M. Whitehead, P. Stevens, A. Rivers, G. Parry, J. S. Roberts, P. Mistry, M. Pate and G. Hill (1988): Effects of well width on the characteristics of GaAs/AlGaAs multiple quantum well electroabsorption modulators, Appl. Phys. Lett. 53, 956

29 M. Whitehead and G. Parry (1989): High contrast reflection modulation at normal incidence in asymmetric multiple quantum well Fabry–Perot structure, Elect. Lett. 25, 566

30 M. Whitehead, A. Rivers, G. Parry, J. S. Roberts, C. Button: A low voltage multiple quantum well reflection modulator with 100:1 on:off ratio, Elect. Lett. to appear in July 1989

31 T. H. Wood, C. A. Burrus, D. A. B. Miller, D. S. Chemla, T. C. Damen, A. C. Gossard and W. Wiegmann (1984): Appl. Phys. Lett. 44, 16

Index

Symbols

Numbers

A

B

C

D

E

F

G

H

I

J

K

L

M

N

Q

T

U

V

W

X

Y

Heterick Memorial Library
3 5111 00399 6547